Flow-Induced Vibration of Power and Process Plant Components

M. K. Au-Yang, Ph.D., PE

Professional Engineering Publishing Limited 2001

**Professional
Engineering
Publishing**

Copyright © 2001 by The American Society of Mechanical Engineers
Three Park Avenue, New York, NY 10016
ISBN: 0-7918-1066-7

Co-published in the UK by Professional Engineering Publishing Limited,
Northgate Avenue, Bury St. Edmunds, Suffolk, IP32 6BW, UK
ISBN: 1-86058-319-9

Library of Congress Cataloging-in-Publication Data

Au-Yang, M. K., 1937 –
Flow-Induced Vibration of Power and Process Plant Components: A Practical Workbook / M. K. Au-Yang.
p. cm.
ISBN 0-7918-1066-7
1. Vibration. 2. Fluid dynamics. 3. Structural dynamics. I. Title.
TA355.A873 2001
624.1'76 — DC21   2001022080

Disclaimer

This book was written for educational purposes. Neither the author, his affiliation, nor ASME Press makes any guarantee that the information contained in this book is 100% accurate. Readers who use the materials contained in this book for any design analysis, diagnostic and troubleshooting should verify the results following their own quality assurance programs.

Acknowledgment

Many of the graphics and drawings contained in this book were reproduced, with cosmetic touch-up and revisions, from the U.S. National Aeronautic and Space Administration, U.S. Department of Energy and other U.S. Government agency reports, and from various publications by the American Society of Mechanical Engineers (ASME). Some are reproduced, with permission, from other professional and private organizations. The author wishes to express his gratitude to Framtome ANP of USA, for permission to use many of the company's drawings in this book. He also thanks the following organizations for permissions to use the specific drawings:

Cambridge University Press, Figure 6.1
Electric Power Research Institute, USA, Figures 2.16 and 11.5
Framatome, France, Figure 7.5
Institution of Mechanical Engineers, UK, Figure 7.2
Nuclear Industry Check Valve Group, Figure 11.11

Finally, the author wishes to express his great appreciation to his wife, Grace, for giving him the time to write this book. Without her moral support, this book would not be possible.

Preface

Unlike stresses, seismic loads and radiation dosages, which are very much regulated and governed by industry codes and standards, flow-induced vibration of power and process plant components is a relatively unregulated technology— and with good reasons. Flow-induced vibration is to a large extent, an operational problem that has relatively little direct impact on public safety. If a component breaks down due to flow-induced vibration, the plant has to be shut down until corrective actions are taken. Very often, this is exactly the approach the power and process industries are taking. Through the years, this kind of unplanned and unforeseen outages can add significantly to the operational costs, as well as tint the public image of an otherwise technologically sound industry.

In the power and process industries at least, flow-induced vibration is often viewed as some kind of "black art"—too mysterious for the average practicing engineers to understand. This is in part due to the vast amount of uncoordinated publications in the open literature, many of which contained undefined notations and terminology. Yet as an engineering science, flow-induced vibration analysis involves no higher mathematics than thermal-hydraulic, seismic or stress analysis. Most of the flow-induced vibration problems observed in the field could have been prevented if only minimum amounts of time were spent on this issue in the design phase of the products.

The purpose of this book is to provide the practicing engineers with a consolidated source to obtain information on the most common flow-induced vibration problems in power and process plant components, as well as the basic equations and charts to enable them to assess the concerns. Realizing that undefined notations and lack of a set of universally accepted terminology constitute a major difficulty for the practicing engineers to follow the published literature, great emphasis is placed on defining every notation used both in the text and at the beginning of each chapter following the summary. Conflicting terminology used by different authors is pointed out, sometimes to the point of redundancy. It is this author's believe that it is better to be repetitious than unclear.

With the exception of the first couple of introductory chapters, each of the chapters in this book follows the same structure. Each one starts with a rather lengthy summary containing the key points and most frequently used equations of the chapter. The intention is that most of time, the analyst should be able to carry out his calculations following just the summaries, with reference to the main text only occasionally. A comprehensive nomenclature follows the summary in each chapter. The main text contains development of the subject in that chapter, extensive work examples and case studies. It is assumed that the readers have access to some minimal computing facility and software, such as personal computers and spreadsheets. Most of the work examples are designed to be solved with pocket calculators or spreadsheets that are virtually part of a modern office setup. Detailed mathematical derivations are given only in cases when such derivations are not readily available in the literature, or when it is the opinion of this author that the particular equation is often mispresented or misused. Emphasis is placed on the physics of the phenomenon. It is the author's believe that, especially in plant component diagnosis and troubleshooting, it is much more important to understand the physics of the phenomenon than the detailed mathematical derivation of the governing equations. Each chapter ends with a reference list. Realizing that too many references frequently intimidate, rather than help, practicing engineers, these reference lists are purposely kept brief. Only the most relevant references are included.

This is a book written by a practicing engineer for practicing engineers. Even though it is neither a research-oriented publication nor a conventional textbook, researchers in flow-induced vibration may find it useful to help them present their results for wider industry acceptances; and students may find this book useful to better prepare themselves for the industry environment. The author is indebted to the many researchers whose publications and friendly discussions have made this book possible. Research results of the following distinguished names in flow-induced vibration form the central part of this book: Drs R. B. Blevins, S. S. Chen, H. J. Connors, F. L. Eisinger, A. Powell, C. E. Taylor; Professors M. P. Paidoussis, M. J. Pettigrew, D. S. Weaver and S. Ziada.

M. K. Au-Yang
Lynchburg, VA.
February 2001

TABLE OF CONTENTS

Chapter 9   Turbulence-Induced Vibration in Cross-Flow   253

CHAPTER 1

UNITS AND DIMENSIONS

Summary

Problems in units account for most errors in dynamic analyses. To avoid this, we must use consistent unit sets (units in **bold** are the main unit sets used in this book):

	Fundamental Units			Derived Units	
	Mass	Length	Time	Force	Stress/Pressure
US Unit	slug	Foot	second	pound (lb)	lb/ft^2
	(lb-s^2/in)	**inch**	**second**	**pound**	**psi (lb/in^2)**
SI Unit	**Kg**	**meter**	**second**	**Newton (N)**	**Pascal (N/m^2)**

Where,
1 slug = 1 lb-s^2/ft = 1 lb mass/32.2
1 slig = 1 (lb-s^2/in) = 1 lb mass/(32.2*12)
1 slug/ft^3 = 1 lb mass per cu.ft./32.2
1 (lb-s^2/in)/in^3 = 1 slig/in^3 = 1 lb mass per in^3/(32.2*12*1728)

As long as we use consistent unit sets, we do not have to carry conversion factors such as g$_c$, and units, such as ft, in and s, in every step of the solutions, and computer programs will not have built-in units. Throughout this book, unless otherwise stated, the two unit systems used are: The SI units and the US units; with inch as the length unit and the entity (lb-s^2/in) as the mass unit. Second is the time unit in both systems. Whenever feasible, examples and solutions are given in both US and SI units. For lack of an official name, some engineers give the name "slig" or "slinch" to the mass unit (lb-s^2/in). While this is very useful in practice, these terms are not used in the remainder of this book. Instead (lb-s^2/in) will be used as an entity for the mass unit in the US unit system.

1.1   Introduction

Problems in units account for most errors in dynamic analyses, especially for US-trained engineers. The reason is the lack of a standard US unit system. While in the SI unit system the time, length and mass units are always second, meter and kilogram; there is no equivalent standard in the US unit system. Structural dynamicist and stress analysts seem to favor using the inch as the length unit, while in fluid dynamics, including the ASME Steam Table, the length unit is often in feet. To make the situation even more confusing, there is no universal agreement as to what is the mass unit in the US unit system. In fact, it is not certain there is a mass unit at all in the US system of units, as will be made apparent in the following paragraphs.

1.2   The Force-Length-Time Unit System

In our everyday lives, we are more concerned with forces rather than masses. Thus, we talk about gaining *weight* instead of gaining mass, even though the weight we gain is a consequence of the mass we gain. Engineers, in particular, are concerned about stresses and pressures, which are force per unit area. Thus, it is not surprising that U.S. engineers prefer to use the force, length and time as the three fundamental units. According to Newton's equation of motion:

Force = mass*acceleration

The mass unit in this unit system is a *derived* unit, and has dimensions given by:

[mass]=force/acceleration

Therefore, the mass unit in a force-length-time unit system is in the form of $lb\text{-}s^2/ft$ or $lb\text{-}s^2/in$. Table 1 gives the units of common variables in this system, which is seldom used outside of the United States.

Table 1.1: The US Force-Length-Time Unit System

Fundamental Units			Derived Units	
Force	Length	Time	Mass	Stress/Pressure
lb	foot	second	$lb\text{-}s^2/ft$	lb/ft^2
lb	inch	second	$lb\text{-}s^2/in$	$psi\ (lb/in^2)$

While this system is more natural in our everyday lives, it is not as convenient in engineering calculations. In addition, it is seldom used outside of the U.S. In this time of world trade and global communication, it is important that we can communicate with the rest of the world even if, due to many practical reasons, it is not yet feasible for us to convert entirely to the SI units. For this reason, the following mass-length-time system will be used throughout this book.

1.3   The Mass-Length-Time Unit System

In the SI unit system, there are three fundamental units: mass in kilogram (Kg), length in meter (m) and time in second (s). The force unit in the SI system is Newton (N) while the stress or pressure unit is Pascal (Pa, 1 Pa=1 N/m^2). These as well as energy, power, viscosity, etc. are all *derived* units. The picture is far less clear in the US unit system. There appears to be no universally accepted mass unit in the US system. In some high school or college textbooks on elementary dynamics, the slug is used as the mass unit to form a mass-length-time unit system. In this system, the three fundamental units are mass in slug, length in feet and time in second. The force unit in this system is pound (lb) while the stress or pressure unit is pound per square foot (lb/ft^2). These are *derived* units. To make the situation confusing, while fluid dynamicists are perfectly at home with the foot as the length unit (with pressure in lb/ft^2), structural dynamicists and stress analysts traditionally use inch as the length unit, with the vibration amplitude expressed in inch and the stress expressed in lb/in^2 (psi). With time in second, length in inch, one must find a corresponding mass unit to complete a mass-length-time unit system. For lack of an official name, some practicing engineers have used the term slig or slinch as the mass unit when the inch is used as the length unit. However, this practice is not followed in this book. Instead ($lb\text{-}s^2/in$) will be used as an entity for the mass unit whenever inch is used as the length unit. One can replace the entity ($lb\text{-}s^2/in$) with slig or slinch as he wishes. Table 2 lists some of the common mass-length-time unit systems used in various disciplines of physical sciences (units in **bold** are the main unit sets used in this book).

Table 1.2: Consistent Unit Sets in Mass-Length-Time Unit Systems

	Fundamental Units			Derived Units	
	Mass	Length	Time	Force	Stress/Pressure
US Unit	slug	foot	second	pound (lb)	lb/ft^2
	($lb\text{-}s^2/in$) or "slig"	**inch**	**second**	**pound (lb)**	**psi (lb/in^2)**
	pound	foot	second	poundal	$poundal/ft^2$
CGS Unit	gm	cm	second	dyne	$dyne/cm^2$
SI Unit	**Kg**	**meter**	**second**	**Newton (N)**	**Pascal (N/m^2)**

Note: 1 slug = 1 $lb\text{-}s^2/ft$ = 1 lb mass/32.2
1 ($lb\text{-}s^2/in$)=1 slig = 1 $lb\text{-}s^2/in$ = 1 lb mass/(32.2*12)
1 $slug/ft^3$ = 1 lb mass per cu.ft./32.2
1 ($lb\text{-}s^2/in$)/in^3 =slig/in^3 = 1 lb mass per in^3/(32.2*12*1,728)

As long as we use consistent unit sets, we do not have to carry conversion factors such as g_c, and units, such as ft, in and s, in every step of the solutions, and computer programs will not have built-in units.

Throughout this book, the two unit systems used are: The SI units and the US units, with inch as the length unit and ($lb\text{-}s^2/in$) as the mass unit. Whenever feasible, examples and solutions to the examples are given in both unit systems. However, since flow-induced vibration analyses often interface with fluid-dynamic calculations (including the ASME Steam Table), which most likely use lb-ft-s as the fundamental units with the lb also used as the mass unit, great care must be exercised to convert the units for different variables into the same unit basis. The following example should make this clear.

Example 1.1: Reynolds Number Calculation

The Reynolds number Re of a circular cylinder subject to cross-flow is given by, $\text{Re} = VD/v$, where

V = cross-flow velocity,
D = diameter of cylinder,
v = kinematic viscosity; v is related to the dynamic viscosity μ by:
$v = \mu / \rho$

where ρ is the fluid mass density. A cylinder is subject to steam cross-flow with the known conditions given in Table 1.3 in both the US and SI unit systems.

Table 1.3: Known parameters in Example 1.1

	US Units	SI Units
Steam pressure	1075 psi	7.42E+06 Pa
Steam temperature	700 F	371 C
Cross flow velocity V	70 ft/s	21.336 m/s
Diameter of cylinder	1.495 in	0.03797 m

What is the Reynolds number of this cylinder in this cross-flow?

Solution

We shall give the solution in both the US and SI unit systems. First, we must find the steam mass density and viscosity at the given temperature and pressure. For that, an electronic version of the ASME Steam Table (1979) is used, which gives the variables in both US and SI units. However, in the US unit system, a non-consistent unit set is used in the ASME Table. In the US unit system, the density of steam is given in lb (mass) per cu.ft. while the dynamic viscosity is given in lb (force) per sq. ft. per second. There are two alternate ways to calculate the Reynolds number in the US unit system. We can convert the steam density to slug/ft^3 and use ft as the length unit in the remaining calculation; or we can convert the steam density to (lb-s^2/in)/in^3 and use inch as the length unit in the rest of the calculations. Because of the inherent consistency in the SI unit system, there is no such problem when we use the SI unit system. The following outlines the steps used in each system (slug and ft are used in the US unit system).

From the ASME Steam Table (1979), we find the following steam properties,

	US Units	SI Units
Density	1.79 lb/ft^3	28.68 Kg/m^3
	=5.559E-2 slug/ft^3	
Dynamic viscosity	4.82E-7 lb/(ft^2-s)	2.31E-5 Pa/s
Kinematic viscosity $v = \mu / \rho$	=8.68E-6 ft^2/s	8.05E-7 m^2/s
$\text{Re} = \dfrac{VD}{v} =$	70*(1.495/12)/8.68E-6	21.336*0.03797/8.05E-7
	=1.0E+6	=1.0E+6

The Reynolds number, being dimensionless, is the same in both unit systems.

1.4 Dimensional Analysis

One reason to use the mass-length-second as the fundamental unit system is that in fluid dynamics, as well as in flow-induced vibration, very often the technique of dimensional analysis is used to derive the *form* of the equation, with several coefficients to be determined by fitting test data. Cornors' equation for fluid-elastic instability (see Chapter 7), as an example, was derived in this way. In dimensional analysis, it is customary to use the square bracket [ ] to mean the "dimension of," and express all the derived units in the dimensions of the three fundamental units: mass, length and time. With this convention:

[mass] $= M$
[length] $= L$
[time] $= T$
[force] = [mass*acceleration]$=ML/T^2$
[pressure] = [force]/[area]$=(ML/T^2)/L^2 = M/LT^2$

and so on. Dimensional analysis is a very powerful tool in the development of empirical equations before a firm equation can be established from theoretical development. It is less important in engineering design and analysis. Therefore, this book will not treat dimensional analysis in great detail. Chapter 1 of Blevins (1990) gives a brief discussion of the application of dimensional analysis to flow-induced vibration. Many books in hydraulics and fluid dynamics discuss dimensional analysis in greater detail. However, very often a simple dimensional check can help the engineer find out if there are any errors in the equations. To be correct, the dimensions on both sides of the equation must be the same.

Example 1.2: Dimensional Check of Bending Moment Equation

The bending moment in a beam vibrating with amplitude $y(x)$ at any point along the beam is given by the equation:

$$\overline{M}(x) = -EI\frac{\partial^2 y}{dx^2}$$

where E, I, x are the Young's modulus, the area moment of inertia and the distance from one end of the beam respectively. Show that the dimensions on both sides of the equation are the same.

Solution:

Since bending moment is in force*length,

$$[LHS] = [\overline{M}] = [force][length] = [mass * acceleration]L = (ML/T^2)(L) = ML^2/T^2$$

On the right hand side (RHS) of the equation, we first note that from the original definition of differential calculus,

$$\frac{\partial^2 y}{\partial x^2} = \lim(\delta x \to 0)\frac{\delta}{\delta x}(\frac{\delta y}{\delta x})$$

i.e., the change in the change of slope of y as a function of x, which therefore has a dimension of L/L^2.

$$[RHS] = [E][I](L/L^2) = [force/area](L^4)(L/L^2) = [mass * acceleration/area]L^3$$
$$= (ML/T^2)(L^{-2})(L^3) = ML^2/T^2$$

The dimensions on both sides of the equation are the same.

References

ASME, 1979, <u>Steam Tables</u>, Fourth Edition, ASME Press, New York.
Blevins, R. D., 1990, <u>Flow-Induced Vibration, Second Edition</u>, Van Nostrand Reinhold, New York.

CHAPTER 2

THE KINEMATICS OF VIBRATION AND ACOUSTICS

Summary

The motion represented by the equation

$$y(t) = a_0 \cos \omega t + b_0 \sin \omega t \tag{2.2}$$

or, in more compact complex variable notation,

$$y(t) = a_0 e^{i\omega t} \tag{2.3}$$

is called simple harmonic motion. Most vibrations and noise we encounter belong to the category of linear vibration and consist of a linear combination of simple harmonic motions of different amplitudes, frequencies and phases. The frequency of a point mass-spring system is given by:

$$\omega = \sqrt{\frac{k}{m}} \quad \text{radian/s} \tag{2.4}$$

The frequency in cycles/s, or Hz, is related to the frequency in radians/s by,

$$f = \frac{\omega}{2\pi} \quad \text{Hz} \tag{2.5}$$

The instantaneous velocity and acceleration of a vibrating point mass is given by:

$$V(t) = \dot{y}(t) = i\omega y(t) \tag{2.7}$$
$$\alpha(t) = \ddot{y}(t) = -\omega^2 y(t) \tag{2.8}$$

The velocity of propagation of a wave is given by:

$$c = f\lambda \tag{2.9}$$

Vibrations can be represented in the time domain (time histories) or in the frequency domain (power spectral densities). Both contain the same information.

Signal 1 is said to be 1.0 bel, or, more commonly, 10 decibels (10 dB) above signal 2 if the log (to base 10) of their ratio is equal to one. That is, if

$$\log_{10}(I_1 / I_2) = 1.0$$

If the energy in one signal I_1 is 20 dB higher than that of another signal I_2, then the vibration amplitude of signal 1 is 10 dB higher than that of signal 2. That is, if

$$\log_{10}(I_1 / I_2) = 2.0 \quad \text{then,} \quad \log_{10}(a_1 / a_2) = 1.0$$

Nomenclature

a	Radius of circular motion
a_0	Vibration amplitude
b_0	Vibration amplitude
c	Velocity of wave propagation
f	Frequency, in Hz
i	$= \sqrt{-1}$
I	Intensity of the sound wave (energy unit)
k	Stiffness
m	Mass
p	Acoustic pressure
s	Second
t	Time
V	Velocity
$y(t)$	Displacement from equilibrium position at time t
α	Acceleration
λ	Wavelength
ω	Angular frequency, radian/s, $= 2\pi f$

2.1   Introduction

It is assumed that the readers are familiar with the fundamentals of engineering dynamics, including the vibration of a point mass-spring system. To be self-contained, the basic equations in the fundamentals of vibration theory and structural dynamics are given in this and the next chapters without derivation to establish the notations and for later references. More details on signal and spectral analysis of vibration data are given in Chapter 13. The readers are referred to the many excellent textbooks in mechanical vibration for the detailed derivation of these basic equations. For applications to engineering analysis, design and, in particular, diagnosis and troubleshooting, it is more important to understand the physics behind the equations than their detailed derivations.

2.2   Free Vibration and Simple Harmonic Motion

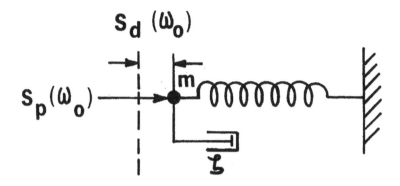

Figure 2.1 Spring-mass system

It is well known that in the absence of any external forces and energy dissipation (damping), a point mass-spring system (Figure 2.1) when disturbed from its equilibrium position will undergo a cyclic motion, with the equation of motion given by:

$$m\ddot{y} + ky = 0 \tag{2.1}$$

where y is the displacement of the point mass m from its equilibrium position, and k is the spring constant. The solution to Equation (2.1) is the well-known "harmonic functions," that is, sine and cosine functions:

$$y(t) = a_0 \cos \omega t + b_0 \sin \omega t \tag{2.2}$$

or, in more compact complex variable notation,

$$y(t) = a_0 e^{i\omega t} \tag{2.3}$$

where

$$\omega = \sqrt{\frac{k}{m}} \tag{2.4}$$

is the natural frequency, in *radians per unit time*, of the cyclic motion, which is more commonly known as *vibration*. While the frequency expressed as radians per unit time (ω) is more convenient for the purpose of mathematical derivations, the term cycles per second, or Hz, is far more convenient for our everyday lives, including engineering applications. The two are related by

$$\omega = 2\pi f \tag{2.5}$$

Throughout this book, both ω and f will be used, with ω used mainly in mathematical derivations. The final equations will be expressed in terms of f. The period of the vibration is given by

$$T = 1/f \tag{2.6}$$

From Equation (2.3), the instantaneous velocity V and acceleration α of the point mass are,

$$V(t) = \dot{y}(t) = i\omega y(t) \tag{2.7}$$
$$\alpha(t) = \ddot{y}(t) = -\omega^2 y(t) \tag{2.8}$$

The motion represented by Equation (2.2) or (2.3) is called simple harmonic motion probably because monotonic musical tones are generated by such motions of the strings in musical instruments—early vibration and acoustic research were closely tied to applications to music. Simple harmonic motion is the most basic unit in vibration theory.

2.3   Linear Vibration and Circular Motion

Before we leave Equation (2.1) and its solution Equation (2.2) or (2.3), we shall spend some time discussing the *kinematics* of vibration. It is amazing how many practical problems in power and process plants can be diagnosed with just some basic knowledge of the kinematics of vibration, without any detailed calculations.

The majority of vibration phenomena we encounter every day can be classified as linear vibrations in which the object undergoes simple harmonic motion: Equation (2.2) or (2.3). Simple harmonic motion can be related to circular motion in the following manner: Imagine a particle moving in a circular orbit with radius a and at a constant angular velocity ω rad/s (Figure 2.2). The time for the particle to complete one revolution is equal to $2\pi/\omega$. This is the period T of the revolution, i.e.,

$$T = 2\pi/\omega = 1/f$$

A particle traveling in a circle of radius a experiences a centrifugal acceleration given by:

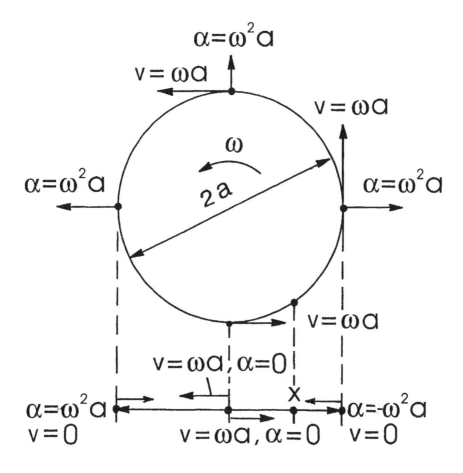

Figure 2.2 Circular motion and simple harmonic motion

$$\alpha = \omega^2 a$$

and this acceleration always points radially outward. Furthermore, the particle's velocity is constant at any moment and is given by:

$$V = \omega a$$

The direction of this velocity is tangential.

Now project the position, velocity and acceleration of the particle onto a straight line, as shown in Figure 2.2. As the particle moves around the circle, its locus moves back and forth along the straight line with a peak-to-peak amplitude of $2a$. At the extreme end points, its velocities are zero while its accelerations are maximum at $\alpha = \omega^2 a$ (and are always inward, which explains the negative sign in Equation (2.7)). At the midpoint of the straight-line path, its velocity is maximum at $V = \omega a$ while its acceleration is zero.

Furthermore, the particle repeats its path every $2\pi/\omega$ seconds, which is the period of the corresponding circular motion. The motion of the locus of the point thus has all the characteristics of simple harmonic motion as described by Equation (2.2).

It can be shown that all linear vibrations are combinations of simple harmonic motions with different periods, frequencies, amplitudes, velocities, and accelerations.

2.4 Vibration Measurement

From the previous discussion, it is seen that a simple harmonic motion is completely defined by its frequency and one of the following parameters: displacement, velocity or acceleration. Thus, the nature of vibration can be determined if the frequency and one of these parameters can be measured and recorded and the data analyzed. This is the principle behind the *accelerometer*, which is by far the most widely used vibration sensor because of its ease in application, ruggedness and economy. The accelerometer measures the instantaneous acceleration as a function of time that its internal parts experience. When the accelerometer is rigidly mounted on a surface, it detects the motion of that surface. However, since an accelerometer detects motion by virtue of movement of its internal parts, any relative movement between an accelerometer and its mounting surface will result in erroneous readings. Thus, it is important to ensure that there is a good bond between the mating surfaces of the accelerometer and the component, the vibration of which we want to measure. When the frequency is very high, e.g., in the 10 KHz range, this will require very rigid attachment methods, such as stiff epoxies, straps or welding.

In spite of its popularity, the accelerometer has its limitations. Its sensitivity decreases with frequency and this sensor is not very useful when the frequency of vibration is below a few cycles per second. In addition, to be effective, the accelerometer must be mounted directly on the vibrating surface. For this reason, it cannot be used to measure the vibration of internal parts of a component without disassembling the component first. It cannot be used to measure the vibration of very slender structures with very low natural frequencies.

Next in popularity among vibration sensors is the *strain gauge*, which again must be bonded directly onto the vibrating surface, making it also an "intrusive" vibration sensor. Furthermore, the strain gauge will not work if the vibrating component is not experiencing any strain, such as the flutter of a valve disk or any rigid-body mode vibration.

The *displacement probe*, both contacting and non-contacting types, is another commonly used vibration sensor. The displacement probe needs a sturdy mounting surface close to the vibrating surface and measures the relative motion between these two surfaces. As such, it is usually an "intrusive" vibration sensor and is more difficult to install compared with the accelerometer. However, it does work even at very low frequencies and when the vibration does not produce any strain. The non-contacting displacement transducer is commonly used to monitor the vibration of pump shafts.

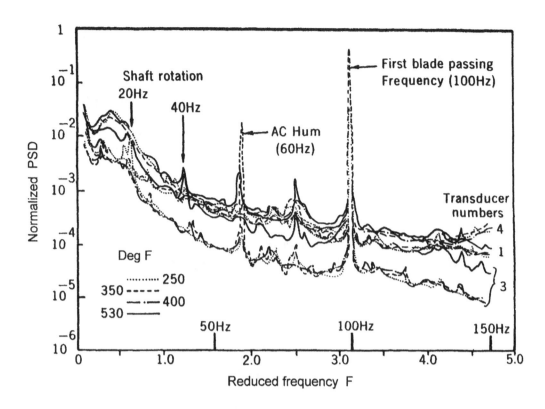

Figure 2.3 Pressure fluctuations inside a nuclear reactor measured by
a dynamics pressure transducer (Au-Yang and Jordan, 1980)

The *dynamic pressure transducer* is a very sensitive instrument used to measure pressure fluctuations in fluids. It can also be used indirectly to measure the vibrations of structures immersed in fluids, because a vibrating structure induces pressure fluctuations in the surrounding fluid. To install pressure transducers, tap holes must often be drilled through the pressure boundary, again making the dynamic pressure transducer an "intrusive" sensor. In addition, dynamics pressure transducers are extremely fragile, and deducing structural vibration amplitudes from the measured pressure fluctuation often requires sophisticated analysis. The dynamics pressure transducer can be used to accurately measure acoustic pressure pulses inside power and process plant components and in the piping systems. Figure 2.3 shows the pressure fluctuations inside a commercial nuclear reactor measured by a dynamic pressure transducer installed inside the reactor vessel. In addition to boundary layer flow turbulence, the acoustic pressure spikes induced by the coolant pump blade passing frequency (100 Hz) and it harmonics are clearly visible.

Low frequency vibrations with fairly large amplitudes can often be visually "measured" either with the naked eye or with the aid of *high-speed photography*. This technique however requires the vibrating component to be accessible and visible.

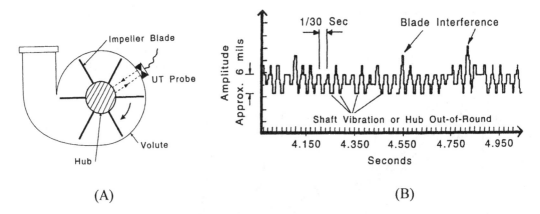

(A)   (B)

Figure 2.4 (A) Measuring the vibration of the impeller hub of a pump using
UT instrument; (B) Its vibration signature (Au-Yang, 1993)

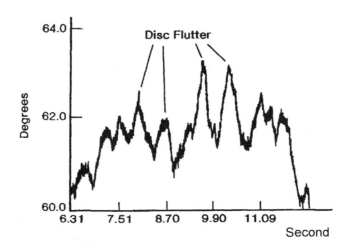

Figure 2.5 Flutter of a check valve disk measured by UT instrument
(Au-Yang, 1993)

The *ultrasonic instrument*, together with the proper signal conditioning electronics and software, presents another option to non-intrusively and quantitatively measure the vibration of internal moving parts of many power plant components. Figure 2.4(A) illustrates an application of the ultrasonic instrument (UT) to measure the vibration of the impeller hub of a small centrifugal pump. In this case, the 0.006-inch (0.15-mm) displacement shown in Figure 2.4(B) was actually caused by a slight out-of-roundness of the hub. This particular pump had no vibration problems. Figure 2.5 shows the vibration signature, as measured by an ultrasonic instrument, of the flutter of a disk inside a check valve.

2.5   Time Domain Representation of Vibration

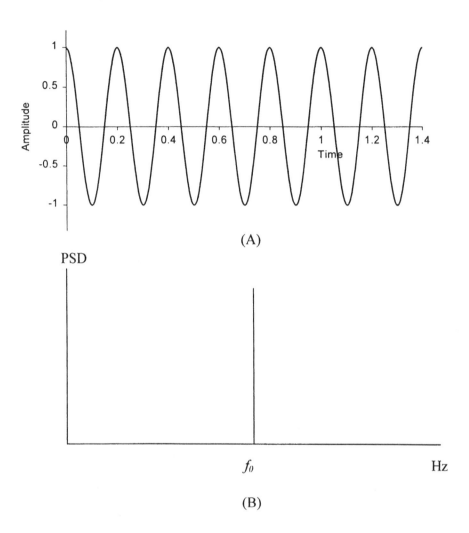

(A)

(B)

Figure 2.6 (A) Time domain representation of a sine wave;
(B) Its corresponding frequency domain representation

Figure 2.6 is a plot of displacement y versus time t from Equation (2.2). This is the simplest example of a time domain representation of vibration. Because of the wavy appearance of the plot, linear vibrations are often associated with wave motions. Figure 2.6 is an example of a sinusoidal (in this particular case a cosine) wave. In the specific example, the period of the wave is 0.2 second; its frequency is 5.0 Hz; and its zero-to-peak amplitude is 1.0 length unit. In musical terms, an acoustic wave, which consists of a single frequency like the one shown above, is called a pure tone. Musical notes are composed of a combination of many pure tones in an orderly manner; while noises are composed of a combination of many pure tones in a disorderly manner.

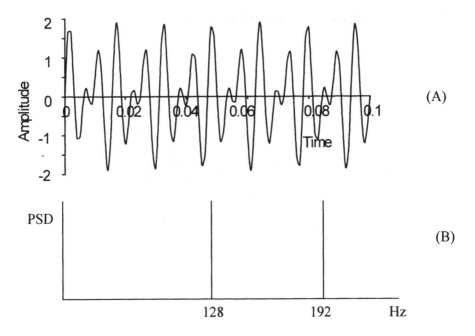

Figure 2.7 Superposition of two sine waves: (A) Time domain; (B) Frequency domain

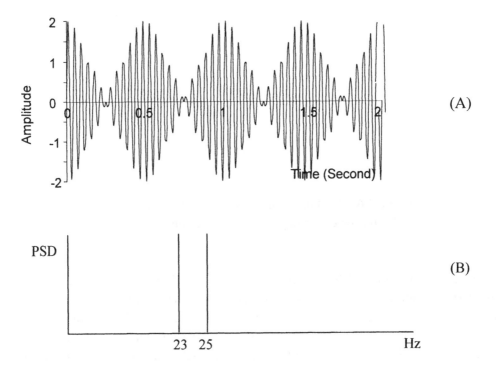

Figure 2.8 Superposition of two sine waves at 23 and 25 Hz showing:
(A) Beating in the time domain representation; (B) Frequency domain representation

2.6   Superposition of Sinusoidal Waves

Figure 2.7(A) is the time history representation of two sinusoidal waves of the same amplitude—one at 128 Hz, the other at 192 Hz. Most of the vibration phenomena, including all musical sounds and most noises, are combinations of sinusoidal waves of different amplitudes, frequencies and phases.

When two sinusoidal waves of the same amplitude and nearly the same frequencies are superimposed onto each other, a phenomenon called beating occurs. This is the familiar modulating sound one hears from a twin-engine airplane flying at a distance, and is caused by the alternate constructive and destructive interference of the sound waves. Figure 2.8(A) shows the resultant wave obtained from the superposition of two sine waves—one at 23 Hz, the other at 25 Hz—both of which have amplitudes equal to 1.0. The resultant wave is also a sinusoidal wave, with a frequency of 24 Hz (the average of the two frequencies), and with an amplitude varying between 0 (the difference between the two amplitudes) and 2.0 (the sum of the two amplitudes). This modulation of the resultant amplitude of several sound waves of nearly the same frequencies is called beating. From Figure 2.8(A), it can be seen that the beating period is 1/(25-23)= 0.5 second. However, this beating period must not be confused with one period of vibration. In the power spectral density plot shown later in this chapter, one would not see a spectral peak at this beating frequency. Instead, there will be two spectral peaks at the frequencies of the constituent waves.

So far, we are considering only sound sources located at the same point in space. In practice, the sound sources usually are distributed in space. In that case the resultant amplitude will modulate in time and also appear to precess in space. Figure 2.9 shows the computer simulation of the resultant space-time pressure distribution from four vibration point sources equally spaced around the circumference of a circle. The frequencies of these four sources are: 145, 146, 147 and 148 Hz. Note that not only does the resultant pressure distribution change with time, but also the resultant force vector appears to precess with time. This phenomenon had been observed in commercial nuclear plants, as will be shown in Case Study 2.1.

2.7   Random Vibration and Noise

We can generalize the above idea to the case of a large number of sinusoidal waves of arbitrary amplitudes, phases and frequencies grouped together. The resulting waveform corresponds to that of random vibration. Figure 2.10(A) is a typical time history representation of a random vibration. Examples of some random vibrations we encounter everyday include the vibration of a car traveling on a rough road; noise from a jet engine; buffeting of an airplane when it hits atmospheric turbulence.

A narrow band random vibration or noise is one in which the component waves all have frequencies around a certain mean value. Figure 2.10(A) is the time history plot of a narrow band random vibration process. The response of a structure to flow turbulence is an example of a narrow band random vibration.

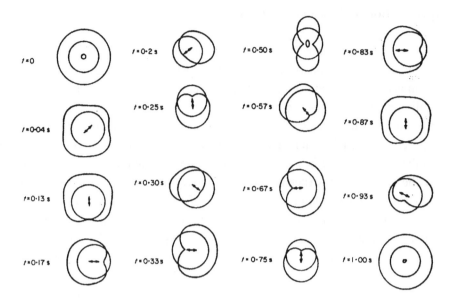

Figure 2.9 Space-time variation of resultant force due to four slightly
out-of-phase acoustic sources. (Au-Yang, 1979)

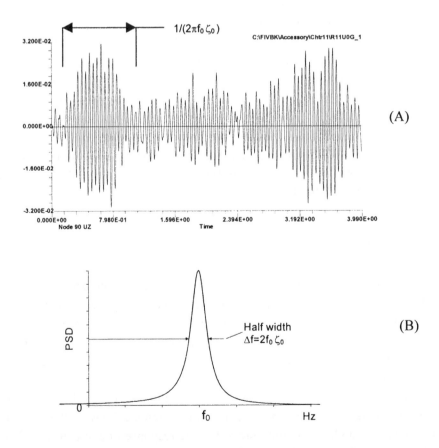

Figure 2.10 Narrow band random vibration signature: (A) Time history; (B) PSD

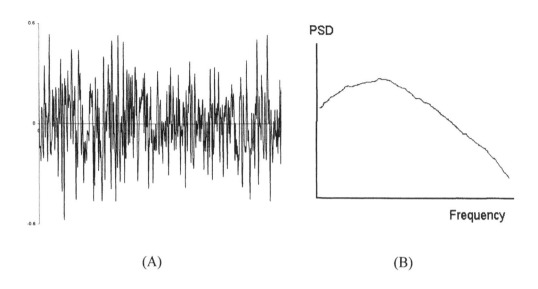

Figure 2.11 Wide band random vibration: (A) Time history; (B) PSD

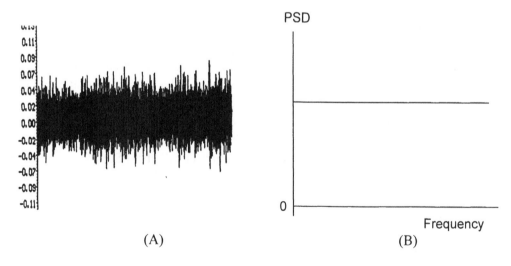

Figure 2.12 White noise: (A) Time history; (B) PSD

A wide band random vibration or noise is one in which the component waves have frequencies spreading over a wide range (Figure 2.11(A)).

When the amplitudes of the component waves are all equal and the frequencies of the components extend from zero to infinity, the resulting noise is called a white noise. Figure 2.12(A) is the time history plot of a white noise.

2.8   Frequency Domain Representation of Vibration

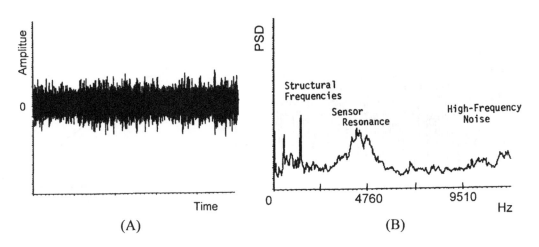

Figure 2.13 (A) Time history of data recorded by an accelerometer mounted on the wall of a check valve; (B) PSD plot of the same data showing valve body natural frequencies, sensor resonant frequencies, and flow noise. (Au-Yang, 1993-(a))

At this point, we should have noticed that except in very simple cases, time domain representation of vibration usually cannot be easily interpreted. This is especially true in the case of random vibrations. For example, take Figure 2.13(A), which is a time history plot of the accelerometer data from a flow test of a check valve. Aside from the fact that the data does not indicate any significant impacts, very little information can be derived from this plot. By a mathematical manipulation called Fourier Transform (see Chapter 13), this graph of amplitude versus time can be "transformed" into a graph of amplitude (squared) versus frequency. The corresponding graph is called a PSD (power spectral density) plot, as shown in Figure 2.13(B). The PSD plot is a graph of the energy distribution of the vibration as a function of frequency. In Figure 2.13(B), two sharp spectral peaks can be observed below 2,000 Hz. These are the structural frequencies of the valve body. A fairly broad frequency peak can be observed around 4,500 Hz. This was traced back to the resonant frequency of the mounted accelerometer at 28,500 Hz, folded back into the lower frequency because of an insufficiently high sampling rate (see Chapter 13). These spectral peaks are superimposed on top of a continuous spectrum extending from 0 Hz to the cut-off frequency of the analyzer (about 10,000 Hz). This continuous spectrum corresponds to vibrations generated by flow turbulence.

Figures 2.5(B) to 2.7(B) and 2.11(B) to 2.14(B) give the PSD plots for typical vibrations, for which we have shown the time history plots earlier. Note, in particular that in Figure 2.8(B), the PSD plots correspond to two wave motions of almost the same frequencies. As discussed before, the resultant amplitude slowly modulates between the sum and difference of the amplitudes of the constituents, and the period of modulation is the inverse of the difference of the constituent frequencies. However, this modulation, or "beating," is not a true vibration. Therefore, no spectral spikes corresponding to the modulation frequencies are observed in Figure 2.8(B). From these examples, we can see that frequency domain representation of vibration can be very useful in revealing information on the constituents of the vibration.

2.9   Traveling Waves

Because of its "wavy" time history plot, vibration motions are often called wave motions. Indeed, waves on the surface of a lake, for example, can be visualized as a kind of vibration as follows: If we choose a fixed point on the surface of the water, and plot the vertical displacement of this point from its mean position as a function of time, we obtain a plot as shown in Figure 2.14. This point on the water surface vibrates (or, as more commonly termed, oscillates,) with a definite amplitude and period. Alternatively, at any given instant of time, we can plot the vertical displacement (from the mean) of different points along a straight line on the surface of the water, and also obtain a plot like Figure 2.14. Figure 2.14 looks identical to Figure 2.6(A), except that in Figure 2.14, the abscissa is a distance (y) rather than time. The curve repeats itself along the y-axis with a definite period λ, which is now a length instead of a time period. This is the wavelength of the periodic motion, and is equal to the distance the wave travels in one period. Returning to the example of surface waves in a lake, we notice that the waves move with a definite speed. The speed of wave motion is given by the product of the frequency f and the wavelength:

$$c = f\lambda \tag{2.9}$$

One can visualize this as similar to a person walking in a straight line. The speed he walks is equal to the product of the number of steps he takes in a given time (frequency), and the length of each step (wavelength).

Return again to the above example of surface wave motion. If we focus on a point on the water surface, we note that this point moves up and down in the vertical direction (the actual motion of the fluid particle is circular; with the plane of the circle in the direction of wave motion). The direction of propagation of the waves, however, is horizontal. This is an example of a transverse wave. In a transverse wave, the direction of particle motion is perpendicular to the direction of wave propagation.

Now consider a row of ball bearings connected by springs (Figure 2.15), initially at rest. If we displace one of the ball bearings from its equilibrium position slightly in the direction of the line joining the bearings, the disturbance will propagate down the line at a definite speed. The motion of this disturbance is again a wave motion. However, contrary to surface waves, the ball bearings move in line with the direction of propagation of the disturbance. This is an example of a longitudinal or compressive, wave.

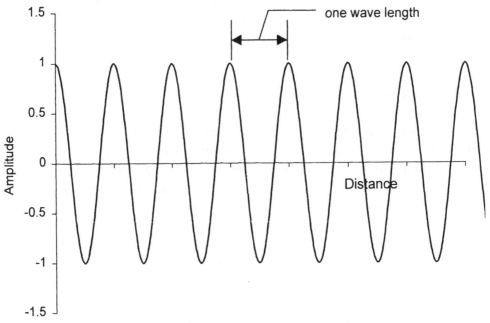

Figure 2.14   Transverse wave motion

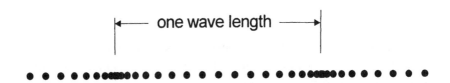

Figure 2.15 Longitudinal wave motion

2.10   Propagation of Sound Waves

Sound and stress waves travel through material media in several waveforms, of which the most important ones are, longitudinal and transversal waves. The latter is more commonly called shear waves in ultrasonic technology. However, bending waves are also a special type of transversal wave which is particularly important in vibration monitoring using accelerometers. Since liquids and gases cannot sustain shear forces, sound waves can propagate through these media only in the form of longitudinal waves. Sound waves can travel through solids in other waveforms, including shear, Rayleigh or Lamb wave. These are important in the application of ultrasonics to flaw detection. For application to vibration monitoring of plant components, the important waveforms are longitudinal, shear and bending waves.

Table 2.1:   Speed of Sound in Water at Various Temperature/Pressure
(Derived from ASME Steam Table, 1979; see Chapter 12)

Pressure		Temperature		Speed of Sound	
psi	PA	Deg. F	Deg. C	ft/s	m/s
15	1.034E+05	80	26.7	4907	1496
		200	93.3	5055	1541
100	6.894E+05	80	26.7	4911	1497
		200	93.3	5060	1542
		300	148.9	4771	1454
500	3.447E+06	80	26.7	4932	1503
		200	93.3	5083	1549
		300	148.9	4801	1463
		400	204.4	4288	1307
1000	6.894E+06	80	26.7	4957	1511
		200	93.3	5111	1558
		300	148.9	4837	1474
		400	204.4	4338	1322
		500	260.0	3604	1099
2000	1.379E+07	80	26.7	5004	1525
		200	93.3	5164	1574
		300	148.9	4907	1496
		400	204.4	4433	1351
		500	260.0	3746	1142
		600	315.6	2766	843

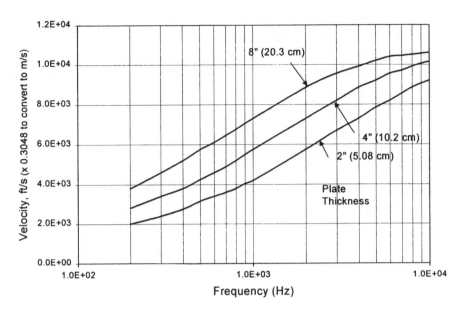

Figure 2.16   Speed of bending waves as a function of pipe wall thickness and frequencies, based on data from Mayo, et al. (Copyright © 1988, EPRI report NP-5743, Electric Power Research Institute, reproduced with permission.)

Sound travels in pipes, plates and shells in longitudinal, shear and bending waveforms. Extensive research by the Electric Power Research Institute (EPRI) in connection with loose part monitoring showed that an accelerometer mounted on the surface of pipe walls, plates and large shell structures responds mainly to the bending waves (Mayo, et al, 1988). The speed of sound wave propagation depends on the material media as well as their ambient conditions. In air at 70°F and one standard atmospheric pressure, the speed is about 1,100 ft/s (335 m/s). In water at room temperature, the speed is about 4,900 ft/s (1,493 m/s). However, at higher temperatures and pressures encountered in power plants, the speed of sound may be less, sometimes less than 3,000 ft/s (914 m/s). Table 2.1 gives the speed of sound in water at several different temperatures and pressures. These values apply to both acoustic and ultrasonic waves.

The speed of sound is generally higher in solids and is different for different modes of propagation. In steel, the speed of sound is about 20,000 ft/s (6,100 m/s) for the longitudinal wave (called L-wave in ultrasonics), and 11,300 ft/s (3,444 m/s) for the shear wave (called S-wave in ultrasonics). Unlike longitudinal and shear waves, which travel at constant speeds irrespective of their frequencies, the speed of bending waves depends on the frequency of the wave, as well as on the thickness of the material (plate, pipe or shell wall thickness) along which the wave travels. Figure 2.16, re-plotted based on data from the above-referenced EPRI report, shows the speed of bending waves in steel as a function of frequency for several material thicknesses commonly encountered in nuclear plants. This information is useful for estimating the distance between the sound source and the accelerometer.

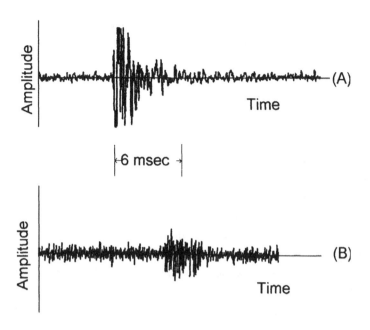

Figure 2.17 (A) A near-field impact wave form; (B) A far-field impact wave form
(Au-Yang, 1993(a))

Because bending waves of different frequencies travel at different speeds, a wave
"packet" such as that generated by an impact, (Figure 2.17), which contains sound waves
of different frequencies, will disperse as it propagates.

The finite and different propagation speeds of sound waves through different material
media, as well as the frequency dependence of the speed of propagation especially in
bending waves, have an effect on the time domain waveforms. Figure 2.17(A) shows a
cavitation wave packet detected by an accelerometer mounted on a valve. The duration
of the event was about 6 ms (milli-second), which is typical of a cavitation.   The
cavitation originated inside the component close to the accelerometer. The resulting
acoustic wave reached the sensor via different paths - some through water, some through
steel in shear, compressive, and bending waveforms, all with difference frequencies. As
a result, these waves reached the accelerometer at different times; thus the accelerometer
received a collection of waves. However, as long as the origin of the cavitation was close
to the sensor, the difference in arrival time was small compared with 6 ms, and the
waveform was not distorted.   Thus, it still had the classic sharp rise, slow decay
waveform of an impact. Figure 2.17(B) shows another cavitation wave packet. Unlike
the one shown in Figure 2.17(A), this wave packet was distorted. Although the overall
duration was still about 6 ms, typical of a cavitation, its wave front was dispersed and it

did not have the sharp rise characteristic of an impact. This cavitation originated some distance away from the accelerometer. The sound waves reached the sensor along different paths. The difference in arrival times in this case was significant compared with the overall duration of the event (6 ms). Thus, the waveform was smeared.

Example 2.1: Tapping Wave Forms From Two Piston Lift Check Valves

Figure 2.18 (A) and (B) show the acceleration time histories of noises picked up by two accelerometers, one mounted on each of two piston lift check valves on two parallel pipe lines about 10 feet (3 meters) apart. The amplitudes and duration of each wave packet is consistent with that caused by valve internal parts hitting the valve body (Au-Yang, 1993). Their regular periods indicate that the tapping was most likely caused by some kind of instability (in this case, leakage flow-induced instability; see Chapter 10). At first it appeared as if both valves were defective and needed to be replaced. More careful examination of the wave forms (Figure 2.19), however, shows that only the signal from the "A" valve has the classic sharp rise, slow exponential decay wave form characteristic of near-field impact waves. The acceleration time history from valve "B," on the other hand, is smeared. Furthermore, it appears that the wave packets from valve B lag behind those from valve A by a small, but finite amount of time. Valve B was actually not defective. The accelerometer on this valve picked up the vibration generated in valve A. These waves had traveled through the pipeline and arrived at the accelerometer on valve B a fraction of a second later. The wave fronts were smeared due to the slight differences in the velocity of propagation among the different frequency components.

2.11   Energy in Sound Waves

Stress and sound waves traveling in solid or fluid media carry energy with them. In the following discussions we shall concentrate on the energy in sound waves in water and in air. As discussed earlier, sound waves travel in fluids with alternate compression and rarefaction both in space and in time. It is this pressure pulsation that excites the ear diaphragm and causes the sensation of sound. In the extreme case, these pressure pulses can cause structural damages. The amount of energy sound waves carry is expressed in terms of energy flux, or power per unit area the wave front is carrying with it. This energy flux is more commonly called sound intensity, or sound pressure level (SPL), in acoustics

$$SPL = p_{rms} c \qquad\qquad (2.10)$$

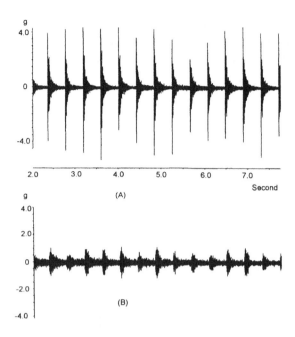

Figure 2.18 Acceleration time histories picked up by two accelerometers
mounted on two piston lift check valves on two parallel pipe lines

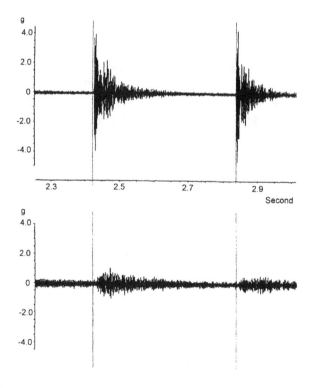

Figure 2.19 "Zoom-in" of the two acceleration traces reveals smearing of
wave front in trace B and difference in arrival times

and has the unit of power per unit area (such as watt/cm^2). Thus, SPL is a measurement of sound intensity, not the pressure pulsation the square of which the sound intensity is proportional to. For the same sound intensity, the pressure pulsation depends on the fluid media. For air, it can be shown that

$$\text{Energy flux} = I = p^2/412 \ \text{ microwatts/cm}^2 \ \text{ in air}$$

$$= p^2/1470000 \ \text{ microwatts/cm}^2 \text{ in water}$$

where p is the rms pressure fluctuation expressed in dyne per sq. cm. For the same sound intensity or energy flux, the corresponding rms pressure fluctuation in water is 60 times larger than that in air. This is why a depth charge exploding under water can cause much bigger damage to marine structures than a bomb exploding in air. Likewise, acoustic waves in piping systems containing water can cause much more damage than in air ducts.

2.12 Threshold of Hearing and Threshold of Pain

Since it is the energy carried in a sound wave that excites our eardrums which in turn causes the sensation of sound, it is reasonable to ask what is the minimum sound intensity that the average "normal" ear can hear. This was a subject of intensive research 120 years or so ago. The answer depends not only on the person, his age and sex, but also on the frequency of the sound. Most human ears are most sensitive to sound waves of about 3,500 Hz. For the sake of establishing a reference standard, it was agreed that the threshold of hearing corresponds to an energy flux of 10^{-16} watt per sq. cm.

Naturally, the next question is what is the maximum sound intensity the normal human ear can tolerate without feeling pain. The answer again depends on as many parameters as the threshold of hearing does. Very approximately, it is 0.1 watt per sq. cm, or a factor of 10^{15} times that at the threshold of hearing.

The rms vibration amplitude of the ear diaphragm at the threshold of hearing can be computed, and was shown to be less than 10^{-8} cm or 4.0 millionth of a mil. This is less than the diameter of a hydrogen molecule. If the ear diaphragm responds linearly to the sound pressure, then at the threshold of pain, the diaphragm would vibrate with a rms amplitude of (remember energy is proportional to the square of amplitude) about 1.0 mm, or about 40 mils. Obviously, the ear diaphragm would not vibrate with such a large amplitude.

2.13 The Logarithmic Scale of Sound Intensity Measurement—The Decibel

The human ear is an extremely sensitive and yet rugged vibration sensor, with a dynamic range far exceeding any instrument ever developed by man. The fact is that the human ear diaphragm does not respond linearly to sound pressure. Instead, it responds more like logarithmically. That is, with every ten-fold increase in the sound intensity, the human ear perceives about a doubling in loudness. Every 100 times increase in the sound intensity, the human ear perceives about 4 times increase in loudness. Because of this and because of the tremendously wide range between the threshold of hearing and the threshold of pain in the human ear, scientists working on the theory of sound introduced, about 100 years ago, the logarithmic scale of measuring sound intensity. One sound intensity I_1 is said to be one *bel* above that of the other I_2 if,

$$\log_{10}(I_1 / I_2) = 1.0$$

where I_1, I_2 are the energy fluxes (intensities) of the two sound waves. In other words, the energies carried in two sound waves are 1.0 bel apart if one is 10 times that of the other. One *decibel (dB)*, naturally enough, is one-tenth of a bel. Thus, I_1 is 10 decibel above I_2 if,

$$\log_{10}(I_1 / I_2) = 1.0$$

As time went on, the unit bel virtually went out of existence. The term 10 decibel is used instead of 1.0 bel. It should be realized that decibel is strictly a relative unit. To use it properly, one should mention what is the reference. It is meaningless to say the electric signal strength is 50 decibel. However, a signal-to-noise ratio of 50 decibel means the energy carried by the electric signal is 100,000 times the energy contained in the background noise ($10\log_{10}(100000)=50$). In acoustics, the reference is taken as the sound intensity at the threshold of audibility, that is, 10^{-16} watt per sq.cm. At the threshold of pain, the sound intensity is 150 dB, corresponding to 0.1 watt/cm^2. In general,

0 dB above   = At the same level
1.0 dB above = 1.26 times
2.0 dB above = 1.59 times
3.0 dB above = 2.00 times
6.0 dB above = 3.98 times
10. dB above = 10.0 times
12. dB above = 10*1.59 = 15.9 times

Table 2.2 gives a very rough estimate of the decibel levels under commonly encountered environment. People exposed to a sustained sound level of 105 dB or above should have hearing protection.

To be specific, the subscript f is sometimes added to dB when we talk about loudness levels, so that it is understood that 0 dB$_f$ refers to a sound energy flux of 1.0^{-16} watt per sq.cm.

Since the power carried in sound waves is proportional to the mean square of pressure fluctuation (and proportional to the square of particle displacement), we have

dB in power = 2*dB in rms pressure = 2*dB in displacement

Table 2.2: Decibel Levels in Common Sound Environment
(Source: Deafness Research Foundation)

Environment	Sound Energy (dB)
Rocket launching pad	180
Threshold of Pain	140
Between reactor coolant pumps	130
Rock concert hall at the loudest	120
Pneumatic drill, garbage truck	100
Lawn mover, subway, motor cycle	90
City traffic, average	80
Noisy restaurant	70
Heated conversation	60
Normal conversation	50
Background noise, inside living room	40
Background noise, quiet library	30
Background noise, country winter night	20
Threshold of hearing	$0 = 10^{-16}$ watt/cm^2

2.14   The Decibel Used in Other Disciplines

Since the decibel scale is just a logarithmic scale to present data that covers a wide range, it was quickly adopted by other technical disciplines such as electrical engineering. Very often, the term dB is used without giving a reference. In the late seventies, it was very common to rate the power output of home stereo amplifiers in terms of dBW, without mentioning what the reference is. In this case, the reference power level was arbitrarily chosen as 1.0 watt rms so that 20 dBW means 100 watt rms while 17.75 dBW means 60 watts rms. This convention quickly died because of lack of popular support.

To make things even more confusing, engineers started using the term dB as amplitude scales such as pressure pulsation, displacement, acceleration (g-level), current or voltage. Thus the g-level in one vibration measurement is 10 dB below that of the other means that the acceleration in one measurement is 10 times smaller than that in another measurement. The current in a circuit is 6 dB more means the current output is four times more. Since power and energy generally vary as the *square* as amplitude (acceleration, displacement, pressure fluctuation, current etc.), we must realize the relationship:

dB in power or energy = 2*dB in amplitude

If the current in the circuit is increased by 3 dB (two times), its power will increase by 6 dB (four times).

In general, it is poor engineering practice to express test data in dB, especially if the reference 0 dB is not explicitly stated. The exception is signal-to-noise ratio, which is by definition a ratio; and in acoustics, for which the dB was originally defined.

2.15   Case Studies

In this section, case studies are presented to show that many of the vibration and noise problems in power and process plants can be diagnosed with just some basic knowledge of the kinematics of vibration and acoustics. More advanced case studies in noise and acoustic problems in power and process plants are given in Chapter 12.

Case Study 2.1:   Forced Vibration of Nuclear Reactor Components by Coolant Pump-Generated Acoustic Load

In a nuclear steam supply system (Figure 2.20), the coolant is circulated by two to four pumps. The coolant enters the nuclear reactor downcomer annulus between the reactor vessel and the core support structure. The inlets of these coolant pumps are distributed around the circumference of the reactor vessel. Acoustic pressure pulses generated by the coolant pump blades propagate along the inlet coolant pipes into the downcomer cavity between the reactor vessel and the core support structure. The frequencies of these acoustic pulses are equal to the coolant pump blade passing frequencies and its higher harmonics. Typical examples are 100 Hz, 200 Hz, 300 Hz etc. (Figure 2.3), with the fundamental and its first harmonic being the most prominent. In an idealized situation, these coolant pumps would have precisely the same rotational speed and the pressure pulses always would be in phase. Under these conditions, these pressure pulses would oppose each other all the time, and there would be no resultant net force on the core support structure. Unfortunately, in practice, due to minute differences in the loading conditions on the pumps, the pump speeds are not always identical. The pump-induced acoustic pressures that are perfectly in phase and balance each other at a certain instant in time, will gradually be out of phase, resulting in a net force vector that slowly precesses and modulates in time, as illustrated in Figure 2.9. This pressure pulses force the reactor core barrel to precess. This type of coolant pump-induced forced vibration motion of a nuclear reactor core barrel was detected by an external monitoring instrument. Figure (2.21), reproduced from Wach and Sunder (1977) shows distinct precessional core motion inside the nuclear reactor vessel. Since the instrumentation used to detect this motion, the ex-core neutron noise detector, could not be used to quantitatively determine the vibration amplitudes of the core, this observed core vibration raised some concern in the early to mid-80s, after some thermal shield restraining bolts were found to be broken in several nuclear plants designed by three different vendors in the United States (Sweeny and Fry, 1986). Extensive analyses were carried out to find the root cause of these bolt

.....

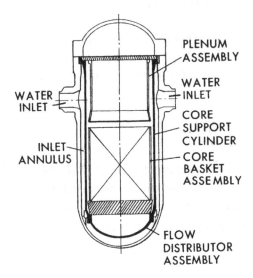

Figure 2.20   Schematic drawing of a commercial pressurized water nuclear reactor showing the narrow water annulus between the core and the outer reactor vessel. (Water driven by external pumps flows down this annulus, turns around at the bottom of the vessel and flow up through the core.)

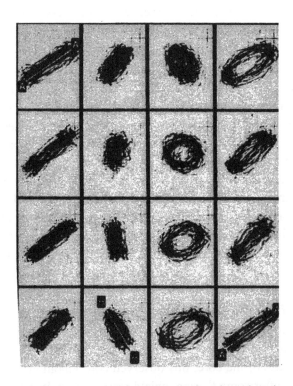

Figure 2.21 Lissajous figure of ex-core neutron noise data showing motion of reactor core in a multi-loop plant (Wach and Sunder, 1977)

failures. After turbulence-induced vibration was ruled out as a probable cause (see Chapters 8 and 9), the attention turned to coolant pump-induced acoustic pulses. After extensive analysis, however, it was found that the core vibration amplitudes caused by these coolant pump induced acoustic pulses, although detectable by the extremely sensitive ex-core neutron noise detector, were far too small to have caused the bolts to fatigue if the material of the bolts had not deteriorated.

By a process of elimination, material degradation was found to be the root cause of these broken thermal shield restraining bolts. They were subsequently replaced with bolts of different kinds of material. There has been no repeated problem after another 15 years of operations.

The two jet pumps in some boiling water reactors also have been found to cause alternate constructive and destructive acoustic pulses that propagate throughout the nuclear reactor. On one occasion these pulses forced the instrument guide tube to vibrate; the motion was subsequently picked up by the loose part monitoring accelerometers (Profitt and Higgins, 1993) and had caused quite a false alarm.

In any power and process plant with multiple pumps operating, this kind of modulating acoustic pulses will propagate through the coolant conduits connecting to these pumps. The components must be designed against these combined acoustic loads. On the other hand, components will not resonate with the beating frequency because, as explained in Section 2.6, the beating frequency is not a real vibration frequency. Still, the components must be designed to withstand the combined acoustic loads even if these pumps are placed opposite to one another.

Case Study 2.2: Detecting Internal Leaks in a Nuclear Plant
(Price and Au-Yang, 1996)

In February 1995, coolant leakage into the quench tank in a nuclear plant was becoming an operational concern. Although it posed no safety concern, this internal leak resulted in measurable increases in both the temperature and level of the quench tank water, and was so severe that, if the trend continued, plant shutdown would be necessary. Preliminary diagnosis based on in-plant instrumentation indicated that any one of 11 valves might be leaking into the quench tank. A decision was made to pinpoint the source of the leaking valve by acoustic signature monitoring using sensitive accelerometers non-intrusively mounted on the outside of these valves. Because of the nature of the problem, data must be collected inside the containment building under very hostile ambient conditions, with the reactor operating at full power. Data collection and in-situ analysis must be rapid to minimize radiation exposure of the personnel. After about one day of testing, the leakage was pinpointed to two of the three reactor shutdown coolant relief valves.

As shown in Figure 2.22(A), there seemed to be an intermittent flow through this valve consistent with the valve plug being lifted off the seat momentarily and then re-seated. The noise level was highest at the seat and apparently originated there. This assessment was confirmed by zooming in around $t=5.0$ seconds and using the cursor on the computer monitor to determine the times of leakage initiation at the three sensor locations.

Figure 2.22(B) is the PSD corresponding to the quiet portion of the time history (2.5 to 6.0 seconds) in Figure 2.22(A). Apart from the broad peak around 4,500 Hz, which was observed in the upstream sensor in several other valves tested and which was not related to seat leakage, there were no high-frequency activities in any of the sensor locations. This confirmed that from 2.5 to 6.0 seconds, the valve was properly seated and there was no through-seat leakage. Figure 2.22(C) is the PSD corresponding to the noisy part of the time history (12.5 to 16.5 second) in figure 2.22(A). High-frequency energy content can be seen at the sensors mounted both near the seat and downstream of the valve. A minute amount of steam-water two-phase mixture had excited the acoustic modes downstream of the valve.

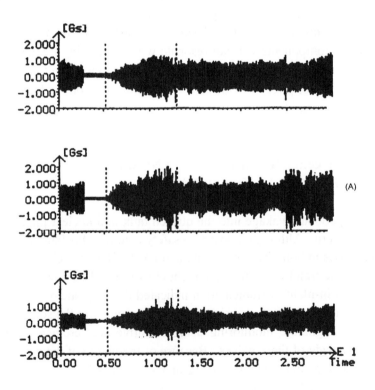

Figure 2.22 (A) Time histories from accelerometers mounted externally on the high-pressure side (top), near the seat (middle) and on the low-pressure side (bottom) of a relief valve showing intermittent leak

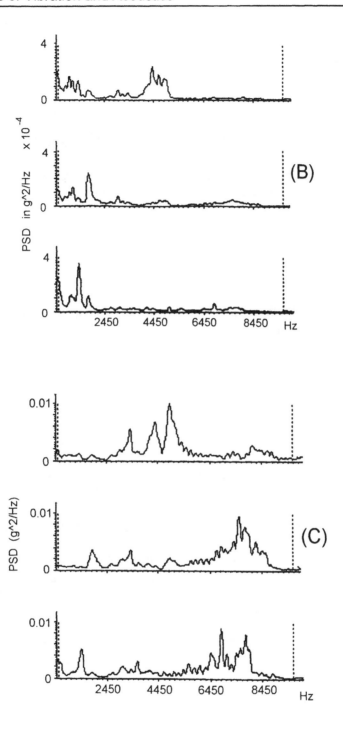

Figure 2.22 (B) PSD plot corresponding to the no-leak portion of the time history showing little high frequency noise picked up by the low pressure side accelerometer; (C) PSD plot corresponding to the high noise portion of the time history showing high frequency energy content picked up by the low pressure accelerometer. (The low-pressure side acoustic modes were excited by the leakage flow through the seat.)

Near Seat

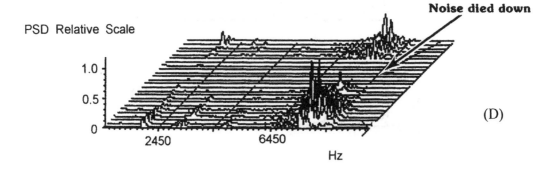

Figure 2.22 (D) Waterfall plots showing noise stopped then gradually
reappeared as the leak resumed (Price and Au-Yang, 1996)

References

ASME, 1979, Steam Tables, Fourth Edition.

ASME Operations and Maintenance Guide Part 23, 2000 Addenda, "In-service Monitoring of Reactor Internal Vibration in PWR Plants."

Au-Yang, M. K., 1979, "Pump-Induced Acoustic Pressure Distribution in an Annular Cavity Bounded by Rigid Walls," Journal of Sound & Vibration, Vol. 62, pp. 557-591.

Au-Yang, M. K.; Jordan, K. B., 1980, "Dynamic Pressure Inside a PWR — A Study Based on Laboratory and Field Test Data," Journal of Nuclear Engineering and Design, Vol. 58, pp. 113-125.

Au-Yang, M. K., 1993(a), "Acoustic and Ultrasonic Signals as Diagnostic Tools for Check Valves," ASME Transaction, Journal of Pressure Vessel Technology, Vol. 115, pp. 135-141.

Au-Yang, M. K., 1993(b), "Application of Ultrasonics to Non-Intrusive Vibration Measurement," ASME Transaction, Journal of Pressure Vessel Technology, Vol. 115, pp. 415-419.

Mayo, C. W., et al, 1988, Loose Part Monitoring System Improvements, EPRI Report NP-5743.

Price, J. E. and Au-Yang, M. K., 1996, "Quench Tank In-Leakage Diagnosis at St. Lucie", Paper presented at the Fourth ASME/NRC Symposium on Valve and Pump Testing, Washington D.C.

Proffitt, R. T. and Higgins, A. F., 1993, "Pump Beating and Loose Part Interaction," Paper presented at the 39th Annual Reactor Noise Technical Meeting and Exposition, Las Vegas.

Sweeney, F. J. and Fry; D. N.; "Thermal Shield Support Degradation in Pressurized Water Reactors," Proceeding of the 1986 PVP Conference, Flow-Induced Vibration, edited by S. S. Chen, J. C. Simonis and Y. S. Shin, PVP Vol. 104, pp. 59-66.

Wach, D. and R. Sunder, R., 1977, "Improved PWR Neutron Noise Interpretation Based on Detailed Vibration Analysis," Paper presented at the Second Specialists' Meeting on Reactor Noise, Tennessee.

CHAPTER 3

FUNDAMENTALS OF STRUCTURAL DYNAMICS

Summary

In the presence of damping but without external forces, the solution to the linear vibration problem takes the form (compare with free vibration in Chapter 2)

$$y = a_0 e^{i\Omega t} e^{-(c/2m)t} \tag{3.2}$$

where Ω, the natural frequency of the damped point-mass spring system, is given by:

$$\Omega = \sqrt{\omega_0^2 - \frac{c^2}{4m^2}} \tag{3.3}$$

When the damping coefficient $c > 2m\omega_0$, energy dissipation is so large that vibration motion cannot be sustained. The motion becomes that of the exponentially decay type instead of vibration. The value of c at which this happens ($c_c = 2m\omega_0$) is called the critical damping. The more commonly used damping ratio is the damping coefficient expressed as a fraction of its value at critical damping

$$\zeta = c/c_c, \quad c_c = 2m\omega_0 \tag{3.4}$$

In terms of the frequency (in Hz) and the damping ratio, the damped natural frequency is given by:

$$f_0' = f_0\sqrt{1-\zeta^2}$$

When the damping ratio is small, the damped natural frequencies are about the same as the un-damped natural frequencies. Under this condition, the damping ratio can be measured in the time domain by the logarithmic decrement method,

$$\zeta = \frac{1}{2\pi n}\log_e(\frac{y_i}{y_{i+n}}) \tag{3.5}$$

where n, i denote n cycles after the ith cycle. Alternatively, the damping ratio can be measured in the frequency domain by the half power point method,

$$\zeta_n = \frac{\Delta f}{2f_n} \tag{3.13}$$

Because the widths of the spectral peaks Δf in power spectral density (PSD, see Chapters 2 and 13) plots are proportional to the damping ratio of the vibrating system, acoustic and electrical signals will appear in PSD plots as spikes with very small or no widths.

An impact is an example of a transient vibration caused by a very short impulsive forcing function. In the time domain, metal-to-metal impact waveforms all have a characteristic sharp rise and slow, exponentially decaying waveform. This is very useful in identifying the presence of loose parts, rattling and cavitation inside plant components.

When the amplitudes are small, vibration problems in continuous structures such as beams, plates and shells can be visualized as infinite assemblies of point mass-spring systems and can be solved by the normal mode analysis method, with the equation of motion for each mode similar to that of a spring-mass system. The total vibration amplitude is obtained by summation over all the important modal contributions,

$$\{y\} = \{a(t)\}^T \{\psi(x)\} = \sum_n a_n(t)\psi_n(x) \tag{3.15}$$

$$a_n(t) = \frac{F_n(t)}{m_n(2\pi f_n)^2[1 - (f/f_n)^2 + i2\zeta_n(f/f_n)]} \tag{3.26}$$

where $a_n(t)$ is the amplitude function and,

$F_n = \int_\ell \psi_n(x)F(x)dx$ is the generalized force

$m_n = \int_\ell \psi_n(x)m(x)\psi_n(x)dx$ is the generalized mass.

$a_n(t)$, F_n and m_n are *not* physically observable quantities. Their values depend on the way the mode shapes are normalized. On the other hand, $y(x)$, being the response at location x, is a physically observable quantity. Its computed value should not depend on the mode shape normalization. The normalization:

$$\int_\ell \psi_m(x)\psi_n(x)dx = \delta_{mn} \tag{3.30}$$

commonly used in applied mathematics and physics is possible only for structures with uniform mass densities. In general, structural mode shape functions are orthogonal only with respect to the mass matrix. The most commonly used mode shape normalization convention in finite-element structural analysis computer programs is therefore:

$$\int_{\ell} \psi_m(x)m(x)\psi_n(x)dx = \delta_{mn} \qquad (3.33)$$

That is, with the generalized mass normalized to unity. In dynamic analyses, great care must be given to the differences in mode shape normalization conventions used in different parts of the calculation.

The amplitude function in Equation (3.26) is often written in the following alternate forms:

$$|a_n(t)|^2 = |H_n(f)|^2 |F_n|^2$$

$$|H_n(f)|^2 = \frac{1}{m_n^2(2\pi f_n)^4\{[1-(f/f_n)^2]^2 + 4\zeta_n^2(f/f_n)^2\}} \qquad (3.27)$$

$H_n(f)$ is called the transfer function and is a commonly used function in vibration measurement. It is also used in the development of turbulence-induced vibration theory. Alternatively, the amplitude function can also be expressed as

$$|a_n(t)| = \frac{A_n(f)|F_n|}{m_n(2\pi f_n)^2}$$

$$A_n(f) = \frac{1}{\{[1-(f/f_n)^2]^2 + 4\zeta_n^2(f/f_n)^2\}^{1/2}} \qquad (3.28)$$

A_n is called the dynamic amplification factor and is useful in estimating the responses of beams using the equivalent static load method.

At resonance, the power dissipated by a vibrating structure is given by:

$$P = 8\pi^3 \zeta_n f_n^3 m_n (y_{0i}^2/\psi_{ni}^2) \qquad (3.40)$$

where (y_{0i}^2/ψ_{ni}^2) is the square of the ratio of the 0-peak vibration amplitude to the mode shape value of the resonant mode at the same point i on the structure. This ratio is, however, a constant and is independent of the particular point selected.

Nomenclature

$a_n(t)$ $= a_0 e^{i\omega t}$, amplitude function of vibration

A_n Dynamic amplification factor

c Damping coefficient

$[c]$	Damping matrix
$[C]$	$= \{\psi\}^T [c]\{\psi\}$, generalized damping matrix
E	Young's modulus
f	Frequency, in Hz
f_0	Un-damped natural frequency, in Hz
f_n	Un-damped natural frequency of mode n, in Hz
f'	Damped natural frequency, in Hz
Δf	Half power point width, in Hz
F	Externally applied force
F_n	$= \int_\ell \psi_n(x) F(x) dx$, generalized force
H_n	Transfer function for mode n
$[H]$	Transfer matrix
i	$= \sqrt{-1}$
I	Moment of inertia
$[k]$	= Stiffness matrix
k_n	$= \int_\ell \psi_n(x) k \psi_n(x) dx$, generalized stiffness for mode n
$[K]$	$= \{\psi\}^T [k]\{\psi\}$, generalized stiffness matrix
L	Length of 1D structure (beam, tubes, etc.)
ℓ	Domain of integration. Over length of 1D structures, surface of 2D structures
m	Linear mass density for 1D structures, surface mass density for 2D structures
m_n	$= \int_\ell \psi_n(x) m(x) \psi_n(x) dx$, generalized mass for the nth mode
$[m]$	Mass matrix
$\overline{M}$	Bending moment
$[M]$	$= \{\psi\}^T [m]\{\psi\}$, matrix of generalized mass
P	Power dissipated by damping
t	Time
T	Period of vibration
w	Equivalent static load per unit length
x	Location on structure
y	Vibration amplitude
δ	Logarithmic decrement
ψ	Mode shape function
$\{\psi\}$	Column vector of mode shape function
σ	Stress
ω	$= 2\pi f$, frequency, in radian/s
ω_0	Natural frequency, in radian/s
Ω	Damped natural frequency, in rad/s

{ }   Column vector
{ }^{T} Row vector

Subscript and Superscripts

$\dot{y} \equiv \partial y / \partial t$

$\ddot{y} \equiv \partial^2 y / \partial t^2$

m, n Modal indices
T Transpose of matrix

3.1   The Equation of Motion with Damping but no External Force

In Chapter 2 we discuss vibrations in which the amplitudes remain constant. In practice, unless there is an external force applied to the mass, its motion will eventually stop due to dissipation of energy through frictional resistance. This energy dissipation mechanism is called damping and is the principal on which the automobile shock absorber works. The equation of motion for a point-mass-spring system with viscous damping but without external forces is written as:

$$m\ddot{y} + c\dot{y} + ky = 0 \tag{3.1}$$

The solution to the above equation consists of two factors: A sinusoidal function just as in the case of free vibration, Equation (2.3) and an exponentially decaying function:

$$y = a_0 e^{i\Omega t} e^{-(c/2m)t} \tag{3.2}$$

where Ω, the natural frequency of the damped point-mass-spring system, is given by

$$\Omega = \sqrt{\omega_0^2 - \frac{c^2}{4m^2}} \tag{3.3}$$

The natural frequency of the damped mass-spring system is smaller than that of the un-damped system. When the damping coefficient c is so large that $(c/2m)$ is larger than ω_0, the quantity inside the square root becomes negative. Cyclic vibrational motion can no longer exist. Instead, the motion is that of the exponentially decaying type. Because of the large energy dissipation, the point-mass cannot maintain its oscillatory motion. Instead, it just "creeps" back to its equilibrium position. The value of c at which the motion changes from oscillatory to creeping is the critical damping coefficient, denoted as c_c. Systems with damping coefficients less than, equal to or larger than c_c is called under-damped, critically damped and overdamped. Figure 3.1 shows the response time history of an under-damped, an overdamped and a critically damped system. A critically damped system will come to its equilibrium position in the shortest amount of time. The

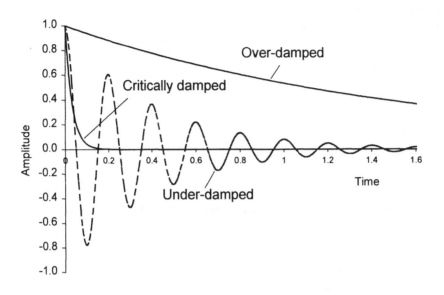

Figure 3.1 Under-damped, over-damped and critically damped vibration

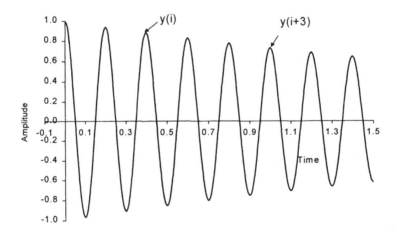

Figure 3.2 Logarithmic decrement of damped vibration in the
absence of external force

commonly referred to damping ratio is the system damping coefficient expressed as a
fraction of its value at critical damping,

$$\zeta = c / c_c, \quad c_c = 2m\omega_0 \tag{3.4}$$

A system with a damping ratio of 100 percent is critically damped and cannot undergo cyclic motion.

When the damping ratio is small, one can show from Equations (3.2), (3.3) and (3.4) that (see also Figure 3.2),

$$\zeta = \frac{1}{2\pi n} \log_e (\frac{y_i}{y_{i+n}}) \tag{3.5}$$

The quantity

$$\delta = \frac{1}{n} \log_e (\frac{y_i}{y_{i+n}}) \tag{3.6}$$

is often referred to as the logarithmic decrement of the damped vibration. Equation (3.5) forms the basis of one of the most reliable and commonly used method to determine the damping ratios of structures. It shows that in the absence of external forces, the vibration amplitude will exponentially decay with the rate of decay governed by the damping *ratio* (and *not* directly by the rate of energy dissipation, which is expressed through c).

3.2   Forced-Damped Vibration and Resonance

In practice all free vibrations will eventually stop due to damping. What maintains a steady-state vibration is an externally applied force. When the energy of the externally applied force just balances the energy dissipated by damping, steady-state vibration is maintained. Return to Equation (3.1): If in addition to viscous damping, there is an externally applied cyclic force F with driving frequency ω acting on the mass, the equation of motion will be

$$m\ddot{y} + c\dot{y} + ky = F_0 \sin \omega t \tag{3.7}$$

The *steady-state* solution to the above equation of motion is

$$y = y_0 \sin(\omega t + \varphi) \quad \text{with} \tag{3.8}$$

$$y_0 = \frac{F_0/k}{\sqrt{(1 - \omega^2/\omega_0^2)^2 + [2(c/c_c)(\omega/\omega_0)]^2}} \tag{3.9}$$

Recalling $c/c_c = \zeta$, the damping ratio, we can re-write the above equation, after replacing ω with the more commonly used frequency in Hz, as follows:

$$y_0 = A(F_0/k) \tag{3.10}$$

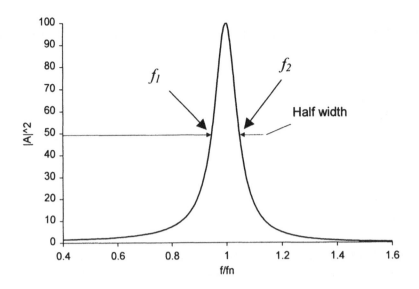

Figure 3.3 The square of response amplitude in the
neighborhood of the resonant frequency

where

$$A_0(f) = \frac{1}{\{[1 - (f/f_0)^2]^2 + 4\zeta^2(f/f_0)^2\}^{1/2}} \qquad (3.11)$$

Since (F_0/k) is the static deflection of the point mass, A_0 is the amplification effect of the dynamic load. For this reason, A_0 is called the amplification factor. When $f = f_0$,

$$A_0(f = f_0) = \frac{1}{2\zeta} \qquad (3.12)$$

Figure 3.3 is a plot of the modulus squared of the amplification factor versus the frequency ratio f/f_n. It is often referred to as the *resonance peak*. In general, the maximum of $|A|$ does not occur exactly at f_0 or f'. In forced vibration, there are three different frequencies:

(1)   The un-damped natural frequency $f_0 = \sqrt{k/m}$.

(2)   The damped natural frequency $f_0' = f_0\sqrt{1 - \zeta^2}$.

(3)   The resonant frequency, or the frequency at which the forced vibration amplitude is maximum.

Fortunately, in almost all of our applications to flow-induced vibration, the damping is light, so the damping ratio is typically less than 0.1. The above three frequencies are about equal. In this case, it can be shown that

$$\zeta_0 = \frac{\Delta f}{2 f_0} \tag{3.13}$$

where $\Delta f = f_2 - f_1$ (see Figure 3.3) is the width of the resonance peak in the frequency space, when $|A|^2 = |A_{max}|^2 / 2$. Δf is often referred to as the half power width of the spectral peak. Figure 3.3 is an example of frequency domain representation of vibration. As stated in Chapter 2, all information that is originally in the time domain vibration data is also in the frequency domain. Equation (3.13) offers an alternate way to measure the damping ratio in the frequency domain by measuring the half power point width and the natural frequency of the structure.

At resonance, the vibration amplitude will grow until the energy dissipated by damping is equal to the energy input by the externally applied force. The system then reaches steady-state vibration.

Since flow turbulence is a combination of forcing functions (to within certain limits) of all possible frequencies, a structure or a fluid cavity exposed to flow turbulence will experience resonant vibration at all of its natural frequencies, each with a resonant peak resembling that in Figure 3.3. By examining the width of the peaks in a PSD plot (see Chapter 2), one can identify the origin of certain vibration components. Electrical hum, for example, frequently appears as spectral peaks at 60 or 120 Hz. Since electrical noise has zero damping, these peaks appear as narrow spikes and can be easily distinguished from other structural vibrations. Similarly, because of their very low damping, acoustic modes also appear in the PSD plots as narrow spikes.

3.3   Transient Vibrations

It was mentioned earlier that unless there is an externally applied force, all vibrations will eventually stop due to damping. This brings in the concept of transient vibrations. For applications to power and process plant components, the most important transient vibration is an impact, and its variants rattling and tapping, which are just other terms describing light impacts. Figure 3.4 is the time history plot of an impact resulting from the seating of the disc of a check valve. Note the characteristic sharp rise, followed by a slow experiential decay of the vibration amplitude. As mentioned before, the rate of amplitude decay is governed by the system damping. The larger the system damping, the more rapid the amplitude decays.

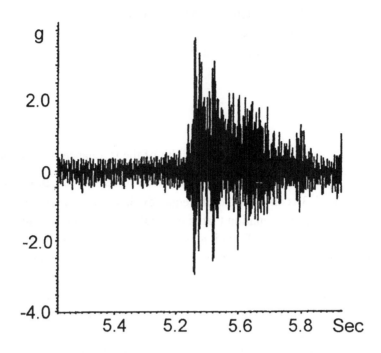

Figure 3.4 An impact waveform in the time domain (Note the sharp rise, exponential decaying characteristic of transient waves caused by metal to metal impacts.)

3.4   Normal Modes

So far we have discussed only a single-degree-of-freedom, point-mass-spring system. Most structural components, such as heat exchanger tubes or pressure vessels, are structures of finite spatial extent, which can be regarded as an infinite number of point-mass-spring systems put together. The entire idea of linear structural dynamic analysis is to reduce these "infinite degree of freedom" systems to a finite, discrete series of single-degree-of-freedom, mass-spring systems.   This is achieved through the technique of modal decomposition and solution in the frequency domain.

Figure 3.5 shows a guitar string, which is a continuous, infinite-degree-of-freedom system. If we pluck the center of the string lightly, it will vibrate in one loop at a definite frequency, emitting a pure tone. If we pluck it harder, it may vibrate with a mixture of single, double or even triple loops, each at a specific frequency. The sound we hear is no longer a pure tone, but a mixture of the fundamental note and its higher harmonics. The single-loop vibration is the fundamental mode of vibration of the guitar string and the specific frequency at which the string vibrates is its fundamental modal frequency. The two-loop vibration is the second modal vibration (called second harmonics) of the guitar string and the frequency at which it vibrates is the second modal frequency, and so on. The fundamental mode, the second mode, third mode and so on are all single-degree-of-freedom systems, like the point mass-spring. As long as the vibration amplitude is not

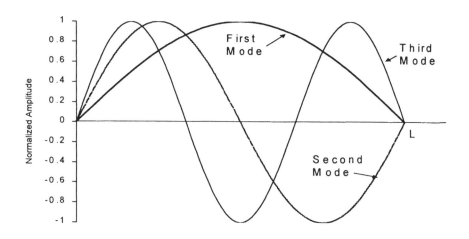

Figure 3.4 Normal modes of a vibrating guitar string

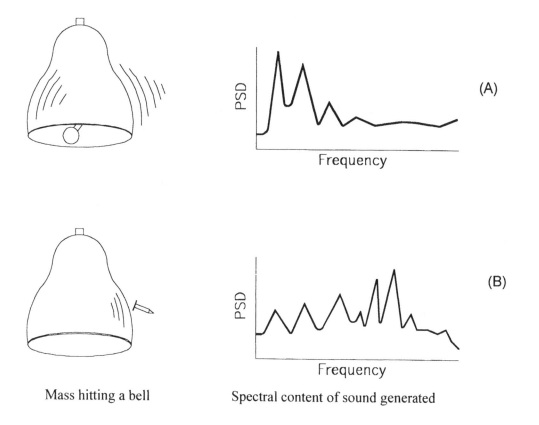

Mass hitting a bell   Spectral content of sound generated

Figure 3.6 (A) A massive moving object excites lower body modes of the component it hits; (B) A small moving object excites higher surface modes of the component it hits

too large that the vibration is in the "linear" domain, then no matter how complex a form the vibrating guitar string appears to be, it is just a combination of many of its normal modes. No matter how complex the guitar string sounds, the notes it generates are combinations of its fundamental note and its higher harmonics. This enables us to solve the vibration problems of continuous structures such as beams, plates and shells by first finding the vibration amplitude of each of its normal modes, and then sum over the contribution from all the normal modes that contribute significantly to the response. This will be treated in the following section.

The above example of a guitar string can be extended to other structures. When struck, the bell, for example, vibrates with a combination of normal modes and emits a characteristic sound, which is a combination of its fundamental note and its higher harmonics. The same is true in fluid-filled cavities, as all of us have probably seen in middle school physics laboratory—a test tube partially filled with water will give a definite tone when we blow air lightly over its open end. In this case the vibrating body is the air column above the water. The organ pipe works on the same principle as a partially water filled test tube.

Closer to power and process plant applications, one can imagine that if, when struck, a bell vibrates with a combination of its normal modes— each at its natural frequency— so will, for example, a check valve when its disk assembly hits the back stop or the seat. This, in short, is the principle behind vibration monitoring. The signals picked up by the vibration sensors are mainly the structural modal vibration frequencies of the component being monitored. When the mass of the moving part is comparable to that of the component being struck, and has sufficient energy to excite the component's body modes, such as when a bell is hit by its clapper (Figure 3.6), the modes excited are predominantly the fundamental and the lower few modes. The characteristic sound a bell generates reflects this combination. However, if the mass of the moving part hitting the component is much smaller than that of the component, and has insufficient energy to excite the component's body modes, such as when the bell is hit by a nail, local deformation will generate many higher modes. The sound it generates will have much more high-frequency components. Similarly, the fluid, whether it is water, steam or gas, filling a valve, for example, forms a resonant cavity just like the air column inside a test tube. When lightly excited, such as by a small leakage flow, one or more of its normal modes may be excited, and the fluid body will vibrate at one or a few of its natural frequencies (see Case Study 2.2 at the end of Chapter 2). When the flow becomes large, these discrete modal frequencies give place to a continuous, turbulent noise. A thorough understanding of the above-mentioned phenomena, together with some experience, will greatly enhance the art of condition monitoring and diagnosis through vibration signal analysis.

3.5   Structural Dynamics

In the following discussion and throughout this book, the word "structure" will be limited to 1D or 2D structures—that is, beams, tubes, rods, plates or shells. "Integrate over the structure" will be used to mean integration over the entire length of the 1D structure, or over the entire surface of the 2D structure. The symbol ℓ is used to denote the domain of integration in general.

For a structure of finite-spatial extent, we can generalize Equation (3.7) and write the response $y(x)$ of the structure at any point x as a collection of Equation (3.1), one for each point on the structure. The resulting equation is a matrix equation (see Hurty and Rubinstein, 1964, for a more detailed explanation):

$$[m]\{\ddot{y}\} + [c]\{\dot{y}\} + [k]\{y\} = \{F\} \tag{3.14}$$

With the advent of inexpensive, high speed electronic computers together with the need to account for non-linear effects, such as the wear of heat exchanger tubes against tube support plates, recently there is an increasingly higher push toward solving Equation (3.14) by direct integration in the time domain. Still, by far the most common method to solve this equation is by modal analysis. We first expand the response vector $\{y\}$ in terms of the normal mode shape functions $\{\psi\}$ and the amplitude function $a(t)$:

$$\{y\} = \{a(t)\}^T \{\psi(x)\} = \sum_n a_n(t)\psi_n(x) \tag{3.15}$$

$$\{\dot{y}\} = \{\dot{a}\}^T \{\psi(x)\} \tag{3.16}$$

$$\{\ddot{y}\} = \{\ddot{a}\}^T \{\psi(x)\} \tag{3.17}$$

Substituting into Equation (3.14), and left multiplying by the transpose of the mode shape vector, we have:

$$([M]\{\ddot{a}(t)\} + [C]\{\dot{a}(t)\} + [K]a(t)) = \{\psi(x)\}^T \{F\} \tag{3.18}$$

where

$$[M] = \{\psi\}^T [m]\{\psi\}$$
$$[C] = \{\psi\}^T [c]\{\psi\}$$
$$[K] = \{\psi\}^T [k]\{\psi\}$$

Based on the theory of structural dynamics, we can simultaneously diagonalize the mass matrix $[M]$ and the stiffness matrix $[K]$. If the damping matrix $[C]$ is also diagonal, then the above equation will be de-coupled with one scalar equation for each mode n. Unfortunately, in practice there is no guarantee that the damping matrix $[C]$ is indeed diagonal. In order to be able to decouple the above equation, it is customary to *assume* that the damping matrix $[C]$ is proportional to either the mass matrix $[M]$, or the stiffness matrix $[K]$, or a linear combination of both, so that it is also diagonal. In this book, we shall assume

$$[C] = [\alpha][M] \tag{3.19}$$

where $[\alpha]$ is a diagonal matrix. With this, we uncouple the equation of structural dynamics into a number of equations, each involving only one mode n:

$$m_n \ddot{a}(t) + \alpha_n m_n \dot{a}(t) + k_n a(t) = F_n \tag{3.20}$$

where

$F_n \qquad = \int_\ell \psi_n(x) F(x) dx$ is the generalized force

$m_n \qquad = \int_\ell \psi_n(x) m(x) \psi_n(x) dx$ is the generalized mass

3.6   Equation for Free Vibration

In the special case when there is no damping and no external force, the resulting equation of motion is that for free vibration:

$$m_n \ddot{a}(t) + k_n a(t) = 0 \tag{3.21}$$

From Chapter 2, the solution to Equation (3.21) is a harmonic function:

$$a(t) = a_0 e^{i\omega t}$$

so that,

$$\dot{a}(t) = i\omega a_0 e^{i\omega t} = i\omega a(t) \tag{3.22}$$

$$\ddot{a}(t) = -\omega^2 a_0 e^{i\omega t} = -\omega^2 a(t) \tag{3.23}$$

Substituting into Equation (3.21) we get

$$k_n = m_n \omega_n^2 \tag{3.24}$$

ω_n is the natural frequency, in radian per second of the nth normal mode of the structure. With this modal frequency defined, we now *assume* the damping matrix $[C]$ is proportional to the mass matrix $[M]$, so that it is diagonal, and with the constant of proportionality given by

$$\alpha_n = 2m_n\omega_n\zeta_n$$

It follows that:

$$C_n = 2m_n\omega_n\zeta_n \tag{3.25}$$

ζ_n is the damping ratio for the nth mode of vibration (compare with Equation (3.4)). In practice, this damping ratio can only be obtained by measurement. Substituting $k_n = m_n\omega_n^2$, $\alpha_n = 2m_n\omega_n\zeta_n$ back into Equation (3.20) and replacing ω_n with the more commonly used engineering parameter, the frequency in Hz, $f_n = \omega_n/(2\pi)$, we get an expression for the amplitude function:

$$a_n(t) = \frac{F_n(t)}{m_n(2\pi f_n)^2[1-(f/f_n)^2+i2\zeta_n(f/f_n)]} \tag{3.26}$$

Notice the assumptions that lead to this simplified de-coupled equation, which forms the building block for all normal mode analysis. We must first of all *assume* that the system is linear so that the response can be obtained by linear superposition. Then we must *assume* that the damping matrix is diagonal and is proportional to the mass matrix (or the stiffness matrix). The constant of proportionality $2m_n\omega_n\zeta_n$ is chosen such that ζ_n has the same physical meaning as the damping ratio in a point-mass-spring system: When $\zeta_n=1.0$, the vibration of that particular mode will stop in the shortest possible time. That is, that mode is critically damped.

Equation (3.26) is often written in the following alternate form:

$$|a_n(t)|^2 = |H_n(f)|^2|F_n|^2, \text{ where} \tag{3.27}$$

$$|H_n(f)|^2 = \frac{1}{m_n^2(2\pi f_n)^4\{[1-(f/f_n)^2]^2+4\zeta_n^2(f/f_n)^2\}}$$

$[H]$ is the diagonal transfer matrix of the dynamic system, with matrix element H_n. The above expression is essential in later developments in turbulence-induced vibration. The equation is also extensively used in dynamic measurement and modal testing. Alternatively, the amplitude function can be written in the following form:

$$|a_n(t)| = \frac{A_n(f)|F_n|}{m_n(2\pi f_n)^2}, \text{ where} \tag{3.28}$$

$$A_n(f) = \frac{1}{\{[1-(f/f_n)^2]^2+4\zeta_n^2(f/f_n)^2\}^{1/2}}$$

A_n is the "dynamic amplification factor" of the nth mode at the forcing frequency f. At resonance, $f=f_n$,

$$A_n(f_n) = 1/(2\zeta_n) \tag{3.29}$$

Recalling

$$\frac{|F_n|}{m_n(2\pi f_n)^2} = \frac{|F_n|}{k_n}$$

is the static deflection of the structure under the load F_n, the dynamic amplification factor gives an estimate of the inertia effect of the load. Equations (3.27) to (3.29) are often used to estimate structural response using the equivalent static load method.

Except in the very simplest case of uniform structures, such as a beam or a plate of uniform cross-sectional properties, subject to uniform loads, Equation (3.14) is now almost always solved with the help of finite-element computer programs. Examples will be given in the following section for a few simple cases for which one can calculate the response by hand or with a spreadsheet.

3.7   Mode Shape Function Normalization

One of the common mistakes in estimating dynamic structural responses by "hand calculation" is caused by mode shape normalization. As shown in Equation (3.15) and used in the subsequent expressions, the mode shape function $\psi_n(x)$ is undefined to within a multiplicative constant. That is to say, if in Equation (3.18), we replace $\psi_n(x)$ with $c\,\psi_n(x)$ where c is a constant, the computed values of y, bending moments, stresses and any physically observable quantities will not change. On the other hand, the generalized mass and generalized force will change. An additional equation is necessary to uniquely define the mode shape function. This is the normalization convention. For structures with uniform cross-sectional properties, the mode shapes are orthogonal to each other:

$$\int_\ell \psi_m(x)\psi_n(x)dx = 0 \quad \text{if } m \neq n$$

Here the integration is over the entire surface (or length) of the structure. For $m=n$, the integral is equal to a constant. Therefore, one way to eliminate the ambiguity in the definition of the mode shape function is to adjust this constant to unity, so that the mode shape functions satisfy the following orthonormal conditions:

$$\int_\ell \psi_m(x)\psi_n(x)dx = \delta_{mn} \tag{3.30}$$

$\delta_{mn} = 1$ if $m{=}n,$ $\delta_{mn} = 0$ if $m \neq n$ $\hfill$ (3.31)

With this orthonormality equation, the mode shape function is uniquely defined, but only to within a sign. If we replace ψ everywhere with $-\psi$, there is still no change in any of the physically observable quantities computed. This has some effect on the cross-acceptance integrals discussed in the chapter on turbulence-induced vibration.

For structures of non-uniform cross-sectional properties, unfortunately the mode shapes functions are no longer orthogonal to each other. That is, the simple Equation (3.30) no longer holds. However, they are still orthogonal to each other with respect to the mass matrix. Instead of Equation (3.30), we have the more complicated equation:

$$\int_{\ell} \psi_m(x)m(x)\psi_n(x)dx = 0 \quad if \quad m \neq n \hfill (3.32)$$

$$= m_n \text{ (a constant) if } m{=}n$$

From Equation (3.18), this constant is called the generalized mass for mode n. Therefore, the generalized mass, the generalized stiffness and force are not physically observable quantities. Their values may change according to the mode shape normalization. Many finite-element computer programs set $m_n =1$ analogous to the normalization convention used in the case of structures with uniform cross-sectional properties:

$$\int_{\ell} \psi_m(x)m(x)\psi_n(x)dx = \delta_{mn} \hfill (3.33)$$

As long as we are consistent, the computed responses, which are physically observable quantities, will not change no matter which mode shape normalization convention we use. However, great care must be exercised when we employ mode shape functions from one source, such as from one computer run or from Blevins' (1979) mode shape tables, and use them with generalized masses or forces computed with another mode shape normalization convention.

3.8   Vibration Amplitudes, Bending Moments and Stresses

After the amplitude functions a_n are computed, the vibration amplitudes at any point x on the structure can be obtained by modal synthesis,

$$y(t,x) = \sum_n a_n(t)\psi_n(x) \hfill (3.34)$$

Very often, only a few modes are needed for convergence in the above sum.

The bending moment in a vibrating structure can be obtained from its mode shape function as follows. From Blevins (1979):

$$\overline{M} = -EI\frac{\partial^2 y}{\partial x^2} \tag{3.35}$$

Substituting into Equation (3.34),

$$\overline{M}_n = -a_n EI\frac{\partial^2 \psi}{\partial x^2} \tag{3.36}$$

If a commercial finite-element structural program is used to solve Equation (3.18), most likely the same program will also calculate the bending moment and stress. If Equation (3.18) is solved by "hand calculation" or by a spreadsheet, there are several alternative ways to calculate the bending moment and the stress after a_n is computed:

(1)   Some commercial finite-element computer programs have capabilities limited to natural frequency and mode shape calculations but do not have the capability to solve for the dynamics responses. These modal analysis computer programs can be used to calculate the mode shapes and their second derivatives $\partial^2 \psi / \partial x^2$. Commercial finite-element computer programs very often normalize the mass matrix to an identity matrix. Great care must be exercised so that the mode shapes used in the calculation of a_n and those used in calculating the bending moment are normalized similarly; otherwise, adjustments will be necessary to put them on an equal basis.

(2)   Use closed form equations to calculate the mode shapes and their second derivative. In the simplest cases of single-span beams of uniform cross-sectional properties, closed form expressions are available for their mode shapes and their natural frequencies. The best known example of this is a single-span beam with uniform cross-sectional properties that is simply-supported at both ends. In this case the mode shape function normalized according to:

$$\int_0^L \psi_m(x)\psi_n(x) = \delta_{mn}$$

are given by,

$$\psi_n(x) = \sqrt{\frac{2}{L}}\sin\frac{n\pi x}{L}$$

and the natural frequencies are given by (see, for example, Blevin, 1979)

$$f_n = \frac{n^2}{2\pi}\sqrt{\frac{EI}{mL^4}}$$

Here m is the mass per unit length of the beam. Uniform, single-span beams with other boundary conditions have similar but considerably more complex expressions (Blevins, 1979). Nonetheless, these equations can be readily implemented onto a spreadsheet. Table 3.1 gives the equations for the natural frequencies and mode shapes for the most commonly encountered beam structures. The mode shapes in Table 3.1 are all normalized in a similar way. For beams with both ends simply-supported or clamped, the normalizations are different from those in Blevins (1979).

(3)   Use numerical mode shape tables. Blevins (1979) not only compiled the numerical values of the mode shapes, but also the first, second and the third derivatives of the mode shapes. These tables can be used to calculate the vibration amplitudes and the bending moments after the amplitude function a_n is computed. Again, great care must be exercised so that the tabulated mode shapes are normalized in the same way as the other parts of the analysis; otherwise, adjustments must be made to put them on an equal basis. The stress can be computed by (Roark, 1989)

$$\sigma = Mc/I \tag{3.37}$$

where c is the distance from the neutral axis.

Example 3.1

Figure 3.7 shows a cylinder subject to a uniformly distributed force F_L. One end of the cylinder can be assumed clamped while the other end is free. Other parameters are given in Table 3.2. Calculate the fundamental modal frequency of this cantilever. Assuming that the only mode that contributes to the response is the fundamental mode, calculate the vibration amplitude at the free end of the cantilever and the maximum dynamic stress.

Table 3.2: Known Parameters for Example 3.1

	US Units	SI Units
OD	1.495 in	0.03797 m
ID	0.896 in	0.02276 m
L	14.5 in	0.368 m
Linear density of material	3.779 lb/ft	5.625 Kg/m
	(=8.15E-4 (lb-s^2/in)/in)	
Young's modulus	25E+6 psi	1.72E+11 Pa
Damping ratio	0.005	0.005
F_L	0.848 lb/in	148.7 N/m
Forcing frequency $f=f_s$	264 Hz	264 Hz

Table 3.1: Natural Frequencies and Mode Shapes for a Few Simple Cases

Single-span beams of uniform cross-sectional properties. All the following mode shapes are normalized according to:

$$\int_0^L \psi_m(x)\psi_n(x)dx = \delta_{mn}, \qquad f_n = \frac{\lambda_n^2}{2\pi L^2}\sqrt{\frac{EI}{m_t}}.$$

Boundary Conditions	λ_n	Normalized Mode Shapes $\psi(x)$	$d^2\psi/dx^2$
Simply-supported at both ends	$n\pi$, $n=1, 2, 3..$	$\psi_n(x) = \sqrt{\frac{2}{L}}\sin\frac{n\pi x}{L}$	$\psi''_n(x) = -(\frac{n\pi}{L})^2\sqrt{\frac{2}{L}}\sin\frac{n\pi x}{L}$
Clamped at both ends	$\lambda_1 = 4.7300$ $\lambda_2 = 7.8532$ $\lambda_3 = 10.9956$	$\psi_n(x) = \frac{1}{\sqrt{L}}\{[\cosh(\frac{\lambda_n x}{L}) - \cos(\frac{\lambda_n x}{L})]$ $- \sigma_n[\sinh(\frac{\lambda_n x}{L}) - \sin(\frac{\lambda_n x}{L})]\}$ $\sigma_{1,2,3} = 0.9825, 1.0008, 1.0000$	$\psi''_n(x) = (\frac{\lambda_n}{L})^2\frac{1}{\sqrt{L}}\{[\cosh(\frac{\lambda_n x}{L}) + \cos(\frac{\lambda_n x}{L})]$ $- \sigma_n[\sinh(\frac{\lambda_n x}{L}) + \sin(\frac{\lambda_n x}{L})]\}$ $\sigma_{1,2,3} = 0.9825, 1.0008, 1.0000$
Clamped-free (cantilever)	$\lambda_1 = 1.8751$ $\lambda_2 = 4.6941$ $\lambda_3 = 7.8548$	$\psi_n(x) = \frac{1}{\sqrt{L}}\{[\cosh(\frac{\lambda_n x}{L}) - \cos(\frac{\lambda_n x}{L})]$ $- \sigma_n[\sinh(\frac{\lambda_n x}{L}) - \sin(\frac{\lambda_n x}{L})]\}$ $\sigma_{1,2,3} = 0.7341, 1.0185, 0.9992$	$\psi''_n(x) = (\frac{\lambda_n}{L})^2\frac{1}{\sqrt{L}}\{[\cosh(\frac{\lambda_n x}{L}) + \cos(\frac{\lambda_n x}{L})]$ $- \sigma_n[\sinh(\frac{\lambda_n x}{L}) + \sin(\frac{\lambda_n x}{L})]\}$ $\sigma_{1,2,3} = 0.7341, 1.0185, 0.9992$

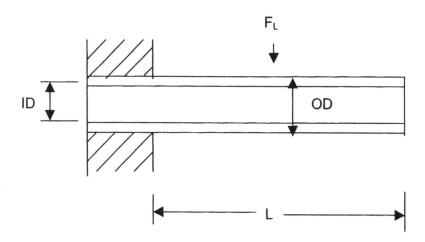

Figure 3.7 Cylinder subject to uniform load

Solution

We can use a finite-element computer program to solve this problem. For a one-span beam of uniform cross-sectional properties subject to a uniform load, it may be beneficial to carry out the calculation by hand occasionally, in order to keep one's memory of the physics of dynamic analysis fresh. One can start implementing the cantilever mode shape function and its second derivative from Table 3.1 onto a spreadsheet, such as Microsoft Excel, and carry out the rest of the calculation with the spreadsheet. Without a spreadsheet, we still have another alternative: Use the tabulated mode shape values for the fundamental mode in Blevins (1979). The following solution is based on this approach. Keeping one mode, the response y is, from Equations (3.15) and (3.28),

$$| y_1(x) | = | a_1 | \psi_1 = \frac{| F_1 | \; \psi(x)}{m_1 (2\pi f_1^2) \; \{[1 - (f_s / f_1)^2]^2 + 4\zeta_1^2 (f_s / f_1)^2\}^{1/2}}$$

We now proceed to evaluate the above equation using the numerical mode shape values tabulated in Blevins (1979). Noting that Blevins normalizes the mode shape for the cantilever according to:

$$\int_0^L \psi^2(x)dx = L$$

we have:

$$| F_1 | = \int_0^L | F_L | \psi_1(x)dx = | F_L | \int_0^L \psi_1(x)dx$$

Similarly,

$m_n = Lm_t$ with $m_t = 3.779/(32.2*12*12) = 8.15\text{E-}4$ (lb-s^2/in). Therefore,

$$| y_1(x = L) |= a_1\psi_1(x = L) = \frac{A_1 | F_L | \psi_1(x = L)}{m_t(2\pi f_1)^2} \frac{1}{L} \int_0^L \psi_1(x)dx$$

From Blevins (1979; Table 8-2b), at the tip of the nozzle, ψ_1 ($x=L$)=2.0. The integral can be obtained from the same table by numerically summing over the 50 mode shape values given at x/L=0.02 intervals:

$$\int_0^L \psi dx = \sum \psi_i \Delta x_i \quad with \quad \Delta x_i = L/50$$

Carrying out the sum we get,

$$\int_0^L \psi dx = 0.8031L \text{, giving,}$$

$$| y_1(x = L) |= \frac{0.8031 A_1 | F_L | \psi_1(x = L)}{m_t(2\pi f_1)^2}$$

and

$y_{max} = 2.90\text{E-}4$ in (7.38E-6 m)

The bending moment at the clamped end of the nozzle is given by:

$$\overline{M}(x = 0) = -EI\frac{\partial^2 y}{\partial x^2} = -a_1 EI\frac{\partial^2 \psi_1}{\partial x^2}$$

From Blevins (1979; Table 8-2b again), for the fundamental mode at x=0,

$$\frac{L^2}{\lambda_1^2}\frac{\partial^2 \psi}{\partial x^2} = 2.0$$

With $\lambda_1 = 1.875$, we get:

	US Unit	SI Unit
$A_I=$	1.48	1.48
$A_1\,\lvert F_L\rvert=$	1.25 lb/in	220 N/m
$a_1=$	1.45E-4	3.69E-6
$y(x{=}L)=$	2.90E-4 in	7.38E-6 m
$\dfrac{\partial^2\psi}{\partial x^2}=$	0.0721 in^{-2}	112 m^{-2}
$\overline{M}_{max}=$	55.9 lb-in	6.32 N-m
$\sigma_{max}=\dfrac{M_{max}(OD/2)}{I}=$	195.5 psi	1.35E+6 Pa

3.9   Equivalent Static Load Method

For simple, one-span beams of uniform cross-sectional properties subject to a uniform, or almost-uniform forcing function, approximate estimates on the structural responses, including the bending moment and stresses, can be obtained by the equivalent static load method using the formulas for stress and strain by Roark (1989). The equivalent static load w, per unit length, that gives the same maximum response is found. The bending moment and the stress are then computed with Roark's equations. Table 3.3 given the equations for a few simple cases.

Table 3.3: Maximum Deflections and Bending Moments for a Few Simple Cases
(Source, Roark, 1989)

Boundary Conditions	Maximum Deflection y_{max}	Maximum Moment M_{max}
Simply-supported both ends	$-\dfrac{5}{384}\dfrac{wL^4}{EI}$ at $x{=}L/2$	$\dfrac{wL^2}{8}$ at $x{=}L/2$
Clamped at both ends	$-\dfrac{1}{384}\dfrac{wL^4}{EI}$ at $x{=}L/2$	$\dfrac{wL^2}{24}$ at $x{=}L/2$
Clamped-free (cantilever)	$-\dfrac{wL^4}{8EI}$ at $x{=}L$	$-\dfrac{wL^2}{2}$ at $x=0$

Example 3.2

Find the maximum bending moment and stress in Example 3.1 using the equivalent static load method.

Solution

At the forcing frequency f_s, the amplification factor, Equation (3.28), is:

$$A_1 = \frac{1}{\{[1-(f_s/f_1)^2]^2 + 4\zeta_1^2(f_s/f_1)^2\}^{1/2}}$$

with $\zeta_1=0.005$. Using $f_s=264$ Hz, we get $f_s/f_1 =264/464=0.569$. Substituting into Equation (3.28) for the amplification factor, we get $A_1=1.47$ (versus 100 if the fundamental frequency coincides with the forcing frequency). Therefore, the equivalent static load on the cylinder is,

	US Units	SI Units
	US Units	_SI Units_
$w=$	1.25 lb/in	220 N/m
$y_{max} = \dfrac{wL^4}{8EI}$	2.7E-4 in	7.1E-6 m
$M_{max} = \dfrac{wL^2}{2}$	61 lb-in	6.92 N-m
$\sigma_{max} = \dfrac{M_{max}(OD/2)}{I}$	214 psi	1.48E+6 Pa

3.10   Power Dissipation in Vibrating Structures

In a vibrating structure, the power dissipated through damping is,

$$P = \int_\ell [\frac{1}{T}\int_0^T c\dot{y}^2(x)dt]dx \tag{3.38}$$

where the integration is over one period T of vibration and over the entire length or surface of the structure. From Equations (3.15) and (3.25),

$$P = \frac{1}{T}\int_0^T \dot{a}(t)\{\psi\}^T[c]\{\psi\}\dot{a}(t)dt = \frac{1}{T}\int_0^T \dot{a}(t)[C]\dot{a}(t)dt = \sum_n \frac{1}{T}\int_0^T 2\zeta_n m_n \omega_n \dot{a}_n^2 dt$$

In the special case when one mode dominates the response, as when the structure is excited to resonance at mode n, one can use Equation (3.15) to reduce the above equation to:

$$P = \frac{1}{T}\int_0^T 2\zeta_n m_n \omega_n (\dot{y}_i / \psi_{ni})^2 \, dt \tag{3.39}$$

In Equation (3.39), since the ratio $\dot{y}_i / \psi_{ni}$ $(= \dot{a}_n(t))$ is a constant independent of the location x on the structure for each mode n, i can be any point on the structure. Furthermore, since the damping ratio is a system parameter, P is the power dissipated by the entire structure, not at one point on the structure. Using

$$y_i = y_{0i} \sin \omega_n t$$
$$\dot{y}_i = \omega y_{0i} \cos \omega_n t$$

From Equation (3.39), we get:

$$P = \frac{1}{T}\int_0^T 2\zeta_n \omega_n m_n (\omega_n y_{0i} \cos \omega_n t / \psi_{ni})^2 \, dt$$

Integrating and reverting back to the more familiar frequency in Hz, we get:

$$P = 8\pi^3 \zeta_n f_n^3 m_n (y_{0i}^2 / \psi_{ni}^2) \tag{3.40}$$

where y_{0i} is the 0-peak amplitude at x_i and f_n is the modal frequency. P can also be expressed in terms of the more commonly measured parameter, the acceleration, as:

$$P = (m_n \zeta_n / 2\pi f_n)(\alpha_{0i}^2 / \psi_{ni}^2) \tag{3.41}$$

where α_{0i} is the measured 0-peak acceleration amplitude at x_i. In the special case of the fundamental mode of a beam simply-supported at both ends, $\psi_1(x) = \sqrt{2/L} \sin(\pi x / L)$ (Table 3.1). We can choose any convenient point to compute the ratio $(\alpha_{0i}^2 / \psi_{ni}^2)$. If we choose x_i as the mid-point of the beam, then

$$y_0^2 / \psi_1^2 = L y_{max}^2 / 2 \quad and \quad m_n = m$$

Equation (3.40) reduces to,

$$P = 16\pi^3 \zeta_n f_n^3 L m \, y_{max}^2 \tag{3.42}$$

where y_{max} is the 0-peak vibration amplitude at the mid-point of the beam. Using $y_{rms} = y_0 / \sqrt{2}$, Equation (3.42) can also be written as,

$$P = 32\pi^3 \zeta_n f_n^3 Lm \ y_{rms}^2 \tag{3.43}$$

where y_{rms} is the rms vibration amplitude at the mid-point of the beam. The above discussion has the following important significance:

- Equation (3.41) offers an alternate way to measure the damping ratio by measuring the power supplied to maintain a steady-state, resonant vibration of the structure, as was shown by Taylor et al (1997). To do this, several point forces may be required to preferentially excite the structure in the mode of interest. Very often, this is the fundamental mode. In addition to the power supplied to maintain the steady, resonant vibration at the nth mode, the 0-peak acceleration at any convenient point on the structure as well as the mode shape has to be measured.

- In heat exchanger tubes with tube-to-support plate clearances, the interaction between these components causes wear between them. This wear is a form of energy dissipation. In service, oxidation and chemical deposits often fill up the tube-support plate clearances and reduce the wear rate between these two components, decreasing the system damping ratio as a result. Incidents have occurred where an otherwise fluid-elastically stable (see Chapter 7) tube bundle suddenly became unstable because of this gradual decrease in the damping ratio of the tube.

- Equation (3.43) offers a simple way to estimate the wear rates in heat exchanger tubes based on measured wear rates in the laboratory or in other operational heat exchangers, together with the rms vibration amplitudes computed with linear vibration theory. This will be shown in Chapter 11.

References:

Blevins, R. D. 1979, Formulas for Natural Frequencies and Mode Shapes, Van Nostand, New York.

Hurty, W. C. and Rubinstein, M. F., 1964, Dynamics of Structures, Prentice-Hall, Englewood Cliffs, New Jersey.

Roark, R. J., 1989, Formulas for Stress and Strain, revised by W. C. Young, McGraw Hill, New York.

Taylor, C. E.; Pettigrew, M. J., Dickinson, T. J. and Currie, I. G., Vidalou, P, 1997, "Vibration Damping in Multispan Heat Exchanger Tubes," 4th International Symposium on Fluid-Structure Interactions, Vol II, edited by M. P. Paidoussis. ASME Special Publication AD-Vol. 53-2, pp. 201-208.

Chapter 4

VIBRATION OF STRUCTURES IN QUIESCENT FLUIDS—I
THE HYDRODYNAMIC MASS

Summary

The general case of coupled fluid-structural dynamic analysis involving the complete set of equations of fluid dynamics and structural dynamics can be prohibitively complicated. In spite of 25 years of intensive research, today many of the coupled fluid-structural dynamic problems are solved based on the "weak coupling" assumption. In this approach, the force induced on the fluid due to structural motion is assumed to be linearly superimposible onto the original forcing function in the fluid. Under this assumption, the effect of fluid-structure interaction can be completely accounted for by an additional mass term, called the hydrodynamic mass, and an additional damping term, called hydrodynamic damping. These hydrodynamic mass and damping terms can then be separately computed and input into standard structural analysis computer programs for dynamic analysis of the coupled fluid-structure system. This approach allows computer programs developed, and data obtained from tests, without taking into account the effect of fluid-structure interaction, to be used in coupled fluid-structural dynamic analyses. Although developed based on coaxial cylindrical shells coupled by an annular fluid gap, the results in this chapter shed light on the physics of fluid-structure interaction and can be used in other geometry in special cases as will be discussed in the following chapter.

As an alternative to the finite-element method, which may be quite complicated in the case of 3D structures, closed-form equations for computing the in-air natural frequencies of cylindrical shells with either both ends simply-supported or both ends clamped, are given in Section 4.2. These equations can be easily solved numerically with today's widely available standard office software.

The weakly coupled fluid-structural dynamics problem can be solved by two alternative formulations of the hydrodynamic mass method: the generalized hydrodynamic mass and the full hydrodynamic mass matrix methods. The first formulation is for estimating the coupled fluid-shell frequencies when the uncoupled, in-air natural frequencies of the individual cylindrical shells are known from measurement, from separate finite-element analysis or by numerical solution of the characteristic equation derived in Section 4.2. The 2x2 effective generalized hydrodynamic mass matrix is given by:

$$\hat{M}^a_{Hmn} = \frac{1}{1+n^{-2}}\sum_\alpha (C^a_{\alpha m})^2 h^a_{\alpha n} \qquad \hat{M}^{ab}_{Hmn} = \frac{1}{1+n^{-2}}\sum_\alpha (C^a_{\alpha m}C^b_{\alpha m})h^{ab}_{\alpha n}$$

$$\hat{M}^{ba}_{Hmn} = \frac{1}{1+n^{-2}}\sum_\alpha (C^a_{\alpha m}C^b_{\alpha m})h^{ba}_{\alpha n} \qquad \hat{M}^b_{Hmn} = \frac{1}{1+n^{-2}}\sum_\alpha (C^b_{\alpha m})^2 h^b_{\alpha n} \qquad (4.50)$$

Both the hydrodynamic mass component h and the Fourier coefficient C play a role in the fluid coupling of the cylinders. The former is the intrinsic fluid pressure each acoustic mode possesses, while the latter is a measure of the compatibility between the structural mode and a particular acoustic mode. Unless there is something in common between these two modes, there will be no hydrodynamic mass coupling between them no matter how large a hydrodynamic mass component the acoustic mode possesses. The Fourier coefficient is given by the projection of the structural mode shape onto the acoustic mode shape:

$$C_{m\alpha} = \int_0^L \psi_m(x)\phi_\alpha(x)dx \tag{4.35}$$

while the hydrodynamic mass components are given by,

$$h_{\alpha n}^a = -\rho(x_\alpha a, x_\alpha b)'_n / [x_\alpha(x_\alpha a, x_\alpha b)''_n]$$

$$h_{\alpha n}^b = \rho(x_\alpha b, x_\alpha a)'_n / [x_\alpha(x_\alpha b, x_\alpha a)''_n]$$

$$h_{\alpha n}^{ab} = -\rho / [x_\alpha^2 a(x_\alpha a, x_\alpha b)''_n$$

$$h_{\alpha n}^{ba} = \rho / [x_\alpha^2 b(x_\alpha b, x_\alpha a)''_n \tag{4.37}$$

where

$'$ = derivative with respect to the argument of the function

$$(x,y)'_n \equiv I_n(x)K'_n(y) - I'_n(y)K_n(x)$$

$$(x,y)''_n \equiv I'_n(x)K'_n(y) - I'_n(y)K'_n(x) = -(y,x)'' \tag{4.38}$$

$$x_\alpha^2 = (\varepsilon\pi/\ell)^2 - (2\pi f/c)^2 \qquad if \quad 2f/c < \varepsilon/\ell$$

$$x_\alpha^2 = (2\pi f/c)^2 - (\varepsilon\pi/\ell)^2 \qquad and \quad J,Y \text{ replace } I,K \text{ if } 2f/c > \varepsilon/\ell$$

In spite of their appearance, these equations can be easily implemented into a computer program or solved with a spreadsheet using its built-in Bessel functions. The following recursive equations for the Bessel functions may be helpful:

$$J'_n(x) = (nJ_n(x) - xJ_{n+1}(x))/x$$

$$Y'_n(x) = (nY_n(x) - xY_{n+1}(x))/x$$

$$I'_n(x) = (nI_n(x) + xI_{n+1}(x))/x$$

$$K'_n(x) = (nK_n(x) - xK_{n+1}(x))/x \tag{4.38}$$

The coupled natural frequencies can be computed directly using the stiffness matrices of the cylindrical shells (Section 4.2) and the effective hydrodynamic mass matrix.

$$\begin{vmatrix} & & 0 & 0 & 0 \\ D(a) & & 0 & 0 & 0 \\ & & 0 & 0 & \hat{M}_H^{ab} \\ 0 & 0 & 0 & & \\ 0 & 0 & 0 & D(b) & \\ 0 & 0 & \hat{M}_H^{ba} & & \end{vmatrix} = 0 \qquad (4.52)$$

where

$$D(a) = \begin{bmatrix} k_{11}^a - \mu_a\,\omega^2 & k_{12}^a & k_{13}^a \\ k_{12}^a & k_{22}^a - \mu_a\,\omega^2 & k_{23}^a \\ k_{13}^a & k_{13}^a & k_{33}^a - (\mu_a + \hat{M}_H^a)\omega^2 \end{bmatrix}$$

The full hydrodynamic mass matrix was formulated for inputting into a finite-element model for subsequent dynamic analyses, using a separate finite-element computer program.

$$[M_H] = \begin{bmatrix} (1/a)\sum_\alpha h_{\alpha n}^a [\phi_\alpha dA_a][\phi_\alpha dA_a]^T & (1/b)\sum_\alpha h_{\alpha n}^{ab}[\phi_\alpha dA_a][\phi_\alpha dA_b]^T \\ (1/a)\sum_\alpha h_{\alpha n}^{ba}[\phi_\alpha dA_b][\phi_\alpha dA_a]^T & (1/b)\sum_\alpha h_{\alpha n}^b[\phi_\alpha dA_b][\phi_\alpha dA_b]^T \end{bmatrix} \qquad (4.55)$$

Since this hydrodynamic mass is originally derived from a pressure, it is effective only in the normal degree-of-freedom. Therefore, when inputting into a finite-element computer program, the hydrodynamic masses given in Equation (4.55) should be associated only with the normal degree-of-freedom.

Nomenclature

a Radius of inner cylinder, also denotes inner cylinder when used as an index
b Radius of outer cylinder, also denotes outer cylinder when used as an index
c Velocity of sound
$[c]$ Damping matrix
C Fourier coefficient, or
           numerical values for computing the stiffness matrix (given in Table 4.1)
dA Element area
dL Element length
$D(a)$ Given by Equation (4.52)
D $=Et^3/[12(1-\sigma^2)]$, or diameter of cylinder
e Energy per unit volume

E	Young's modulus, or total energy
f	Frequency, in Hz
$\bar{f}$	In-water frequency
F	External force
h	Hydrodynamic mass components
I	Bessel function of second kind
J	Bessel function of first kind
$[k]$	Stiffness matrix
K	Modified Bessel function of second kind
L	Length of cylinder
m	Axial modal number for structure
M_H	Hydrodynamic mass
$\hat{M}_H$	Effective hydrodynamic mass
$[m]$	Mass matrix
n	Circumferential modal number, acoustic or structural
N, N'	Amplitude function for acoustic pressure
p	Pressure
p_0	Incident pressure
p'	Induced pressure
P'_{mn}	Generalized pressure
$\{q\}$	$=\{u, v, w\}$ Displacement vector on surface of cylindrical shell
$\{Q\}$	$=\{U, V, W\}$ Generalized displacement vector
r	Radial position
$\vec{r}$	Position vector
R	Radius of cylinder, or radial function of acoustic pressure
t	Time, or thickness of shell
u	Axial displacement component on the surface of cylindrical shell
U	Axial amplitude function
v	Tangential displacement components on the surface of cylindrical shell
V	Tangential amplitude function, or velocity of fluid flow
w	Normal displacement component on the surface of cylindrical shell
W	Normal amplitude function
x	Position vector
y	Displacement
Y	Modified Bessel function of first kind
α, β	Acoustic axial modal number
$[\delta]$	$= \begin{bmatrix} 0 & 0 & 0 \\ 0 & 0 & 0 \\ 0 & 0 & 1 \end{bmatrix}$
γ	Numerical values for computing stiffness matrix (given in Table 4.1)

δ_{mr}	Kronecker delta
ε	$=\alpha$, or $(2\alpha-1)/2$ depending on end condition of fluid gap
θ	Angular coordinate
κ	$=Et/(1-\sigma^2)$
μ	Physical mass surface density
ρ	Fluid mass density
σ	Poison ratio
ϕ	Acoustic axial mode shape function
χ_α^2	$=|(2f/c)^2-(\varepsilon/L)^2|$
ψ	Axial structural mode shape function
ω	Angular frequency, rad/s
∇	Gradient operator
∇^2	Laplacian operator
Θ	Circumferential mode shape function, structural and acoustic
[ ]	Denotes matrix
{ }	Denotes column vector
$\dot{y}$	$\equiv \partial y / \partial t$
$\ddot{y}$	$\equiv \partial^2 y / \partial t^2$
$\vec{V}$	Denotes vector quantity

Subscripts

a	Denotes outer cylinder
b	Denotes inner cylinder
n	Circumferential modal index, or denotes normal component
m, r	Axial modal index for structure
α, β	Axial acoustic modal index

4.1   Introduction

In a broad sense, fluid-structure interaction covers such diversified subjects as aeroelasticity, hydroelasticity and flow-induced vibration. In the power and process industries, this term seems to apply specifically to a class of dynamics problems that involves fluids, quiescent or flowing, confined by structural boundaries. Unlike in aeroelasticity, the structures involved are usually stationary except for their vibratory motions. Furthermore, the velocities of fluid flow are usually much lower than those encountered in aeroelasticity. Therefore, fluid-structure interaction problems in the power and process industries belong to a unique field of engineering science, which, in a broader sense, includes flow-induced vibration. However, flow-induced vibration is such a specialized subject that it has evolved as a separate field of study, except for the estimates of added mass and damping. The last two dynamic variables are often calculated using methods developed for fluid-structure interaction problems.

The history of fluid-structure interaction problems, as we understand it today, can be traced back to 1843, when Stoke studied the uniform acceleration of an infinite cylinder in an infinite fluid medium. He concluded that the only effect of the fluid on the motion of the cylinder is to increase its effective mass by an amount equal to the mass of fluid it displaces, and hence the term "added mass."

Stoke's equation applies only to an infinite cylinder moving in an infinite fluid medium. In the 1960s, designers of nuclear steam supply systems found that in a confined fluid medium, the added mass due to fluid-structure interaction was much larger than the mass of fluid displaced by the structure.

Fluid-structure interaction, in the form known to the power and process plant structural engineers today, apparently began in the 1950s. The application was not to nuclear plants, but to large liquid fuel tanks in aerospace vehicles (Abramson and Kana, 1967). The earliest documented study of fluid-structure interaction specifically for power plant application appears to be the work of Fritz and Kiss (1966) at the Knolls Atomic Power Laboratory. Since Fritz and Kiss addressed only the rocking motion of coaxial cylinders, it was a fluid-solid interaction study except that the fluid medium is finite, and the acceleration of the solid is not uniform. From the early 1970s to the early 1980s, in order to address safety-related issues associated with the early stage of commercial nuclear power, a large number of technical papers on the dynamics of coupled fluid-elastic shell systems was published. Among some of these landmark papers are: Fritz (1972), Krajcinovic (1974), Chen and Rosenberg (1975), Horvay and Bowers (1975), Au-Yang (1976, 1977), Yeh and Chen (1977), Scavuzzo et al (1979), Au-Yang and Galford (1981, 1982). In 1998, ASME International published in its Boiler Code Sec III Appendix N-1400, an official guide on dynamic analyses of couple fluid-shell systems. This guide also contains extensive bibliography on technical papers and special bound volume publications on fluid-structure dynamic analysis up to the late 1980s.

Strongly Coupled Fluid-Structure Systems

In common with the science of interacting fields, a fluid-structure system can be classified as strongly or weakly coupled. A strongly coupled fluid-structure system is one in which the flow field induced by the structural motion (called the induced field) and the original flow field (called the incident field) cannot be linearly superimposed upon each other. This is usually caused by large structural displacement, resulting in large induced fluid velocity and completely distorted incident flow field. Vortex-induced vibration (Chapter 6), aeroelasticity and fluid-elastic instability of heat exchanger tube banks (Chapter 7) are all examples of strongly coupled fluid-structure systems. Even for a structure in a non-viscous, non-conductive and single-phase fluid without internal heat generation, the governing equations of motion will include the following (Note: these equations are presented to show why we follow the much simpler hydrodynamic mass and hydrodynamic damping approach. The readers do not have to memorize these equations):

Equation of Fluid Dynamics:

Continuity: $\partial\rho/\partial t + \nabla(\rho\vec{V}) = 0$ (4.1)

Momentum: $\partial\vec{V}/\partial t + \vec{V}.\nabla\vec{V} + \nabla p/\rho - \vec{F} = 0$ (4.2)

Energy: $\partial E/\partial t + \nabla.(E\vec{V} + p\vec{V}) - \rho\vec{F}.\vec{V} = 0$ (4.3)

$E = \rho e + \rho\vec{V}.\vec{V}/2$ (4.4)

Equation of Structural Dynamics: From Chapter 3, the equation of structural dynamics is, in matrix notation,

$$[m]\{\ddot{y}\} + [c]\{\dot{y}\} + [k]\{y\} = \{F\}$$ (3.14)

The equations of fluid dynamics (4.1 to 4.3) and structural dynamics (3.14) are coupled by the requirement that at the fluid-structure interface, the fluid velocity normal to the structural surface must be equal to the normal component of the structural velocity:

$$\dot{y}_n = V_n$$ (4.5)

Equations (4.1 to 4.5), which have been written in compact vector and matrix notations, are much more complex than they appear to be. In nuclear safety applications, these equations are further coupled with heat transfer equations and equations of neutronics that causes internal heat generation in the fluid. This is the reason why today, after more than a quarter of a century of intensive research, there is still no satisfactory analytical method to solve the heat exchanger tube bank instability, as well as many other strongly coupled fluid-structural dynamic problems.

Weekly Coupled Fluid-Structure Systems

By contrast, a weakly coupled fluid-structure system is one in which the flow field induced by the structural motion can be regarded as a small perturbation of, and can be linearly superimposed onto, the incident flow field. That is, we assume that,

$$p = p_0 + p'$$ (4.6)

where p, p_0, and p' are the total, incident and induced pressures on the shells. Furthermore, because the effect of fluid-structure coupling can be accounted for in a quiescent fluid, one can start with the much simpler equation of acoustics instead of the full set of equations of fluid dynamics, Equations (4.1) to (4.4). Perhaps the best know example of weakly coupled fluid-structure systems is turbulence-induced vibration (Chapters 8 and 9). The standard approach to solving this problem is to assume the structural motion does not affect the fluctuating pressure in turbulent flow. The latter can therefore be either measured or calculated independently of any structural parameters.

The Hydrodynamics Mass and Damping Method

In a particularly attractive technique of solving the weakly coupled fluid-structure problem, the entire effect of fluid-structure interaction is accounted for by an added mass (called hydrodynamics mass, except in tube bundles in the remaining part of this book) term. As a result, neither the basic structural analysis computer software nor the computer software developed to generate the forcing functions needs to be extensively revised. Test data obtained without taking into account the effect of fluid-structure coupling (such as the turbulence-induced forcing function) remain valid even in the presence of fluid-structure interaction.

In spite of over 20 years of intensive development effort, today much of the flow or transient fluid-induced and almost all seismic structural dynamic problems are solved by the hydrodynamic mass approach. Some commercial finite-element structural analysis computer programs have incorporated, in various forms of sophistication, equations for computing the hydrodynamic mass matrix.

Likewise, the effect of fluid viscosity can be completely accounted for by a hydrodynamic damping term. Hydrodynamic damping will be discussed in Chapter 5. It will be seen from the examples given in Chapter 5 that unlike hydrodynamic mass, hydrodynamic damping is relatively insignificant even for large structures with narrow entrained fluid gaps.

In the following sections the vibration of two coaxial circular cylindrical shells coupled by an inviscid fluid gap will be discussed. Since the subject matter involves not only acoustics but also the dynamics of thin shells, considerable details and mathematics will be involved. It is important to understand the hydrodynamic mass concept thoroughly to avoid mistakes in its applications.

4.2   Free Vibration of Thin Circular Cylindrical Shells in Air

The subject of hydrodynamic mass and hydrodynamic damping for thin-walled cylindrical shells with uniform surface density was extensively studied in the late 1970s and early 1980s as part of a combined regulatory agency and industry effort to ensure the safe operation of nuclear reactors. Although they were derived based on a highly idealized geometry, the results shed light on the dramatic hydrodynamic mass effect on structures with entrained narrow fluid gaps; a phenomenon not well understood before the 1970s. To establish the notation and convey the basic idea, it is necessary to begin with the fundamentals of thin shell theory. Much of the following derivations are based on the fundamentals of structural dynamics in Chapter 3.

   Figure 4.1 shows the vibration mode of a finite, thin cylindrical shell. Viewed from the ends, the vibration of the cylinder may consist of any number of waves distributed around the circumference. The number of circumferential waves is denoted by n, with $n = 1$ being the beam mode and $n=0$ the symmetric, breathing mode[1]. Following the common notations in cylindrical shell theory, we denote the displacement at a point on the surface of the cylindrical shell by q and the surface mass density (assumed constant) by μ. The force on the right-hand side of Equation (3.14) is now the pressure induced by the motion of the cylindrical shell. Instead of Equation (3.14), we have, in the present case of thin circular cylindrical shells of uniform thickness and surface mass density,

$$[\mu]\{\ddot{q}\} + [c]\{\dot{q}\} + [k]\{q\} = \{p_0\} + \{p'\}$$ (4.7)

Here p_0 is the "incident" pressure, that is, any external forcing function that is not caused by the motion of the cylinder itself, and p' is the pressure induced by the motion of the shells on themselves.

   Resolving the displacement q into three orthogonal directions u, v, w in the axial, tangential and normal directions (Figure 4.2):

$$\{q\} = \{u, v, w\}$$ (4.8)

and expanding the displacement in each of these directions as a modal sum (see Chapter 3), we get:

$$u = \sum_{mn} U_{mn}\psi_m\Theta_n \, , \qquad v = \sum_{mn} V_{mn}\psi_m\Theta_n \, , \qquad w = \sum_{mn} W_{mn}\psi_m\Theta_n$$ (4.9)

U, V, W are the amplitude functions in the axial, tangential and normal direction while ψ, Θ are the axial and circumferential mode shape functions (compare with Chapter 3). Throughout this chapter, it is assumed that the both the axial and circumferential mode shapes are normalized to unity, which this is possible, because the shells have uniform surface densities (Chapter 3).

$$\int_0^L \psi_m(x)\psi_r(x)dx = \delta_{mr}$$ (4.10)

$$\int_0^{2\pi} \Theta_n\Theta_s(\theta)d\theta = \delta_{ns}$$ (4.11)

---

[1] The $n=0$ or symmetric breathing mode is usually of much higher frequency and does not contribute much to structural response. This mode is not considered in the following discussion.

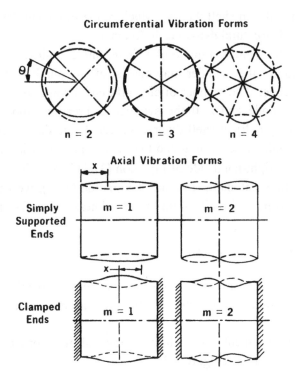

Figure 4.1 Normal modes of a thin cylindrical shell

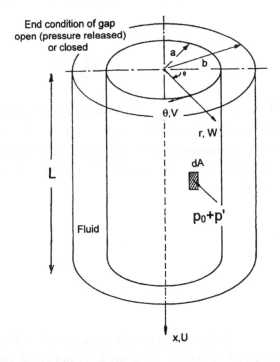

Figure 4.2 Cylindrical coordinate system u, v, w

When viewed from the side, the deformation of the cylinder consists of a number of waves distributed along the length of a generator. The number of half-waves along a generator is denoted by m, $m = 1, 2, 3, \ldots$ As shown in Figure 4.1, the axial wave forms depend on the end conditions of the cylinder.

The equation of motion for the free vibration of the shell is (compare with Section 3.6):

$$[\mu]\{\ddot{q}\} + [k]\{q\} = 0$$

In long-hand notation,

$$\mu\ddot{u} + k_{11}u + k_{12}v + k_{13}w = 0$$
$$\mu\ddot{v} + k_{21}u + k_{22}v + k_{23}w = 0 \qquad\qquad (4.12)$$
$$\mu\ddot{w} + k_{31}u + k_{32}v + k_{33}w = 0$$

The natural frequencies of the cylindrical shell are given by the characteristic equation:

$$\begin{vmatrix} k_{11} - \mu\omega_{mn}^2 & k_{12} & k_{13} \\ k_{21} & k_{22} - \mu\omega_{mn}^2 & k_{23} \\ k_{31} & k_{32} & k_{33} - \mu\omega_{mn}^2 \end{vmatrix} = 0 \qquad (4.13)$$

Those with access to finite-element structural analysis computer programs will most likely find the in-air natural frequencies with finite-element models. For uniform cylindrical shells with either both ends simply-supported or clamped, analytical solutions for the stiffness elements have been derived based on thin shell theory, $(h/R)^2 \ll 1$ and $(h/L)^2 \ll 1$:

Shells with both ends simply-supported (Arnold and Warburton, 1952):

$$k_{11} = (m\pi/L)^2\kappa + n^2\kappa(1-\sigma)/(2R^2)$$
$$k_{12} = -mn\pi\kappa(1+\sigma)/(2LR)$$
$$k_{13} = m\pi\kappa\sigma/(LR)$$
$$k_{22} = n^2\kappa/R^2 + \kappa(1-\sigma)(m\pi/L)^2/2 \qquad (4.14)$$
$$k_{23} = -n\kappa/R^2$$
$$k_{33} = \kappa/R^2 + D[n^2/R^2 + (m\pi/L)^2]^2$$

where

$$\kappa = Et/(1-\sigma^2)$$

$$D = Et^3/[12(1-\sigma^2)]$$

(4.15)

For shells with clamped ends (Au-Yang, 1978, 1986)[1]:

$$k_{11} = \frac{\kappa}{C_m^2}\left(\frac{\gamma_m}{L}\right)^2 + \frac{n^2\kappa(1-\sigma)}{2R^2}$$

$$k_{12} = -C_m\frac{n\kappa}{R}\frac{\gamma_m}{L}\frac{1+\sigma}{2}$$

$$k_{13} = C_m\frac{\kappa\sigma}{R}\frac{\gamma_m}{L}$$

(4.16)

$$k_{22} = \frac{\kappa n^2}{R^2} + C_m^2\frac{\kappa(1-\sigma)}{2}\left(\frac{\gamma_m}{L}\right)^2$$

$$k_{23} = -\frac{\kappa n}{R^2}$$

$$k_{33} = \frac{\kappa}{R^2} + D\left(\frac{n^4}{R^4} + 2C_m^2(\frac{\gamma_m}{L})^2(\frac{n}{R})^2 + (\frac{\gamma_m}{L})^4\right)$$

The constants C and γ are given in the Table 4.1 for the first six axial modes:

Table 4.1: Constants for the Stiffness Matrix Elements

m	C	γ
1	0.7416	4.7300
2	-0.8642	7.8532
3	0.9045	10.9956
4	-0.9266	14.1372
5	0.9403	17.2788
6	-0.9498	20.4203

---

[1] Note: Equation (4.14) has been considerably simplified from the expressions in Au-Yang's (1978) original publication, based on a hint by Warburton (1978). The constants C_m here are related to the constants C_1, C_2, C_{13} etc by $C=C_1/C_2$ for $m=1, 3, 5 \ldots$ and $C=-C_3/C_1$ for $m=2, 4, 6. \ldots$ Also in Equation (4.14), the signs in k_{12} and k_{13} are opposite to those in Au-Yang's (1978) paper in order for the expressions to be more like the corresponding ones in the simply-supported cylinder. The final computed frequencies do not change as a result of changing the signs for both matrix elements at the same time. These expressions are also much simpler.

The natural frequencies of the shell can be obtained by substituting k into the characteristic Equation (4.13). There will be one set of k and three roots for each pair of mode numbers (m, n). Two of these frequencies are usually much higher than the third one. These are associated with vibrations predominantly in the axial and tangential directions. The lowest frequency is associated with vibrations predominantly in the normal direction. This is the flexural mode and is the mode of interest in structural dynamic analysis.

Equation (4.14) offers a simple alternate way to the finite-element method of computing the natural frequencies of cylindrical shells with clamped ends. It can be easily implemented into a spreadsheet and the natural frequencies can be obtained by numerically searching the zeros of the determinant $|M\text{-}\mu\omega^2|$. Furthermore, as will be seen later, it can easily be adopted to find the natural frequencies of the cylindrical shells with fluid-structure coupling. Application of Equation (4.14) to thin shells of aspect ratios commonly encountered in the power and process industries gives results that agree well with experimental results in a much shorter time compared with the finite-element or other "exact methods." An example of using Equation (4.14) will be given at the end of Subsection 4.4.

4.3   Acoustic Modes of a Fluid Annulus Bounded by Rigid Walls (Figure 4.2)

After we establish the dynamics of the cylindrical shells, we must turn our attention to the dynamics of the fluid between the two cylinders. As mentioned before, under our weak coupling assumption, the equations of fluid dynamics in the annular cavity are replaced by the much simpler acoustic equations. To distinguish the acoustic modes from the structural modes, we shall denote the acoustic mode numbers in the axial and circumferential directions by $\alpha, \beta = 1, 2, 3 \ldots$ The acoustic pressure (p') distribution in an annular cavity bounded by rigid walls can be written as a product of three functions, one function in each of the cylindrical coordinate variables:

$$p'(\vec{r}) = \sum_{\alpha\beta} R_{\alpha\beta}(\vec{r})\phi_\alpha(x)\Theta_\beta \tag{4.17}$$

where R is the radial function, ϕ is the axial (acoustic) mode shape function and Θ is the circumferential mode shape function. Depending on the end conditions of the annular cavity, the axial shape function takes the form:

$$\phi_\alpha(x) = \sqrt{\frac{2}{L}}\sin\frac{\alpha\pi x}{L} \quad \text{if both ends are open (pressure released, or } p=0 \text{ at } x=0, L) \tag{4.18}$$

$$\phi_\alpha(x) = \sqrt{\frac{2}{L}}\cos\frac{\alpha\pi x}{L} \quad \text{with} \quad \phi_0 = \frac{1}{\sqrt{L}} \tag{4.19}$$

if both ends are closed (p=max at x=0, L):

$$\phi_\alpha(x) = \sqrt{\frac{2}{L}} \sin\frac{(2\alpha-1)\pi x}{2L} \quad \text{or} \quad \phi_\alpha(x) = \sqrt{\frac{2}{L}} \cos\frac{(2\alpha-1)\pi x}{2L} \tag{4.20}$$

if one end is open while the other end is closed. The circumferential mode shape function is given by

$$\Theta_\beta = \cos\beta\theta / \sqrt{\pi} \tag{4.21}$$

As with the structural mode shape functions, the acoustic mode shape function must also be properly normalized:

$$\int_0^L \phi_\alpha(x)\phi_\beta(x)dx = \delta_{\alpha\beta} \tag{4.22}$$

$$\frac{1}{\pi}\int_0^{2\pi} \cos\alpha\theta \cos\beta\theta d\theta = \delta_{\alpha\beta} \tag{4.23}$$

In the small displacement limit, the pressure distribution $p'(r)$ must satisfy the wave equation:

$$(\nabla^2 - \frac{1}{c^2}\frac{\partial^2}{\partial t^2})p'(\vec{r}) = 0 \tag{4.24}$$

where ∇^2 is the Laplacian in cylindrical coordinates. Just as a combination of sine and cosine functions is the most general solution for the wave equation in Cartesian coordinates, the general solution for the wave equation in cylindrical coordinates is given by a linear combination of Bessel functions and modified Bessel functions J, Y, I and K. On the surface of the inner rigid wall, the radial function R must satisfy the boundary condition that the normal component of the acceleration must be zero. That is,

$$\left|\frac{\partial p'}{\partial r}\right|_{r=a} = 0 \tag{4.25}$$

The radial function that satisfies the Laplace Equation (4.24) and the boundary condition Equation (4.25) is given by (Au-Yang, 1976):

$$R_{\alpha\beta}(r) = N_{\alpha\beta}[J_\beta(\chi_\alpha r) - J'_\beta(\chi_\alpha a)Y_\beta(\chi_\alpha r)/Y'_\beta(\chi_\alpha a)] \tag{4.26}$$

for

$$\chi_\alpha^2 = (2f/c)^2 - (\varepsilon/L)^2 > 0$$

and

$$R_{\alpha\beta}(r) = N'_{\alpha\beta}\,[I_\beta(\chi_\alpha r) - I'_\beta(\chi_\alpha a)K_\beta(\chi_\alpha r)/K'_\beta(\chi_\alpha a)] \tag{4.27}$$

for

$$\chi_\alpha^2 = (\varepsilon/L)^2 - (2f/c)^2 > 0$$

where

c = speed of sound

$\varepsilon = \alpha$ for fluid annulus with either both ends open ($p = 0$) or both ends closed ($p = p_{max}$),
   $= (2\alpha-1)/2$ if one end is open and the other is closed.

$N_{\alpha\beta}$, $N'_{\alpha\beta}$ are amplitude constants (compared with the amplitude constants A in Chapter 3), and J, Y, I, K are Bessel functions and modified Bessel functions of the first and second kinds.

So far we have imposed only the boundary conditions at the surface of the inner wall. The above equations, therefore, apply to the case of a single cylinder vibrating in an infinite fluid medium. Depending on whether

$$(2f/c)^2 - (\varepsilon/L)^2 > 0, \quad \text{or}$$
$$(2f/c)^2 - (\varepsilon/L)^2 < 0$$

the pressure can be in the form representing energy radiated from the vibrating cylinder, or an inertia effect. The frequency at which:

$$f = \varepsilon c/2L \tag{4.28}$$

is known in acoustics as the coincidence frequency. Above the coincidence frequency, energy is radiated away from the vibrating structure. The effect of fluid-structure interaction on the structure is an additional damping due to this energy radiation. Below the coincidence frequency, energy cannot be radiated, and the net effect of fluid-structure interaction on the structure is an "added mass" term. This will be called hydrodynamic mass in this book, although the term added mass is also commonly used in the literature.

The case under consideration in this chapter is that of an annular fluid gap bounded by rigid walls. Since there cannot be any energy loss (assuming inviscid fluid) in an enclosed fluid cavity, we know from the physics of the problem that, in this case, the effect of fluid-structure interaction on the structure can only be that of an added mass.

Imposing the rigid wall boundary condition on the outer radius, we have

$$\left|\frac{\partial p'}{\partial r}\right|_{r=b} = 0 \tag{4.29}$$

Just as the boundary conditions at the two ends of a simply-supported beam together with the fundamental equation of structural dynamics gives rise to the characteristic equation from which we can calculate the discrete natural frequencies of a vibrating beam, these rigid wall boundary conditions, Equations (4.25) and (4.29), lead to the following characteristic equation from which we can calculate the natural frequencies of the acoustic modes inside this annulus:

$$J'_\beta(\chi_\alpha a)Y'_\beta(\chi_\beta b) = J'_\beta(\chi_\alpha b)Y'_\beta(\chi_\beta a) \tag{4.30}$$

if $\quad (2f/c)^2 - (\varepsilon/L)^2 > 0 \quad$ and

$$I'_\beta(\chi_\alpha a)K'_\beta(\chi_\beta b) = I'_\beta(\chi_\alpha b)K'_\beta(\chi_\beta a) \tag{4.31}$$

if $\quad (2f/c)^2 - (\varepsilon\pi/L)^2 < 0$

Equation (4.31) has no real roots. All the natural frequencies of the fluid annulus are given by Equation (4.30), which can be solved numerically with today's computer software. From the above discussion, we see that acoustic modes in the annular cavity exist only if the frequency is higher than the coincidence frequency, so that:

$$f > \varepsilon c/2L$$

Below this frequency, there are no standing waves in the cavity. We shall return to Equation (4.30) in Chapter 12.

The Ripple Approximation

For many applications, the fluid annulus is thin compared with its radius, so that the condition $a/(b-a) >> 1$ is satisfied. Under this condition, the following approximate equation can be used to calculate the natural frequencies of the acoustic mode:

$$f_{\alpha\beta} = \frac{c}{2\pi}\left[\left(\frac{\varepsilon\pi}{L}\right)^2 + \left(\frac{2\beta}{a+b}\right)^2\right]^{1/2} \tag{4.32}$$

This is known as the ripple approximation. In this approximation, i.e., when the annular gap is thin, the curvature effect disappears. Equation (4.32) can therefore be used in rectangular fluid channels.

4.4   Vibration of a Coupled Fluid-Cylindrical Shell System

In Section 4.3 we discussed the acoustic modes in a cylindrical annulus of fluid bounded by two rigid walls. Now suppose these bounding walls are co-axial flexible cylinders—a situation often encountered in power and process plant components—then instead of boundary conditions Equations (4.25) and (4.29), we have, on the surfaces of the inner and outer cylinders (see Figure 4.2):

$$\left|\frac{\partial p'}{\partial r}\right|_{r=a} = -\rho \ddot{w}_a \quad and \quad \left|\frac{\partial p'}{\partial r}\right|_{r=b} = +\rho \ddot{w}_b \tag{4.33}$$

That is, on the surfaces of the inner and outer walls, the acceleration of the fluid particles $\pm (1/\rho)(\partial p'/\partial r)$ must be equal to the accelerations of the normal components w of the cylinders. These boundary conditions couple the fluid vibration modes in the annular gap with those of the cylindrical shells.

It was shown (Au-Yang, 1976) that, because of the orthogonality properties of the cosine functions, which are the common circumferential mode shape functions for both the cylindrical shell modes and the acoustic modes, there will be no cross-circumferential modal coupling between the cylindrical shell modes and the acoustic modes. We shall use a common index n to denote the circumferential mode of both cylinders and the fluid annulus. The normalized circumferential mode shapes for both the cylinders and the fluid gap are given by Equation (4.21), $\Theta(\theta) = \cos n\theta / \sqrt{\pi}$. In the special case when the annular gap has two open ends ($p=0$ at both ends) and both cylinders are simply-supported, then both the fluid and the cylinder axial modes are given by the sine function, and there will be no cross-modal coupling between the axial modes. In the general case, there will be cross-modal coupling between the cylinder and acoustic axial modes. We shall denote the structural axial mode shape function by $\psi_m(x)$ and acoustic axial mode shape function by $\phi_\alpha(x)$; both are normalized to unity.

There are two alternate methods of calculating the coupled fluid-shell natural frequencies: the 2 x 2 generalized hydrodynamic mass matrix, or the full hydrodynamic mass matrix. The first approach is a "hand calculation" using pocket calculators or personal computers with only standard office software such as Microsoft Office. It may start with in-air natural frequencies of the individual cylinders obtained by measurement, by solving the characteristic Equation (4.13), or from a finite-element computer program. The second approach is based on the finite-element method, with a structural analysis computer program and a separately computed full hydrodynamic mass matrix. This will be discussed in the following two sections.

4.5   The 2 x 2 Generalized Hydrodynamics Mass Matrix

Very often, the in-air natural frequencies of the individual cylinders are known either from measurement, from separate calculations using a finite-element computer program, or by solving the characteristic Equation (4.13). In this case the coupled frequencies of the fluid-shell system can be obtained using the "generalized hydrodynamic mass matrix" method. The generalized hydrodynamic masses for each of the cylindrical shells as a whole, as well as the cross coupling between the two shells are computed. The resulting matrix is 2 x 2. The following derivation is based on Au-Yang (1977) using a series expansion method. The structural mode shapes are expanded in terms of the acoustic mode shapes:

$$\psi_m(x) = \sum_\alpha C_{m\alpha}\phi_\alpha(x) \tag{4.34}$$

Since the acoustic modes shapes are sine or cosine functions (Equations (4.18) to (4.21)), the above is a Fourier series and the coefficients are Fourier coefficients given by:

$$C_{m\alpha} = \int_0^L \psi_m(x)\phi_\alpha(x)dx \tag{4.35}$$

Following this approach, it was shown that, as long as the vibration amplitudes of the cylinders are small compared with the dimensions of the annular fluid gap, the mth axial modal generalized pressure induced by the motion of cylinders on the surfaces of cylinders are proportional to the normal component of acceleration of the cylindrical shells:

$$p'_a = \sum_\alpha (C^a_{m\alpha} h^a_\alpha \ddot{w}_a + C^b_{m\alpha} h^{ab}_\alpha \ddot{w}_b)\phi_a \Theta_n$$

$$p'_b = \sum_\alpha (C^a_{m\alpha} h^{ba}_\alpha \ddot{w}_a + C^b_{m\alpha} h^b_\alpha \ddot{w}_b)\phi_a \Theta_n \tag{4.36}$$

h^a_α is the contribution of the αth acoustic mode to the pressure induced on cylinder a due to its own motion; h^{ab}_α is the contribution of the αth acoustic mode to the pressure on cylinder a due to the motion of cylinder b. h^b_α and h^{ba}_α have similar meanings for cylinder b. The *hydrodynamic mass components h* are given by

$$h^a_{\alpha n} = -\rho(x_\alpha a, x_\alpha b)'_n / [x_\alpha(x_\alpha a, x_\alpha b)''_n]$$

$$h^b_{\alpha n} = \rho(x_\alpha b, x_\alpha a)'_n / [x_\alpha(x_\alpha b, x_\alpha a)''_n]$$

$$h^{ab}_{\alpha n} = -\rho / [x^2_\alpha a(x_\alpha a, x_\alpha b)''_n] \tag{4.37}$$

$$h^{ba}_{\alpha n} = \rho / [x^2_\alpha b(x_\alpha b, x_\alpha a)''_n]$$

where

' denotes derivative with respect to the argument of the function

$$x_\alpha^2 = (\varepsilon\pi/\ell)^2 - (2\pi f/c)^2 \qquad if \quad 2f/c < \varepsilon/\ell \quad and$$

$$(x,y)'_n \equiv I_n(x)K'_n(y) - I'_n(y)K_n(x)$$
$$(x,y)''_n \equiv I'_n(x)K'_n(y) - I'_n(y)K'_n(x) = -(y,x)''$$

$$x_\alpha^2 = (2\pi f/c)^2 - (\varepsilon\pi/\ell)^2 \qquad if \quad 2f/c > \varepsilon/\ell \quad and$$

$$(x,y)'_n \equiv J_n(x)Y'_n(y) - J'_n(y)Y_n(x)$$
$$(x,y)''_n \equiv J'_n(x)Y'_n(y) - J'_n(y)Y'_n(x) = -(y,x)''$$

In spite of its formidable appearance, Equation (4.37) can be easily implemented on a spreadsheet or programmed in a standard computer language. The following formulas for the derivatives of the Bessel functions may help.

$$J'_n(x) = (nJ_n(x) - xJ_{n+1}(x))/x$$
$$Y'_n(x) = (nY_n(x) - xY_{n+1}(x))/x$$
$$I'_n(x) = (nI_n(x) + xI_{n+1}(x))/x \qquad\qquad (4.38)$$
$$K'_n(x) = (nK_n(x) - xK_{n+1}(x))/x$$

It should also be noted that in general, h^a and h^b should come out positive while h^{ab} and h^{ba} should come out negative, and that $bh^{ab} = ah^{ba}$. The generalized pressure (with respect to each cylindrical shell) is given by:

$$P'_{mna} = \int_0^L \psi_a p'_a \, dx \qquad\qquad (4.39)$$

with a similar expression for P'_b. From Equations (4.36) and (4.9),

$$P'_{amn} = (\sum_\alpha (C^a_{m\alpha})^2 h^a_{\alpha n}) \ddot{W}_{amn} + (\sum_\alpha C^a_{m\alpha} C^b_{m\alpha} h^{ab}_{\alpha n}) \ddot{W}_{bmn}$$
$$P'_{bmn} = (\sum_\alpha C^a_{m\alpha} C^b_{m\alpha} h^{ba}_{\alpha n}) \ddot{W}_{amn} + (\sum_\alpha (C^b_{m\alpha})^2 h^b_{\alpha n}) \ddot{W}_{bmn} \qquad (4.40)$$

This can be written as

$$\begin{Bmatrix} P'_{amn} \\ P'_{bmn} \end{Bmatrix} = \begin{bmatrix} M^a_{Hmn} & M^{ab}_{Hmn} \\ M^{ba}_{Hmn} & M^b_{Hmn} \end{bmatrix} \begin{Bmatrix} \ddot{W}_{amn} \\ \ddot{W}_{bmn} \end{Bmatrix} \tag{4.41}$$

or in matrix notation,

$$\{P'\} = [M_H]\{\ddot{W}\} \quad \text{for each set of mode number } (m, n) \tag{4.42}$$

where

$$M^a_{Hmn} = \sum_\alpha (C^a_{\alpha m})^2 h^a_{\alpha n} \qquad M^{ab}_{Hmn} = \sum_\alpha (C^a_{\alpha m} C^b_{\alpha m}) h^{ab}_{\alpha n}$$

$$M^{ba}_{Hmn} = \sum_\alpha (C^a_{\alpha m} C^b_{\alpha m}) h^{ba}_{\alpha n} \qquad M^b_{Hmn} = \sum_\alpha (C^b_{\alpha m})^2 h^b_{\alpha n} \tag{4.43}$$

Equation (4.42) shows that the net effect of the pressure induced on the cylinders is that of added mass terms. This should be expected from discussions in Section 4.3. Since the fluid annulus is completely enclosed, there can be no energy loss from the system. The net effect of fluid-structure coupling can only be an inertia term as is pointed out in Section 4.3. For this reason, the 2 x 2 matrix

$$[M_H]_{mn} = \begin{bmatrix} M^a_H & M^{ab}_H \\ M^{ba}_H & M^b_H \end{bmatrix}_{mn} \tag{4.44}$$

has been given the name "generalized added mass" or "generalized hydrodynamic mass". The latter terminology will be used in this book. It is "generalized" because there is one of this matrix per each set of mode number (m, n); and it is a mass term because the pressure is proportional to the normal component of the acceleration $\ddot{W}$. However, unlike the physical mass, the hydrodynamic mass is effective only in the normal component of the acceleration. In terms of the hydrodynamic mass matrix, the modal equation of motion for each cylinder becomes,

$$([\mu] + \{Q\}^T [M_H][\delta_{33}]\{Q\})(-\omega^2 + i2\omega\omega_{mn} + \omega^2_{mn})Q^a_{mn} = P_0 \tag{4.45}$$

where the matrix

$$[\delta_{33}] = \begin{bmatrix} 0 & 0 & 0 \\ 0 & 0 & 0 \\ 0 & 0 & 1 \end{bmatrix} \tag{4.46}$$

denotes the fact that the hydrodynamic mass matrix, being originally a pressure, is effective only in the normal component W of the motion of the shell. In the finite-element approach to be discussed in the next section, these hydrodynamic masses should be associated only with the normal degree-of-freedom in cylindrical coordinates. In the present approach, to be able to treat the hydrodynamic mass on the same basis as the physical mass density matrix $[\mu]$, we carry out the matrix multiplication and define the *effective* hydrodynamic mass matrix as

$$\hat{M}_{Hn} = \{Q\}^T [M_H][\delta_{33}]\{Q\}$$

$$= (U_n, V_n, W_n)\begin{bmatrix} 0 & 0 & 0 \\ 0 & 0 & 0 \\ 0 & 0 & M_{Hn}^a \end{bmatrix}\begin{pmatrix} U_n \\ V_n \\ W_n \end{pmatrix} = \frac{M_{Hn}}{1 + (U_n/W_n)^2 + (V_n/W_n)^2} \tag{4.47}$$

From the theory of thin circular cylindrical shells,

$$1 + (U_n/W_n)^2 + (V_n/W_n)^2 \approx \frac{1}{1 + n^{-2}} \tag{4.48}$$

It follows that,

$$\hat{M}_H = \frac{M_H}{1 + n^{-2}} \tag{4.49}$$

Physically this means that because the hydrodynamic mass is effective only in the normal direction, its inertia effect is not the same as that of a physical mass that is effective in all three degree-of-freedom. In the beam mode ($n=1$), its inertia effect is reduced by a factor of 2 compared with the physical mass. In higher shell modes when n is large, its inertia effect is almost the same as the physical mass. From Equations (4.43) and (4.49),

$$\hat{M}_{Hmn}^a = \frac{1}{1+n^{-2}}\sum_\alpha (C_{\alpha m}^a)^2 h_{\alpha n}^a \qquad \hat{M}_{Hmn}^{ab} = \frac{1}{1+n^{-2}}\sum_\alpha (C_{\alpha m}^a C_{\alpha m}^b)h_{\alpha n}^{ab}$$

$$\hat{M}_{Hmn}^{ba} = \frac{1}{1+n^{-2}}\sum_\alpha (C_{\alpha m}^a C_{\alpha m}^b)h_{\alpha n}^{ba} \qquad \hat{M}_{Hmn}^b = \frac{1}{1+n^{-2}}\sum_\alpha (C_{\alpha m}^b)^2 h_{\alpha n}^b \tag{4.50}$$

The modal equation of motion for the cylinders can now be written as (suppressing the modal index m, n for brevity),

$$\left\{-\omega^2\left(\begin{bmatrix} \mu_a & 0 \\ 0 & \mu_b \end{bmatrix} + \begin{bmatrix} M_a\delta_{33} & M_{ab}\delta_{33} \\ M_{ba}\delta_{33} & M_b\delta_{33} \end{bmatrix}\right) + \begin{bmatrix} k_a & 0 \\ 0 & k_b \end{bmatrix}\right\}\begin{Bmatrix} Q_a \\ Q_b \end{Bmatrix} = \begin{Bmatrix} P_{0a} \\ P_{0b} \end{Bmatrix} \tag{4.51}$$

The equations of motion for the cylinders are now coupled together through the off-diagonal element of the hydrodynamic mass matrix. It must be emphasized, however, that coupling is only through the normal degrees-of-freedom of the shells, as denoted by the δ_{33} matrix in Equation (4.51). The coupled natural frequencies can be computed directly using the stiffness matrices and the *effective* hydrodynamic mass matrix:

$$
\begin{vmatrix}
 & & 0 & 0 & 0 \\
 D(a) & & 0 & 0 & 0 \\
 & & 0 & 0 & \hat{M}_H^{ab} \\
 0 & 0 & 0 & & \\
 0 & 0 & 0 & D(b) & \\
 0 & 0 & \hat{M}_H^{ba} & &
\end{vmatrix} = 0
\tag{4.52}
$$

where

$$
D(a) = \begin{bmatrix}
 k_{11}^a - \mu_a\,\omega^2 & k_{12}^a & k_{13}^a \\
 k_{12}^a & k_{22}^a - \mu_a\,\omega^2 & k_{23}^a \\
 k_{13}^a & k_{13}^a & k_{33}^a - (\mu_a + \hat{M}_H^a)\omega^2
\end{bmatrix}
$$

In spite of its formidable appearance, Equation (4.52) can be easily solved numerically with a spreadsheet and its built-in mathematical functions, as illustrated in the Examples 4.1 and 4.2. Notice that the off-diagonal hydrodynamic elements $\hat{M}_H^{ab}$, $\hat{M}_H^{ba}$ are numerically negative.

The Physics of Fluid-Structure Coupling

Equation (4.50) indicates that the effective hydrodynamic mass depends not only on h, which can be regarded as the hydrodynamic mass components, but also the Fourier coefficients C. From its definition, Equation (4.35), the Fourier coefficient $C_{m\alpha}$ is a measure of the compatibility of structure mode m and acoustic mode α. If the two modes have nothing in common, i.e., if these two modes are orthogonal to each other, then no matter how large the acoustic pressure, (i.e., h), it cannot couple with the structure. As a result, that particular acoustic mode cannot contribute to the hydrodynamic mass $\hat{M}^a$, because $C_{m\alpha}$ is zero (Figure 4.3). Failure to include the Fourier coefficients into the hydrodynamic mass formulation of fluid-structure coupling has lead to many apparently paradoxical conclusions. This will be discussed in Chapter 5.

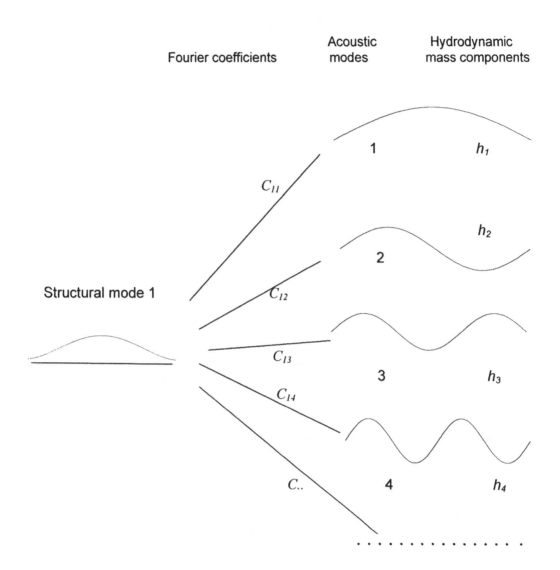

Figure 4.3 The physics of fluid-structure coupling

Example 4.1: Coupled Shell-Mode Vibration

It is suggested that the next generation of nuclear reactors should be of low-pressure design, operating at moderate temperatures so that they are intrinsically safe. In a hypothetical design, the structure supporting the core and the outer reactor vessel can be modeled as two coaxial cylindrical shells, with water filling the gap between them (Figure 4.4). The dimensions of these two cylinders and the pressure and temperature of the water between the two cylinders are given in Table 4.2. To illustrate the dramatic effect of the water in the annular gap on the natural frequencies of the shells, we assume that there is no water inside the inner shell.

Table 4.2: Data for Coaxial Cylindrical Shells

	US Unit	SI Unit
Length L	300 in	7.620 m
Inner shell outer radius	78.5 in	1.994 m
Outer shell inner radius	88.5 in	2.248 m
Thickness of inner shell, t	1.5 in	0.0254 m
Thickness of outer shell, t	1.5 in	0.0254 m
Boundary condition of shell:	Clamped both top and bottom	
Boundary condition of water annular gap:	Closed at the top and pressure released at the bottom	
Density of material, both shells	0.2831 lb/in^3 (7.327E-4 (lb-s^2/in) per in^3)	7863 Kg/m^3
Poison ratio	0.29	0.29
Young's modulus	2.96E+7 psi	2.041E+11 Pa
Water between cylinders:		
Water temperature	300 deg. F	148.9 deg. C
Water pressure	80 psi	5.516E+5 Pa
Assume no water inside the inner shell		
Surface density $\mu=t\rho$ both shells	0.001099 (lb-s^2/in) per in^2	298.6 Kg/m^3

From ASME Steam Table (1979), we find the following:

	US Unit	SI Unit
Density of water	57.3 lbm/ft^3 (8.582E-5 (lb-s^2/in) per in^3)	917.9 Kg/m^3
Velocity of sound	4769 ft/s (57230 in/s)	1453.6 m/s

Solution

Following our standard practice, we convert the given mass density into a consistent unit in the US unit system. This is given in the parentheses in Table 4.2, from this and the given thickness of the shells, we can calculate the surface mass densities of the shells. These are given in Table 4.4. From the ASME Steam Table (1979), we can find the density of water at the given temperature and pressure. There are two alternate ways to estimate the velocity of sound in water at the given ambient conditions: The first is by interpolating the given values in Table 2.1. The second is by calculation based on a method outlined in Chapter 12, Example 12.2. The values given in the last row in Table 4.2 were obtained from the latter method. The following paragraphs outline the steps for the rest of the solution.

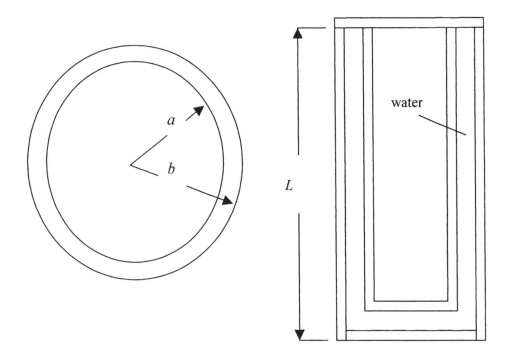

Figure 4.4: Coupled fluid-shell system

The Stiffness Matrix and the Natural Frequencies

The first step is to calculate the stiffness matrix from Equation (4.15), after which the in-air frequencies of each shell can be obtained from Equation (4.13). The following stiffness matrices were computed with Microsoft Excel. A Fortran program can also be used. Since the stiffness matrices are given for the purposes of illustration, for brevity only the matrix elements for the $(m, n)=(1, 3)$ mode in the US unit system are shown. Notice that as long as we use consistent units within each unit system, we do not have to carry along the units in each step of the calculation. Mean radii of 77.75 in. and 89.25 in. are used in Equation (4.15) in calculating the following stiffness matrix elements.

$(m, n) = (1, 3)$ k matrix in US units:

Inner cylinder k^a

	u	v	w
u	47533	-14107	2114
v	-14107	74526	-24058
w	2114	-24058	8044

Outer cylinder k^b

	u	v	w
u	41356	-12289	1842
v	-12289	57125	-18257
w	1842	-18257	6101

The in-air (1,3) modal frequencies for cylinder a is given by, from Equation (4.13),

$$\left| [k_a] - \mu_a (2\pi f)^2 [I] \right| = 0 \quad \text{with } \mu_a = 0.0010991 \text{ (lb-s}^2/\text{in) per in}^2.$$

and similarly for cylinder b. The above determinant equation can be easily solved for the unknown f numerically by trial and error, using a spreadsheet. There will be three roots for each cylinder for each pair of modal indices (m, n). Of these, the one with the lowest values is the mode with motion predominantly in the radial direction. This is the flexural mode and is the only mode of interest in the subsequent dynamic analysis. Therefore, in searching for the zeros of the above equation, it is important to start from low values of f. The following table gives the values of the in-air modal frequencies for the two cylinders obtained with the MDETERM function of Microsoft Excel. Note that the (1,4) mode has the lowest in-air natural frequency. It is not uncommon in shells that the mode with n larger than 1 has the lowest modal frequency.

Table 4.3: In-air natural frequencies (Hz) of cylinders

Mode No. (m,n)-->	(1,1)	(1,2)	(1,3)	(1,4)	(1,5)
Cylinder					
Inner (a)	151.3	84.7	54.8	51.0	64.6
Outer (b)	151.7	87.7	56.3	46.4	52.6

Computation of the Fourier Coefficients

The first step in calculating the hydrodynamic mass matrices is to calculate the Fourier coefficients, Equation (4.35). As discussed earlier in this section, the Fourier coefficient is a measure of the compatibility between a structural and an acoustic mode.

$$C_{ma} = \int_0^L \psi_m(x)\phi_\alpha(x)dx = \sum \psi_m(x_i)\phi_\alpha(x_i)\Delta x_i$$

From Equation (4.20), for a water annular gap closed at the top and pressure released at the bottom, the acoustic mode shape function is:

$$\phi_\alpha(x) = \sqrt{\frac{2}{L}} \cos\frac{(2\alpha - 1)\pi x}{2L}$$

while the m=1 axial structural mode shape for a shell with both ends clamped are from Table 3.1, Chapter 3,

$$\psi_1(x) = \frac{1}{\sqrt{L}}\{[\cosh(\frac{\lambda_1 x}{L}) - \cos(\frac{\lambda_1 x}{L})] - \sigma_1[\sinh(\frac{\lambda_1 x}{L}) - \sin(\frac{\lambda_1 x}{L})]\}$$

with $\lambda_1 = 4.73$, $\sigma_1 = 0.9825$

The Fourier coefficients are computed with the above axial mode shape functions and Equation (4.35) by numerical integration with a spreadsheet, using $\Delta x_i = L/30$, with the following results:

Table 4.4: Fourier Coefficients for the Cylinders

Shell Boundary Conditions	C_{11}	C_{12}	C_{13}	C_{14}	C_{15}
Both ends clamped	0.7958	-0.5447	-0.2393	0.02843	-0.03010

Since both cylinders have the same boundary conditions and the same lengths, the Fourier coefficients are the same for both cylinders.

Computation of the Hydrodynamic Masses

The next step is to calculate the hydrodynamic component h, using Equation (4.37). Since the hydrodynamic mass is very sensitive to the annular gap width, when calculating the hydrodynamic mass components h, it is important that the radii of the cylinders be chosen such that the true width (10 in. in the present case) of the gap is preserved. A spreadsheet can be used to calculate h. The following is computed with a small Fortran program. Because the hydrodynamic masses are frequency-dependent, we must initially compute these at a "guessed" frequency. The following are computed at a frequency of 15 Hz, with radii of 78.5 and 88.5 in. for the inner and outer shell, respectively. For brevity only the hydrodynamic masses for $n=3$ and in US units are given (notice the off-diagonal matrix elements are negative).

Table 4.5: Hydrodynamic mass components for $n=3$ (in (lb-s^2/in) per in^2)

α	h^{aa}	h^{ab}	h^{bb}	h^{ba}
1	6.3930E-03	-6.7570E-03	7.2070E-03	-5.9930E-03
2	5.5200E-03	-5.7710E-03	6.2180E-03	-5.1190E-03
3	4.3560E-03	-4.4570E-03	4.8990E-03	-3.9530E-03
4	3.3370E-03	-3.3070E-03	3.7440E-03	-2.9330E-03
5	2.5720E-03	-2.4450E-03	2.8770E-03	-2.1690E-03

The effective hydrodynamic masses are, from Equation (4.50), equal to (we now suppress the modal indices (m, n), with the understanding that this is for the $(1,3)$ mode),

$$\hat{M}_H^a = \frac{1}{1+n^{-2}} \sum_\alpha (C_\alpha^a)^2 h_\alpha^a , \quad n=3$$

with similar expressions for the $M_H^{ab}, M_H^{ba}, M_H^b$. The results are given below:

$$\left[\hat{M}_H\right] = \begin{bmatrix} 5.400E-3 & -5.682E-3 \\ -5.040E-3 & 6.086E-3 \end{bmatrix} \quad \text{(lb-s}^2/\text{in)/in}^2 \text{ for } (m,n)=(1,3)$$

The "In-Water" and Coupled Frequencies

We now proceed to calculate three separate "in-water" frequencies—that of cylinder a *assuming* cylinder b is rigid; that of cylinder b, *assuming* cylinder a is rigid; and finally, the coupled frequency of both when they are flexible. Again, in each case we search only for the lowest modal frequency.

In-water (1,3) frequency of cylinder a when cylinder b is rigid:

The flexural $(1,3)$ modal frequency is given by lowest root of the determinant equation:

$$\left| [k_a] - (\mu_a + \hat{M}_H^a)(2\pi f)^2 [I] \right| = 0, \text{ or}$$

$$\left| \begin{bmatrix} 47533 & -14107 & 2114 \\ -14107 & 74526 & -24058 \\ 2114 & -24058 & 8044 \end{bmatrix} - (2\pi f)^2 \begin{bmatrix} 1.0991E-3 & 0 & 0 \\ 0 & 1.0991E-3 & 0 \\ 0 & 0 & 6.4992E-3 \end{bmatrix} \right| = 0$$

This equation can be easily solved by trial and error, using, for example, the MDETERM function in Microsoft Excel. For the flexural mode, we get,

$$\bar{f}_{13}^a = 23.6 \text{ Hz, similarly,}$$
$$\bar{f}_{13}^b = 23.0 \text{ Hz}$$

compared with in-air frequencies of 54.8 and 56.3, respectively. Finally, the coupled frequencies of the shells are, from Equation (4.51), given by the roots of the characteristic equation:

$$
\begin{bmatrix}
47533 & -14107 & 2114 & 0 & 0 & 0 \\
-14107 & 74526 & -24058 & 0 & 0 & 0 \\
2114 & -24058 & 8044 & 0 & 0 & 0 \\
0 & 0 & 0 & 41356 & -12289 & 1842 \\
0 & 0 & 0 & -12289 & 57125 & -18257 \\
0 & 0 & 0 & 1842 & -18257 & 6101
\end{bmatrix} - (2\pi f)^2
$$

$$
\begin{bmatrix}
1.0991\text{E-}03 & 0 & 0 & 0 & 0 & 0 \\
0 & 1.0991\text{E-}03 & 0 & 0 & 0 & 0 \\
0 & 0 & 6.4992\text{E-}03 & 0 & 0 & -5.6821\text{E-}03 \\
0 & 0 & 0 & 1.0991\text{E-}03 & 0 & 0 \\
0 & 0 & 0 & 0 & 1.0991\text{E-}03 & 0 \\
0 & 0 & -5.0398\text{E-}03 & 0 & 0 & 7.1848\text{E-}03
\end{bmatrix} = 0
$$

Again this equation can be solved numerically by trial and error, using, for example, the MDETERM function in MicroSoft Excel, except now we have *two* (1,3) flexural modal frequencies: With the lower one representing the two cylinders vibrating out-of-phase, while the higher one represents the two cylinders vibrating in-phase, we find

Coupled out-of-phase (1,3) frequency = 17.5 Hz.
Coupled in-phase (1,3) frequency = 48.5 Hz

The dramatic effect of fluid-structure coupling on the natural frequencies of large-scale structures can now be appreciated. Notice that so far we purposely assume there is no water inside the inner cylinder. All this fluid-loading effect is due to the small amount of water in the narrow annular gap. This phenomenon has been observed during pre-operational tests of nuclear reactors.

Example 4.2: Beam Mode Vibration inside an Annular Fluid Gap

A typical pressurized water nuclear reactor can be modeled as two concentric cylinders with the annular gap between them filled with water (see Figures 2.20 and 4.5). The core of the reactor (the inner cylinder) is supported at the top by a flange with an equivalent rotational spring constant. The core can be assumed to vibrate in a rigid body, compound pendulum mode. The "in-air" frequency and the equivalent rotational spring constant have been computed separately. The diameters of the two cylinders and other relevant information are given in Table 4.6. Find the effective hydrodynamic and the "in-water" natural frequency for the beam ($n=1$) mode vibration of the core inside the stationary reactor vessel. It can be assumed that the mass of the core is evenly distributed along its entire length.

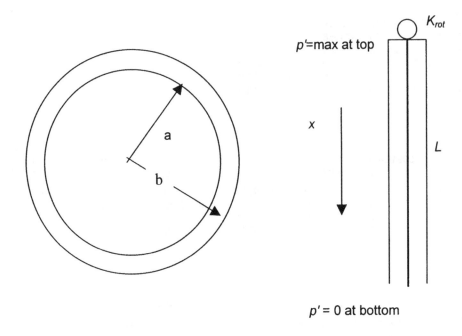

Figure 4.5 Schematic of the model

Table 4.6: Data for a Pressurized Water Reactor

	US Unit	IS Unit
Core (inner cylinder) OD	157 in	3.988 m
Vessel (outer cylinder) ID	177 in	4.496 m
Length L	300 in	7.62 m
Water temperature	575 deg. F	301.7 C
Pressure	2200 psia	1.517E+7 Pa
Total core weight (including water)	728000 lb	330215 Kg
	(22608 slugs; 1884 (lb-s^2/in))	
"In-air" beam mode frequency	21.9 Hz	21.9 Hz
Equi. spring constant at support	1.07E+12 lb-in/rad	1.209E+11 N-m/rad

This example is presented in a form problems of this type may be posed to the practical engineers. Very often the "in-air" frequencies, that is, frequencies without fluid-structure coupling, are computed separately with a finite-element model. The water contained in the inner cylinder is included in the calculation. However, the hydrodynamic mass effect due to the water in the annular gap between the inner and outer cylinders has not been included. The engineer is now asked to assess the effect of fluid-structure coupling.

Solution

U.S. engineers must be ready to handle "non-consistent" units that are frequently given to them. In dynamic calculations, it is important to convert all these units into consistent units (such as mass unit in (lb-s^2/in), see Chapter 1). After the engineer becomes proficient enough, he may be able to skip some of these, as in the case when the masses are involved in simple calculation of ratios. Even though it is not absolutely necessary in this particular example, the mass units are nevertheless changed to the consistent unit of (lb-s^2 /in) in the US unit system in order to form a habit. From ASME Steam Table (1979):

	US Unit	SI Unit
Water density, ρ	45.1 lb/ft^3	723 Kg/m^3
	(1.401 slug/ft^3 or 6.7545E-5 (lb-s^2/in)/in^3)	
Dynamic viscosity μ	1.834E-6 lbf-s/ft^2	8.778E-5 Pa-s
Kinematic viscosity $v = \mu/\rho$ [1]	1.876E-4 in^2 -s	1.214E-7 m^2 -s
Velocity of sound, c	3099 ft/s	944.5 m/s
Inner radius, a	78.5 in	1.999 m
Outer radius, b	88.5 in	2.248 m
Gap width, g =b-a	10 in	0.254 m

---

[1] This value is computed here for use in an example in Chapter 5.

The velocity of sound is obtained either by interpolating the values given in Table 2.1, or by calculation based on a method outlined in Example 12.2 in Chapter 12. In computing the kinematic viscosity in US units, we must first convert the density into the proper mass per unit volume unit (see Chapter 1), in this case slug/cu.ft. The computed kinematic viscosity will be in ft^2 /s. We then multiply the number by 144 to convert it into sq.in./s unit that is the correct unit for structural analysis. The following shows the step-by-step calculations.

<u>The Acoustic and Structural Mode Shape Functions</u>

The reactor core (see Figure 2.20) is supported at the top by a flange. Therefore, the top (x=0) of the water gap is closed (p'=maximum). The bottom of the annular gap is open to the lower plenum. Therefore, p'=0 at x=L. From Equation (4.20), the *normalized* acoustic mode shapes are given by,

$$\phi_\alpha(x) = \sqrt{\frac{2}{L}} \cos\frac{(2\alpha - 1)\pi x}{2L} \quad \alpha=1, 2, 3 \ldots$$

The *normalized* mode shape for a beam vibrating in compound pendulum mode is

$$\psi_1(x) = \sqrt{\frac{3}{L^3}}\; x$$

The subscript 1 emphasizes that we are considering the first beam mode. The reader may want to prove for himself that the above mode shapes is normalized to unity, Equation (4.10).

The Fourier Coefficients

The Fourier coefficients are given by (Equation (4.35)),

$$C_{1\alpha} = \int_0^L \psi_1 \phi_1 dx = \frac{\sqrt{6}\Delta x_i}{L^2} \sum x_i \cos\frac{(2\alpha - 1)\pi x_i}{2L}$$

The above summation can be easily carried out with today's widely available software. The following are computed with a spreadsheet with 30 integration steps. These coefficients are dimensionless and are therefore the same in both unit systems. They are given in Table 4.7 for $\alpha = 1$ to 8—it can be seen that they converge quite satisfactorily.

Table 4.7: Fourier Coefficients and Hydrodynamic Mass Components

α	Fourier Coeff. C	h ((lb-s^2/in)/in^2)	h(Kg/m^2)
1	0.567	0.04055	11017
2	-0.631	0.01708	4640
3	0.273	0.00832	2180
4	-0.244	0.00453	1230
5	0.163	0.00292	795
6	-0.152	0.00207	562
7	0.117	0.00155	422
8	-0.111	0.00122	330

The Hydrodynamic Mass Components h

Though Equations (4.37), (4.38) and the recursive equations for the derivatives of the Bessel functions look formidable, they can be easily coded in any standard computer language or on a spreadsheet, from which h can be computed in the general case. The values given in Table 4.7 were computed with a small Fortran program. It should be cautioned here that since the hydrodynamics mass is extremely sensitive to the gap width, it is important to preserve the true dimension of the annular gap width (10 in. or 0.254 m). Using these values of the hydrodynamic mass components together with the respective Fourier coefficients, we find from Equations (4.43), (4.49) and (4.52):

	US Unit	SI Unit
$\sum_{\alpha=1}^{8} C_{1\alpha}^{\prime 2} h_{\alpha}^{a} =$	2.086E-2 (lb-s^2/in)/in^2	5666 Kg/m^2
$\hat{M}^{a}(\pi a L) =$	1543 (lb-s^2/in)	2.705E+5 Kg
$f_{1}^{water} =$	16.2 Hz	16.2 Hz

By comparison, the physical mass of the water in the annular gap is only 106 lb-s^2/in. The hydrodynamic mass of the inner cylinder is almost 15 times the mass of water in the annular gap.

4.6   Extension to Double Annular Gaps

The development in Section 4.5 can be extended to cylindrical shells surrounded by an inner and an outer annular gap. In this case the diagonal elements of the hydrodynamic mass matrix will be the sum of the diagonal elements of the hydrodynamic mass matrix due to the inner and the outer fluid gaps; while the off-diagonal element remains the same as those between the shells coupled by the fluid gap. Equation (4.52) can be generalized to:

$$
\begin{vmatrix}
 & & & 0 & 0 & 0 & 0 & 0 & 0 \\
 & D(a) & & 0 & 0 & 0 & 0 & 0 & 0 \\
 & & & 0 & 0 & \hat{M}_{H}^{ab} & 0 & 0 & 0 \\
0 & 0 & 0 & & & & 0 & 0 & 0 \\
0 & 0 & 0 & & D(b) & & 0 & 0 & 0 \\
0 & 0 & \hat{M}_{H}^{ba} & & & & 0 & 0 & \hat{M}_{H}^{bc} \\
0 & 0 & 0 & 0 & 0 & 0 & & & \\
0 & 0 & 0 & 0 & 0 & 0 & & D(c) & \\
0 & 0 & 0 & 0 & 0 & \hat{M}_{H}^{cb} & & &
\end{vmatrix} = 0
$$

Thus if in Example 4.1, the inside of cylinder a is filled with water as is most likely the case in practice, then the effective hydrodynamic mass due to the water contained in the inner cylinder must be calculated with equations derived in chapter 5 and added to the total effective hydrodynamics mass $\hat{M}_H^a$.

4.7   The Full Hydrodynamic Mass Matrix

Very often it is desirable to implement the hydrodynamic mass matrix into a finite-element model and compute the coupled response of the cylinders directly. In this case we must derived the *full* hydrodynamic mass matrix, instead of the 2 x 2 generalized hydrodynamic mass matrix given in Section 4.5. Following the same method as described in Section 4.5, Au-Yang and Galford (1981) showed that the pressure induced by the motion of the cylinders on themselves is given by

$$\{p'\} = [M_H]\{\ddot{w}\} \tag{4.54}$$

where

$$\{p'\} = \begin{Bmatrix} p'_a \\ p'_b \end{Bmatrix}$$

$$\{\ddot{w}\} = \begin{Bmatrix} \ddot{w}_a \\ \ddot{w}_b \end{Bmatrix}$$

$$[M_H] = \begin{bmatrix} (1/a)\sum_\alpha h_{\alpha n}^a [\phi_\alpha dA_a][\phi_\alpha dA_a]^T & (1/b)\sum_\alpha h_{\alpha n}^{ab} [\phi_\alpha dA_a][\phi_\alpha dA_b]^T \\ (1/a)\sum_\alpha h_{\alpha n}^{ba} [\phi_\alpha dA_b][\phi_\alpha dA_a]^T & (1/b)\sum_\alpha h_{\alpha n}^b [\phi_\alpha dA_b][\phi_\alpha dA_b]^T \end{bmatrix} \tag{4.55}$$

Since h^{ab} and h^{ba} are negative, the two off-diagonal sub-matrices in the above matrix have negative matrix elements. Note that from Equation (4.42), since the hydrodynamic mass is actually an induced pressure, it is effective only in the normal component w in the equation of motion for the shells. Therefore, when inputting into the mass nodes in a finite-element model, *the hydrodynamic masses should be associated with only the normal degree-of-freedom* (Figure 4.6). Most commercial finite-element computer programs offer this option.

Equation (4.55) indicates that in addition to the hydrodynamic mass at every structural node that is immersed in the fluid, each of these nodes is coupled (through the off-diagonal terms) to each of the other nodes immersed in the fluid, as shown in Figure 4.7.

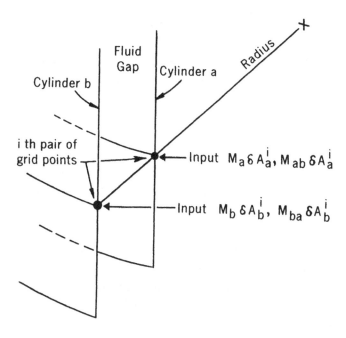

Figure 4.6 Direct input of hydrodynamic masses into grid point of a finite-element model

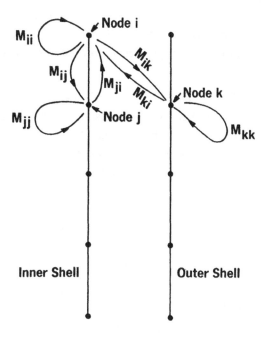

Figure 4.7 Hydrodynamic mass coupling

Example 4.3: Beam Mode Vibration inside a Fluid Annulus—Full Matrix Method

Return to Example 4.2: Now suppose the reactor vessel (the outer shell) is partially constrained. Calculate the full hydrodynamic mass matrix for the coupled beam mode vibration of these two shells.

Solution

This situation often exists in practice. Instead of two coaxial cylinders vibrating freely, very often the system is elastically restrained to the building. A finite-element structural analysis computer program will be necessary to calculate the responses. Under these conditions, it is much easier to calculate the coupled frequencies of the coaxial cylinders in Example 4.2 using the full hydrodynamic mass approach. There are two points we have to remember for an efficient computation: First, unlike physical masses or structural properties that may contain discontinuities, the hydrodynamic masses are in general well-behaved functions of the spatial variable. Considerably fewer nodes are required to account for the effect of fluid coupling than for other parts of structural analysis. This means that we do not have to calculate the hydrodynamic masses at every structural node point. Second, in Equation (4.55), the summation over h is constant for each of the sub-matrices and thus needs to be done only four times. Once the hydro-dynamic mass components h are computed by using Equations (4.37) and (4.38) or one of the simplified methods discussed in Chapter 5, the full hydrodynamic mass can be computed easily with a spreadsheet. As an illustration, we choose a model with only 10 elements axially for each of the core (cylinder a) and the vessel (cylinder b), as shown in Figure 4.8.

The hydrodynamic mass components h_α^a have been computed in the last example. Using the same method, i.e., either by a spreadsheet or a Fortran program, the other hydrodynamic mass components can be computed. The values in Table 4.8 were computed with a small Fortran program. For brevity, only values in US units ((lb-s^2/in)/in^2) are given. For the purpose of illustration, only five acoustic modes are used in the sum in this example:

Table 4.8: Hydrodynamics Mass Components (in (lb-s^2/in)/sq.in.)

α	h^a	h^{ab}	h^b	h^{ba}
1	0.040577	-0.045388	0.045743	-0.040259
2	0.010708	-0.018890	0.019237	-0.016755
3	0.008010	-0.008662	0.009006	-0.007684
4	0.004532	-0.004741	0.005083	-0.004205
5	0.002922	-0.002930	0.003269	-0.002598

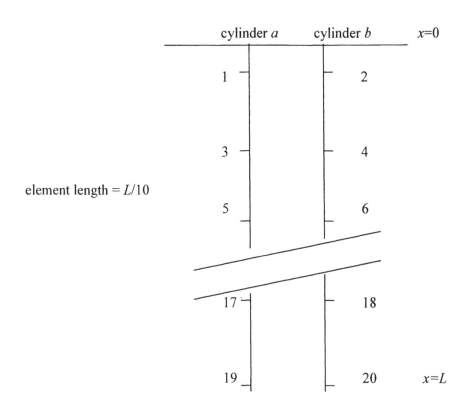

Figure 4.8 Finite-element model for coupled beam mode vibration

The element areas are:

$$dA_a = 2\pi a dL \quad and \quad dA_b = 2\pi b dL$$

respectively, for cylinders a and b. From Equation (4.55), the effective hydrodynamic mass for the beam mode ($n=1$) for each pair of nodes i, j are:

$$\hat{M}^a_{ij} = \pi a \sum_\alpha h^a_\alpha \phi_\alpha(x^a_i)\phi_\alpha(x^a_j)dL^2 \qquad \hat{M}^{ab}_{ij} = \pi a \sum_\alpha h^{ab}_\alpha \phi_\alpha(x^a_i)\phi_\alpha(x^b_j)dL^2$$

$$\hat{M}^{ba}_{ij} = \pi b \sum_\alpha h^{ba}_\alpha \phi_\alpha(x^b_i)\phi_\alpha(x^a_j)dL^2 \qquad \hat{M}^b_{ij} = \pi b \sum_\alpha h^b_\alpha \phi_\alpha(x^b_i)\phi_\alpha(x^b_j)dL^2$$

Note that these are "effective" (because they have been divided by the factor $1+n^{-2}$) but not "generalized" hydrodynamic mass, the latter is given by $\{\Psi\}^T[M_H]\{\Psi\}$ and is therefore dependent on the mode shapes of the two components.

Using the hydrodynamic mass components given in Table 4.8, as well as the acoustic mode shape for closed (at $x=0$) and opened (at $x=L$) boundary conditions,

$$\phi_\alpha(x) = \sqrt{\frac{2}{L}} \cos\frac{(2\alpha - 1)\pi x}{2L}$$

we can compute the full hydrodynamic mass matrix, using either a spreadsheet or a short computer program. The values in Table 4.9 were computed with a short Fortran program. For brevity only elements in alternate nodes are given. The following should be noted:

- The hydrodynamic mass is symmetric
- In each sub-matrix, the hydrodynamic mass elements decrease as their distance from the diagonal increase, showing that mass coupling decreases for nodes that are far apart.
- The matrix elements in sub-matrices $[M^a]$ and $[M^b]$ are all positive while those in sub-matrices $[M^{ab}]$ and $[M^{ba}]$ are all negative.

Using a finite-element computer program with a lump-mass structural model in which the physical mass of the water in the annular gap is added to the nodes, the coupled out-of-phase fundamental modal frequency was found to be 21.9 Hz. Using the hydrodynamic mass approach with the above full hydrodynamic mass added to the corresponding structural nodes, the coupled natural frequency of the out-of-phase mode was found to be 18.8 Hz. The frequency reduction is not as much as that in Example 4.2 because the outer cylinder is elastically constrained. It is of interest to find what the effective hydrodynamic mass would be when the core/vessel system moves as a whole in the horizontal direction without flexural deformation. In this case the mode shape vector would be $\{\psi\} = \{1 \quad 1 \quad 1.....1 \quad 1\}$. The effective hydrodynamic mass for the *system* will be:

$$\hat{M}_H^{system} = \{1 \quad 1 \quad 1.......1\}[M_H]\begin{Bmatrix} 1 \\ 1 \\ . \\ . \\ . \\ 1 \end{Bmatrix}$$

=summation over all the hydrodynamic mass elements. Carrying out the summation (note that not all the nodes are shown in Table 4.9), we get

$$\hat{M}_H^{system} = 93 \qquad \text{(lb-s}^2\text{/in)} \qquad \text{(16300 Kg)}$$

At 45.1 lb/cu.ft., the total mass of water in the annular gap is 106 (lb-s^2/in) or 18,578 Kg. Thus in seismic response, the net effect of fluid-structure coupling reduces approximately to that of the mass of water contained in the annular gap. It will be shown in Chapter 5 that in the limit of a slender cylinder, this relationship is exact.

Table 4.9: The full hydrodynamic mass matrix for the beam mode ($n=1$)
(For brevity, only values at alternate nodes are shown.)

$$\begin{bmatrix} [\hat{M}_H^a] \,|\, [\hat{M}_H^{ab}] \\ --- \,|\, --- \\ [\hat{M}_H^{ba}] \,|\, [\hat{M}_H^b] \end{bmatrix}$$

<------- [Maa] ----------->   <------------ [Mab] ------------->

	1	5	9	13	17	2	6	10	14	18
1	101.1	51.16	25.12	11.69	4.14	-111.2	-57.70	-28.33	-13.18	-4.67
5		68.21	34.57	16.11	5.70		-74.25	-39.00	-18.17	-6.43
9			59.20	28.59	10.13			-64.09	-32.25	-11.42
13	symmetric			53.22	19.58	symmetric			-57.34	-22.09
17					36.63					-38.64
2						128.3	65.07	31.95	14.87	5.26
6							86.52	43.98	20.50	7.25
10	symmetric about							75.06	36.36	12.88
14	main diagonal						symmetric		6.75	24.90
18										46.35

<------------ [Mba] ----------->   <------------ [Mbb] ------------>

4.8   Forced Response of Coupled Fluid-Shells

We now return to the equation of motion for the coupled fluid-shell system. Substituting Equation (4.54) into Equation (4.7), we get

$$[\mu]\{\ddot{q}\} + [c]\{\dot{q}\} + [k]\{q\} = \{p_0\} + [M_H][\delta_{33}]\{\ddot{q}\}$$

Rearranging,

$$([\mu] + [M_H][\delta_{33}])\{\ddot{q}\} + [c]\{\dot{q}\} + [k]\{q\} = \{p_0\} \tag{4.56}$$

Equation (4.56) is exactly the same as the equation of motion for a structure in the absence of fluid-coupling, with the exception that an additional term $[M_H][\delta_{33}]$ has been added to the original physical mass matrix $[\mu]$. This term accounts for all the effect of fluid-structure coupling and with one exception, behaves in every aspect like a physical mass. The one exception is that being originally a pressure, this added mass is effective only in the normal degree-of-freedom (w-direction) of the shell. The matrix δ_{33} is carefully included with $[M_H]$ to emphasize this fact. As discussed in Section 4.5, this has a very important effect on the calculation of the effective hydrodynamic masses.

Knowing the values of the acoustic mode shape functions ϕ at each node point, equations (4.37) and (4.55) can be used to calculate the hydrodynamic mass matrix, which can then be used together with a finite-element structural analysis computer program for the responses analysis. In general, the hydrodynamic mass matrix is a full matrix (see Example 4.3). Since it is not uncommon to find, in a finite-element model for two fluid- coupled coaxial cylindrical shells, one hundred elements in each of the inner and outer shells, this would result in a 100 x 100 hydrodynamic mass matrix. Experience shows, however, that elements separated by several nodes can be ignored without significantly affecting the results. Thus, reasonable results can be obtained by keeping five, or even three, of the diagonal elements in the full added mass matrix.

A much bigger complication arrives from the frequency-dependence of the hydrodynamic masses. At low frequencies and in small-scale structures, the hydrodynamic masses are relatively constant so frequency-dependence can be ignored. Otherwise these hydrodynamic masses must be computed by iteration starting with a "guessed" frequency, usually a frequency lower than the in-air frequency. Because there is one set of hydrodynamic masses for each mode, this approach can be tedious if more than two to three modes are required in the modal sum.

It is not the objective of this chapter or indeed this book, to get into the details of dynamic responses of coupled fluid-shell system. Specific areas of application, such as turbulence-induced vibration, will be treated in detail in Chapters 8 and 9. In the following section, we briefly review the effect of fluid-structure coupling on the responses of the structure to three different kinds of forcing functions—turbulence, rapid depressurization due to a pipe break in a pressurized nuclear reactor, and seismic excitation.

(A) Turbulence-Induced Vibration

Turbulence-induced vibration is addressed in detail in Chapters 8 and 9. Here we review the justification of using the weakly coupled fluid-structure system approach to solve for the turbulence-induced responses of shells. As discussed before, this approach is justified if the response is small compared with the dimension of the flow channels bounded by these structures, and if the pressure induced by the motion of the shells is linearly superimposable onto the original turbulence forcing function. The former is confirmed from operational experience, so we need to confirm only the superposition assumption.

Figure 4.9 shows an experimental setup (Au-Yang et al, 1995) to verify that in turbulence-induced vibration, as long as the vibration amplitudes are small, the pressure induced by the vibration of the structure just linearly superimposes onto the original forcing function due to turbulence. In this test, the outer shell was a thick-walled shell further stiffened by rings to make sure it was non-responsive. In the first test, the inner shell was also thick-walled and non-responsive. Water flowed through the annular gap between the shells and the turbulent forcing function was measured by groups of dynamic pressure transducers mounted on the wall of the outer shell. The thin line in Figure 4.10 shows the measured power spectrum at one location. The test was then repeated, with the

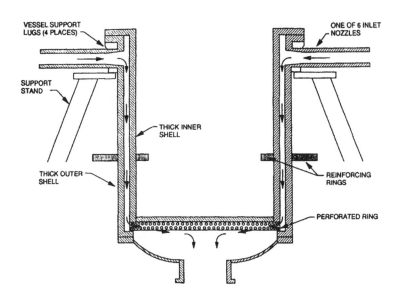

Figure 4.9 Test setup to verify the "linear superposition" assumption
(Au-Yang et al, 1995)

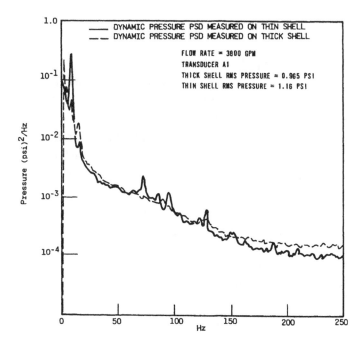

Figure 4.10 Test data showing that in turbulence-induced vibration, the pressure induced
by the motion of the structure just linearly superimposes upon the original fluctuating
pressure power spectrum (Au-Yang et al, 1995).

inner shell replaced by a flexible, responsive one. Before the flow test, the shell mode frequencies of this shell were measured in-air. The shell was then coaxially placed inside the thick outer shell and the annular gap was filled with water. The shell mode frequencies were again measured. Water was then allowed to flow through the annular gap at the same rate as in test 1, and the turbulent forcing function was measured. The thick line in Figure 4.10 shows the dynamics pressure spectrum at the same location as in test 1. From the two curves, it is obvious that the dynamic pressures in the two tests were essentially the same. The only difference is that in the second test, there were discrete pressure spikes linearly superimposed onto the original turbulence-induced forcing function as measured with the thick-walled, non-responsive shell in place. Comparison with the measured "in-water" shell frequencies showed that these discrete frequencies corresponded to the "in-water" modal frequencies of the inner shell. These discrete pressure components were induced by the motion of the inner shell, thus confirming the linear superposition assumption. The most important effect of fluid-structure coupling in turbulence-induced vibration is that it lowers the natural frequencies of the structure, causing it to resonate with the more dominant spectrum of the forcing function.

Turbulence-induced vibration will be discussed extensively in Chapters 8 and 9, using the weakly coupled fluid-structure system approach. This greatly simplifies the problem as the turbulence forcing function can be measured, or computed numerically, without taking into account the effect of fluid-structure coupling.

(B) The Lost-of-Coolant Accident (LOCA) Problem

Figures 2.20 and 4.11 show simplified schematic drawings of a pressurized water nuclear reactor. One of the safety issues concerning this type of nuclear reactors is that should the inlet pipe break, a rarefaction wave would propagate through the coolant inlet duct into the downcomer (the annular gap between the core and the outer vessel), resulting in a transient asymmetric pressure distribution around the circumference of the core. This asymmetric pressure distribution gives rise to a net force, which would then cause transient responses of the entire reactor system. This is known in the nuclear industry as a loss-of-coolant-accident (LOCA; Figure 4.11).

Figure 4.12 shows a typical integrated (around the circumference) LOCA-induced force on the core of a pressurized water reactor obtained from computational fluid dynamics without taking into account the effect of structural motion. It shows that the force usually lasts only a fraction of a second. Fluid-dynamics analysis shows that up to 0.2 seconds after the pipe breaks, the pressure in the downcomer still exceeds 1,700 psi (1.72 E+7 Pa), while the temperature is about 560 deg. F (293 deg. C). Under these circumstances, the coolant is still in a sub-cooled state. In order not to exceed the stress limit during a LOCA or earthquake, the core's response is limited by guide blocks to less than 0.5 inch (1.27 cm) inside an annular gap of fluid about 10 in. (25.4 cm). As a result, the LOCA problem is also a weekly coupled, fluid-structure interaction problem. Since the force is generated in the annular gap between the two components; it favors the out-of-phase response of the two components. This problem was analyzed by Au-Yang and

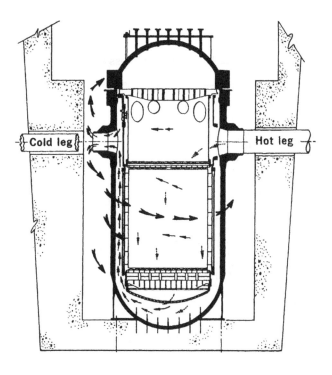

Figure 4.11 Fluid flow during a hypothetical Loss-of-Coolant Accident (LOCA)

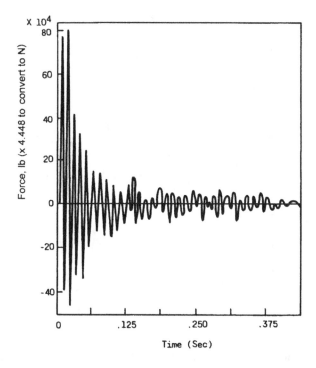

Figure 4.12 Integrated force acting on the core of a nuclear reactor in a LOCA

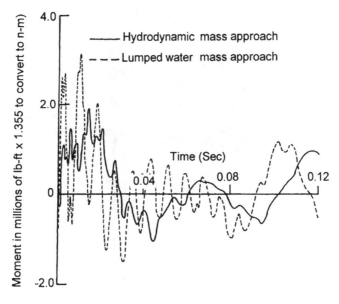

Figure 13 Induced bending moment at the top flange computed by two
different methods (Au-Yang and Galford, 1981)

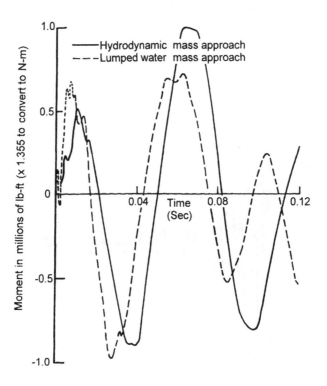

Figure 4.14 Induced bending moment at the nozzle support calculated by
two different methods (Au-Yang and Galford, 1981)

Galford (1981) using a weekly-coupled fluid-structure interaction approach. Example 4.3 shows the hydrodynamic mass matrix computed separately and used with a finite-element structural model in which the core and the vessel are joined together at the top flange. The system is supported at the nozzle. To see the effect of fluid-structure interaction, the problem was solved twice: First, by ignoring the effect of fluid-structure coupling, the physical mass of the water in the downcomer was lumped to the core nodes in the structural model. Then the effect of fluid-structure coupling was accounted for by implementing the full hydrodynamic mass computed in Example 4.4 into the structural model. Figure 4.13 shows the bending moments at the top flange computed by these two different methods; Figure 4.14 shows the bending moment at the nozzle support computed with these two different methods. As expected, fluid-structure coupling significantly reduces the bending moment induced a the top flange but does not have a large effect on the nozzle support. The former is induced by the out-of-phase mode of the two components, while the latter is induced by the in-phase mode.

(C) Seismic Response

The forcing function in seismic events arises from base motion, which tends to excite the in-phase mode of structural components. As shown in Example 4.3 and also in (B), above, the net effect of fluid-structure coupling in in-phase modal responses often reduced back to the net effect of the physical mass of the water. Thus, fluid-structure interaction has little effect on seismic response except in sloshing problems.

References

ASME International Boiler Code Appendix N-1400 Series, 1998.

ASME Steam Tables, 1979, Fourth Edition.

Abramson, H. N. and Kana, D. D., 1967, "Some Recent Research on the Vibration of Elastic Shells Containing Liquids," in Proceeding Symposium on the Theory of Shells, University of Houston, Houston.

Arnold R. H. and Warburton, G. B., 1952, "The Flexural Vibration of Thin Cylinders," Proceeding of the Institute of Mechanical Engineers, Vol. 167, pp. 62-74.

Au-Yang, M. K., 1976, "Free Vibration of Fluid-Coupled Coaxial Cylindrical Shells of Different Lengths," ASME Transaction, Journal of Applied Mechanics, Vol. 43, pp. 480-484.

Au-Yang, M. K., 1977, "Generalized Hydrodynamic Mass for Beam Mode Vibration of Cylinders Coupled by Fluid Gap," ASME Transaction, Journal of Applied Mechanics, Vol. 44, pp. 172-173.

Au-Yang, M. K., 1978, "Natural Frequencies of Cylindrical Shells and Panels in Vacuum and in a Fluid," Journal of Sound and Vibration, Vol. 57(3), pp. 341-355.

Au-Yang, M. K. and Galford, G. E., 1981, "A Structural Priority Approach to Fluid-Structure Interaction Problems," ASME Transaction, Journal of Pressure Vessel Technology, Vol. 103, pp. 142-150.

Au-Yang, M. K. and Galford, G. E., 1982, "Fluid-Structure Interaction - A Survey with Emphasis on Its Application to Nuclear Steam System Design," Journal Nuclear Engineering and Design, Vol. 70, pp. 387-399.

Au-Yang, M. K., 1986, "Dynamics of Coupled Fluid-Shells," ASME Transaction, Journal of Vibrations, Acoustics, Stress, Reliability in Design, Vol.108, pp. 339-347.

Au-Yang, M. K.; Brenneman, B. and Raj, D., 1995, "Flow-Induced Vibration Test of an Advanced Water Reactor Model-Part I: Turbulence-Induced Forcing Function," Journal Nuclear Engineering and Design, Vol. 157, pp. 93-109.

Chen, S. S., Rosenberg, G. S., 1975, "Dynamics of a Coupled Fluid-Shell System," Journal Nuclear Engineering and Design, Vol. 32, pp. 302-310.

Fritz, R. J., Kiss, E., 1966, The Vibration Response of a Cantilever Cylinder Surrounded by an Annular Fluid, Report KAPL M-6539, Knolls Atomic Power Laboratory.

Fritz, R. J., 1972, "The Effect of Liquids on the Dynamic Motion of Solids," ASME Transaction, Journal of Engineering for Industry, pp. 167-173.

Horvay, G. and Bowers, G., 1975, "Influence of Entrained Water Mass on the Vibration of a Shell," ASME Paper No. 75-FE-E.

Krajcinovic, D., 1974, "Vibration of Two Co-Axial Cylindrical Shells Containing Fluid," Journal of Nuclear Engineering and Design, Vol. 30, pp. 242-248.

Scavuzzo, R. J., Stokey, W. F. and Radke, E. F., 1979, "Fluid-Structure Interaction of Rectangular Modules in Rectangular Pools" in Dynamics of Fluid-Structure Systems in the Energy Industry, ASME Special Publication PVP-39, edited by M. K. Au-Yang and S. J. Brown. pp. 77-86.

Warburton, G. B., 1978, "Comments on Natural Frequencies of Cylindrical Shells and Panels in Vacuum and in a Fluid by M. K. Au-Yang," Journal of Sound and Vibration, Vol. 60(3), pp. 465-469.

Yeh, T. T. and Chen, S. S., 1977, "Dynamics of a Cylindrical Shell System Coupled by a Viscous Fluid," Journal of Acoustical Society of America, Vol. 62(2), pp. 262-270.

CHAPTER 5

VIBRATION OF STRUCTURES IN QUIESCENT FLUIDS— II SIMPLIFIED METHODS

Summary

In the special case when only one of the cylinders is flexible, the "in-water" natural frequencies of the shell can be obtained from the "in-air" frequencies by simple ratioing:

$$\bar{f}_n = f_n^{air} \sqrt{\frac{\mu}{\mu + \hat{M}_H}} \tag{5.1}$$

As shown in Chapter 4, one of the major tasks of calculating the hydrodynamic mass is the calculation of the hydrodynamic mass component h. In this chapter, simplified expressions for calculating h in special cases are given. Of these, the most commonly used in the industry is the "slender cylinder" approximation:

$$h_{an}^a = \frac{\rho a}{n} \frac{b^{2n} + a^{2n}}{b^{2n} - a^{2n}} \qquad\qquad h_{an}^{ab} = -\frac{2\rho b}{n} \frac{b^{2n} + a^{2n}}{b^{2n} - a^{2n}}$$

$$h_{an}^{ba} = -\frac{2\rho a}{n} \frac{b^{2n} + a^{2n}}{b^{2n} - a^{2n}} \qquad\qquad h_{an}^b = \frac{\rho b}{n} \frac{b^{2n} + a^{2n}}{b^{2n} - a^{2n}} \tag{5.2}$$

However, for application to large-shell structures commonly encountered in the power and process industries, Equation (5.2) often overestimates the hydrodynamic masses by as much as a factor of two or more. In addition, these equations give the hydrodynamic mass *components* in the form of surface densities. The effective hydrodynamic mass still has to be computed based on Equation (4.50). Failure to observe this has lead to inconsistencies in the hydrodynamic mass formulation of coupled fluid-shell problems.

When the thickness of the fluid gap is small compared with the diameter of the shells, curvature effect is negligible. The hydrodynamic mass components can be obtained from the simplified equation:

$$h_{an}^a = h_{an}^b = \rho / [\lambda_{an} \tanh(\lambda_{an} g)]$$
$$h_{an}^{ab} = h_{an}^{ba} = -\rho / [\lambda_{an} \sinh(\lambda_{an} g)] \tag{5.6}$$

where $g = b - a$ = annular gap thickness and

$$\lambda_{\alpha n}^2 = (\frac{\varepsilon\pi}{L})^2 + (\frac{2n}{a+b})^2 - (\frac{2\pi f}{c})^2$$

This is the "ripple approximation." For application to most power and process plant components, Equation (5.6) gives results that agree closely with those obtained from the exact equations. Equation (5.6) can be used in square cylindrical shells or flat rectangular channels provided that the fluid gap thickness is small compared with the characteristic dimension of the structure.

The hydrodynamic mass components for a single cylinder containing fluid can be obtained from the general expression, Equation (4.37), and are given by:

$$
\begin{aligned}
h_{\alpha n} &= \frac{\rho I_n(\chi_\alpha R)}{\chi_\alpha I'_n(\chi_\alpha R)} && \text{for} && f < \frac{\varepsilon c}{2L} \\
&= \frac{\rho J_n(\chi_\alpha R)}{\chi_\alpha J'_n(\chi_\alpha R)} && \text{for} && f > \frac{\varepsilon c}{2L} \\
&= \frac{\rho R}{n} && \text{for} && f = \frac{\varepsilon c}{2L}
\end{aligned}
$$
(5.7)

Likewise, for the special case when a single cylinder is immersed in an infinite fluid medium,

$$
\begin{aligned}
h_{\alpha n} &= \frac{\rho K_n(\chi_\alpha R)}{\chi_\alpha K'_n(\chi_\alpha R)} && \text{for} && f < \frac{\varepsilon c}{2L} \\
&= \frac{\rho Y_n(\chi_\alpha R)}{\chi_\alpha Y'_n(\chi_\alpha R)} && \text{for} && f > \frac{\varepsilon c}{2L} \\
&= \frac{\rho R}{n} && \text{for} && f = \frac{\varepsilon c}{2L}
\end{aligned}
$$
(5.8)

If in addition, the cylinders are slender, then in both cases:

$$h_{\alpha n} = \frac{\rho R}{n} \quad \text{at all frequencies}$$
(5.9)

Because of their simple forms, Equations (5.7) to (5.9) are most suitable for studying the physics and consistency of the hydrodynamic mass formulation of fluid-structure interaction problems.

The hydrodynamic mass for a tube surrounded by neighboring *rigid* tubes is given by the simple equation,

$$\textit{Added mass coefficient} = \left[\frac{(D_e/D)^2 + 1}{(D_e/D)^2 - 1}\right]$$
(5.11)

where

$$D_e / D = (1 + \frac{P}{2D})(\frac{P}{D})$$

and the added mass coefficient is defined as

$$Added\ Mass\ per\ Unit\ Length = Added\ mass\ cofficient \times \frac{\rho \pi D^2}{4} \tag{5.10}$$

Although Equation (5.11) is commonly used in tube bundle dynamic analysis, it does not agree well with Chen's (1985) much more complicated equation when the surrounding neighboring tubes are also flexible.

Fluid-structure interaction adds very little to the damping ratio of the coupled systems, even with very small entrained fluid gaps. In virtually all the situations encountered in the power and process industries, hydrodynamic damping due to entrained viscous fluids can be ignored.

Nomenclature

See also nomenclature in Chapter 4

g	$= b-a$, annular gap width
H	Function given Figure 5.8
S	$= \omega a^2 / v$
ζ	Damping ratio
ζ_H	Hydrodynamic damping ratio
v	Kinematic viscosity

5.1   Introduction

The examples in Chapter 4 show that one of the major tasks of computing the coupled fluid-structural frequencies is the calculation of the hydrodynamic mass component h. Except in the incompressible fluid limit, h is dependent on the frequency. As a result, numerical or iterative techniques are often necessary to solve the coupled fluid-structural dynamic problem. This chapter offers some simplified equations for calculating the hydrodynamic mass in some special cases. Because of their simplicity, these special cases can often shed light on the physics of fluid-structure interaction and the consistency of the hydrodynamic mass formulation of fluid-structure interaction problems.

5.2   One Cylinder Flexible

In many practical applications, a flexible cylindrical is coaxially placed inside an outer vessel of slightly larger diameter but with a much thicker wall, and the annular gap between the two cylinders are filled with a liquid. The vibration of the inner shell only is of interest. In this special case, the only effect of fluid coupling is the term $\hat{M}^a$. It is often assumed that in this case, the "in-water" frequency $\bar{f}$ can be obtained by the following simple equation,

$$\bar{f}_n = f_n^{air}\sqrt{\frac{\mu}{\mu + \hat{M}_H}} \tag{5.1}$$

This approach is only approximate, as it ignores the cross (structural) coupling to the tangential and axial directions of the shell. For example, suppose that in Example 4.2, we calculate the "in-water" (1,3) modal frequencies of the inner and outer shells, each assuming the other one is rigid, using Equation (4.52) with the hydrodynamic masses $\hat{M}_H^a, \hat{M}_H^b$ and the in-air frequencies of 54.8 and 56.3 Hz computed in that example, we would get:

$$\bar{f}_{13}^a = 22.5\ Hz, \quad \bar{f}_{13}^b = 22.0\ Hz$$

compared with 23.6 Hz and 22.9 Hz obtained by solving the determinant equation as in Example 4.2.

Equation (5.1) can be used to approximately estimate the in-water frequency *if* the hydrodynamic mass is known. Unfortunately, except in the incompressible fluid limit when $c \to \infty$, the hydrodynamic mass is frequency-dependent because χ_α in Equation (4.37) is frequency-dependent. At low frequencies, $\hat{M}^a$ is relatively insensitive to the frequency. At the (hard-wall) acoustic modal frequencies, $\hat{M}^a$ asymptotically approaches $\pm\infty$ (Figure 5.1). Very often only the lowest modal frequency is necessary, the hydrodynamic mass and the coupled frequencies can be found by iteration starting with a "guessed", usually the in-air frequency. Alternatively, the coupled frequencies can be found by a graphical method. If we plot the functions

$$g_1 = \frac{f}{f_{mn}^{air}}$$

and

$$g_2 = \sqrt{1 + \hat{M}_{mn}(f)/\mu}$$

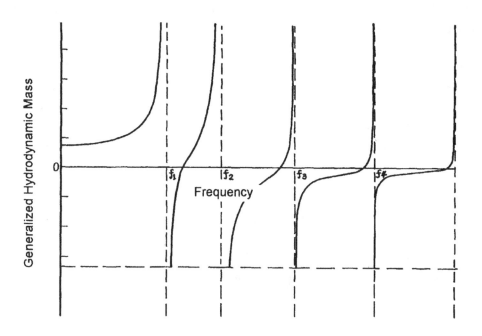

Figure 5.1 Frequency dependence of the hydrodynamics mass $\hat{M}^a$

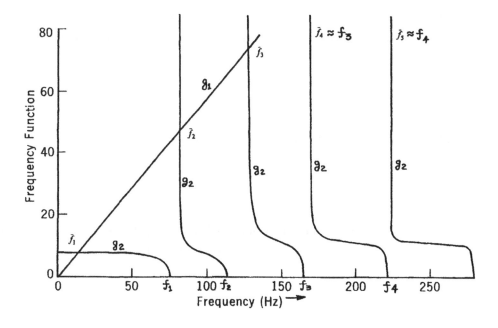

Figure 5.2 Graphical solution for the coupled modal frequencies

against f (Figure 5.2), the points at which the two curves intersect give the coupled frequencies of the fluid-shell system. Notice that there are many solutions for each set of (m, n). Of these, the lowest frequency is normally regarded as the "in-water" frequency of the shell, while the higher frequencies are the acoustic modal frequencies of the annular cavity bounded by flexible walls. Because of the asymptotic behavior of $\hat{M}^a$ near the acoustic modal frequencies, it can be readily seen from Figures 5.1 and 5.2 that while the in-air frequency of the shell may be significantly lowered by fluid coupling, the acoustic modal frequencies are not significantly affected by the flexibility of the bounding walls.

5.3   The Slender Cylinder Approximation

When the conditions $|\varepsilon a/L - 2fa/c|$ and $|\varepsilon b/L - 2fb/c| \ll 1$ are simultaneously satisfied, Equation (4.37) reduces to (Au-Yang, 1986):

$$h_{an}^a = \frac{\rho a}{n} \frac{b^{2n} + a^{2n}}{b^{2n} - a^{2n}} \qquad h_{an}^{ab} = -\frac{2\rho b}{n} \frac{b^{2n} + a^{2n}}{b^{2n} - a^{2n}}$$

$$h_{an}^{ba} = -\frac{2\rho a}{n} \frac{b^{2n} + a^{2n}}{b^{2n} - a^{2n}} \qquad h_{an}^b = \frac{\rho b}{n} \frac{b^{2n} + a^{2n}}{b^{2n} - a^{2n}} \tag{5.2}$$

In the slender cylinder approximation, the hydrodynamic mass components are independent of the axial mode number. Note that $\hat{M}$ and h are both in units of surface mass *density,* and

$$ah_{an}^{ab} = bh_{an}^{ba} \tag{5.3}$$

so that the hydrodynamic mass matrix on the two cylinders are symmetrical.

Example 5.1

Return to the nuclear reactor model in Example 4.2; calculate the beam mode frequency of the core using the slender cylinder approximation, Equation (5.2).

Solution

Since the outer vessel is held stationary, we need to compute only h^a. The Fourier coefficients are the same as those computed in Example 4.2 and are reproduced here in Table 5.1. Using the data given in Table 4.6 in Example 4.2 with $n=1$ in Equation (5.2), we get:

Table 5.1: Fourier Coefficients and Hydrodynamic mass Components
in the Slender Cylinder Approximation.

$\alpha=$	1	2	3	4	5	6	7	8
$C=$	0.567	-0.631	0.273	-0.244	0.163	-0.152	0.117	-0.111
h ((lb-s²/in)/in²)	0.04436	0.04436	0.04436	0.04436	0.04436	0.04436	0.04436	0.04436
h(Kg/m²)	12051	12051	12051	12051	12051	12051	12051	12051

$$h_{1\alpha}^a = \frac{\rho a}{1} \frac{[(b/a)^2 + 1]}{[(b/a)^2 - 1]} = 8.38 \rho a \quad \text{for all } \alpha$$

In the slender cylinder approximation, the hydrodynamics mass components are independent of the acoustic mode number. The numerical values of h, in both US and SI units, are given in rows 2 and 3 in Table 5.1. The total *effective* hydrodynamic mass is

$$\hat{M}_{H1}^a = \frac{2\pi a}{1 + n^{-2}} \sum C_{1\alpha}^2 h_{1\alpha}^a$$

Using these mass densities, the Fourier coefficients and h given above, we find for $n=1$,

$$(\hat{M}_{H1}^a)_{total} = 3048 \, (\text{lb - s}^2/\text{in}) \quad \text{or} \quad 5.342\text{E+5} \ \ \text{Kg}$$

This is almost twice as large as the corresponding values computed with the exact equations. This shows that in the case of relatively short shells, as in this example, using the slender cylinder approximation can greatly overestimate the effect of hydrodynamic mass coupling. Since the outer cylinder is stationary in the present example, this is the only added mass we need. The "in-water" frequency is, from Equation (5.1),

$$f_1^{water} = f_1^{air} \sqrt{\frac{m_{phy}}{m_{phy} + (\hat{M}_{H1}^a)_{total}}}$$

where m_{phy} is the total physical mass (1,884 lb-s²/in or 330,215 Kg). Using the in-air frequency of 21.9 Hz, we get:

$$f_1^{water} = 13.5 \ \text{Hz}$$

5.4   Incompressible Fluid Approximation

When $2f/c \ll \varepsilon/L$, the term $2\pi f/c$ can be dropped from the argument of the Bessel function in Equation (4.37); the resulting hydrodynamic masses are independent of frequency. Since this condition is satisfied when the speed of sound is infinite, this is called the incompressible fluid assumption. It should be cautioned that in application to large-scale structures, the incompressible fluid assumption should be used with caution. For a structure 300 in. or 8 m long vibrating in water at 625 deg. F (330 deg. C), typical of a nuclear reactor core support structure, the incompressible fluid assumption generally introduces more than 10% error in the hydrodynamic mass matrix elements for frequencies above 20 Hz. Therefore, it should not be used in analyses involving rapid transients with high frequency contributions to the responses.

5.5   Equation of Fritz and Kiss

In the special case of beam mode (n=1) vibration of a cylinder coaxially inside another cylinder of slightly larger diameter, a semi-empirical equation was given by Fritz and Kiss (1966):

$$h_a = \frac{\rho a^2}{b-a} \frac{1}{1+12a^2/L^2} \quad \text{(surface density)} \tag{5.4}$$

$$\hat{M}_H = \frac{\rho \pi a^2}{(b-a)/a} \frac{1}{1+12a^2/L^2} \quad \text{(linear density)} \tag{5.5}$$

Here h is a surface density and is effective only in the normal direction, while $\hat{M}_H$ is an effective linear mass density (compare with Equation (4.50)).

Example 5.2

Return to the nuclear reactor model, Example 4.2 again, and calculate the effective hydrodynamic mass of the core using Equation (5.5).

Solution

From Equation (5.5), we find from straightforward calculation:

$$\hat{M}_{H1}^a = 1690 \text{ (lb-s}^2/\text{in)} \quad (2.96\text{E}+5 \text{ Kg})$$

In spite of its simplicity, the equation of Fritz and Kiss gives a reasonable estimate of the beam mode hydrodynamic mass. It over-estimates the hydrodynamic mass because Equation (5.5) assumes full coupling between the acoustic and structural modes (i.e., $C_{11}^2 = 1$, while all other Cs are zero). This is true if the annular gap has open ends and the cylinder is simply-supported at both ends, so that the structural and acoustic modes shapes are both given by the sine function. In the present case, however, difference in the mode shape functions causes imperfect coupling between the structural and acoustic mode shapes. As a result, the hydrodynamic mass of cylinder a is slightly less than what is predicted by Equation (5.5).

5.6   The Ripple Approximation

When the ratio *(b-a)/(b+a)* is small, curvature effect becomes unimportant. The annular gap can be treated as a rectangular cavity. The expressions for the hydrodynamic masses can be simplified to

$$h_{an}^a = h_{an}^b = \rho /[\lambda_{an} \tanh(\lambda_{an} g)]$$
$$h_{an}^{ab} = h_{an}^{ba} = -\rho /[\lambda_{an} \sinh(\lambda_{an} g)] \qquad (5.6)$$

where $g=b-a=$annular gap thickness and

$$\lambda_{an}^2 = (\frac{\varepsilon\pi}{L})^2 + (\frac{2n}{a+b})^2 - (\frac{2\pi f}{c})^2$$

Example 5.3: PWR Beam Mode Vibration

Return once again to the nuclear reactor model of Example 4.2; calculate the total effective hydrodynamic mass on the core, and its "in-water" pendulum mode natural frequency, using the ripple approximation.

Solution

As in the exact solution in Example 4.2, we need to calculate the Fourier coefficients. These, however, are the same as those in Example 4.2 and are reproduced here in Table 5.2.

Using the ripple approximation Equation (5.6), we can compute h as a function of α. These are also given in Table 5.2. Note that unlike in the slender cylinder approximation, in the ripple approximation, h decreases rapidly with α. Multiplying by the square of the Fourier coefficients C and sum over the eight modes, we get

$$\hat{M}^a_{H1} = 1643 \ (\text{lb-s}^2/\text{in}) \quad (2.882\text{E}+5 \ \text{Kg})$$

This is within 7% of that calculated with Equation (4.37) in Example 4.2. Using this value in Equation (5.1), we get

$$f_1^{water} = 16.0 \ \text{Hz}$$

Table 5.2: Fourier Coefficients and Hydrodynamic Mass Components
in the Ripple Approximation

$\alpha=$	1	2	3	4	5	6	7	8
$C=$	0.567	-0.631	0.273	-0.244	0.163	-0.152	0.117	-0.111
h ((lb-s^2/in)/in^2)	0.0432	0.0182	0.00851	0.00481	0.0031	0.00218	0.00164	0.001
h(Kg/m^2)	11748	4939	2314	1307	842	593	445	350

5.7   Single Cylinder Containing Fluid

The case when there is only one cylinder containing fluid can be obtained from Equation (4.38) by letting $a \to 0$ and then replacing b with R, the radius of the cylinder,

$$
\begin{aligned}
h_{\alpha n} &= \frac{\rho I_n(\chi_\alpha R)}{\chi_\alpha I'_n(\chi_\alpha R)} & \text{for} & \quad f < \frac{\varepsilon c}{2L} \\
&= \frac{\rho J_n(\chi_\alpha R)}{\chi_\alpha J'_n(\chi_\alpha R)} & \text{for} & \quad f > \frac{\varepsilon c}{2L} \\
&= \frac{\rho R}{n} & \text{for} & \quad f = \frac{\varepsilon c}{2L}
\end{aligned}
\quad (5.7)
$$

5.8   Single Cylinder in Infinite Fluid

The case when there is only one cylinder in an infinite fluid medium can likewise be reduced to,

$$h_{\alpha n} = \frac{\rho K_n(\chi_\alpha R)}{\chi_\alpha K'_n(\chi_\alpha R)} \qquad for \qquad f < \frac{\varepsilon c}{2L}$$

$$= \frac{\rho Y_n(\chi_\alpha R)}{\chi_\alpha Y'_n(\chi_\alpha R)} \qquad for \qquad f > \frac{\varepsilon c}{2L} \qquad\qquad (5.8)$$

$$= \frac{\rho R}{n} \qquad\qquad for \qquad f = \frac{\varepsilon c}{2L}$$

If, in addition, the cylinder is infinitely long, then

$$h_{\alpha n} = \frac{\rho R}{n} \quad \text{at all frequencies} \qquad\qquad (5.9)$$

Equation (5.9) holds true both for a single cylinder in a finite fluid, or one containing fluid.

5.9   Consistency of the Hydrodynamics Mass Formulation

In the slender cylinder limit when the end effects are negligible, the hydrodynamic mass formulation must reduce to what we observe physically. Because of its simplicity, Equation (5.2) and (5.7) to (5.9) are most suitable for checking the consistency of the hydrodynamic mass formulation. We investigate the following three special cases.

Single Infinite Cylinder Containing Liquid Undergoing Rectilinear Motion

The effective hydrodynamics mass of an infinite cylinder containing liquid and in rectilinear motion can be obtained from Equation (5.9) by putting $n=1$, $h = \rho R$ per unit surface area of the cylinder. The effective hydrodynamic mass due to the water contained inside the cylinder will be considered in two separate cases—with both top and bottom ends of the annular gap closed (p'=maximum) and with the top opened (p'=0) while the bottom is closed. In both cases, the *normalized* structural mode shape for rectilinear motion is $\psi_1(x) = 1/\sqrt{L}$. With both ends closed, the normalized acoustic mode shape is given by Equation (4.19). In particular, $\phi_0(x) = 1/\sqrt{L}$.

The Fourier coefficients are $C_0 = 1.0$, while all the others are 0 because of the symmetry of the cosine function. That is, only the 0-th acoustic mode contributes the hydrodynamic mass to the cylinder. From Equation (4.62), the effective hydrodynamic mass is:

$$\hat{M}_H = \sum C_{1\alpha}^2 \frac{2\pi\rho r^2 L}{1 + 1^{-2}} = \pi\rho r^2 L$$

This is equal to the mass of liquid contained in the cylinder, as it should be.

Now consider the case when one end is pressure-released (p'=0) and the other end is closed (p'=maximum). The acoustic mode shapes are given by Equation (4.20). The Fourier coefficients are:

$$C_{11} = \int_0^L \frac{1}{\sqrt{L}} \sqrt{\frac{2}{L}} \sin \frac{\pi x}{L} \, dx = \frac{\sqrt{8}}{\pi}$$

$$C_{12} = \int_0^L \frac{1}{\sqrt{L}} \sqrt{\frac{2}{L}} \sin \frac{3\pi x}{L} \, dx = \frac{\sqrt{8}}{3\pi}$$

$$C_{13} = \frac{\sqrt{8}}{5\pi}, \quad \ldots\ldots$$

The effective hydrodynamic mass on the cylinder is

$$\hat{M}_H = \sum C_{1\alpha}^2 \frac{2\pi\rho b^2 L}{1 + 1^{-2}} = \pi\rho b^2 L (1 + \frac{1}{3^2} + \frac{1}{5^2} + ...) \frac{8}{\pi^2}$$

The infinite series inside the parenthesis sums to $\pi^2/8$. Thus, again $\hat{M}_H = \pi\rho b^2 L =$ mass of water contained in the cylinder, as it should be.

Inertia Load due to Hydrodynamics Mass (Figure 5.3)

Equation (5.2) and Examples 4.1 and 4.2 show that in the case of very narrow fluid gaps, such as in a pressurized water nuclear reactor, the effective hydrodynamic mass can be much larger than the physical mass of water contained in the annulus between the reactor vessel and the reactor core (see Figure 2.20). The concern is that during an earthquake, this hydrodynamic mass may give rise to large added inertia loads that the reactors are not designed for. In Example 4.3, we show that in the particular case under study, such added inertia load does not exist. More insight on this "added inertia load paradox" can be gained by studying the case of two infinitely long coaxial cylinders with fluid filling the annular gap between them. Let us compute the effective generalized hydrodynamic mass of this system when it is undergoing rectilinear motion, as it would experience during an earthquake. Since we are considering rigid body motion, each of the inner and outer cylinders can be represented by one node. The generalized hydrodynamic mass for the *system* is given by

$$\hat{M}_{Hsystem} = \{\Psi\}^T [\hat{M}_H]\{\Psi\}$$

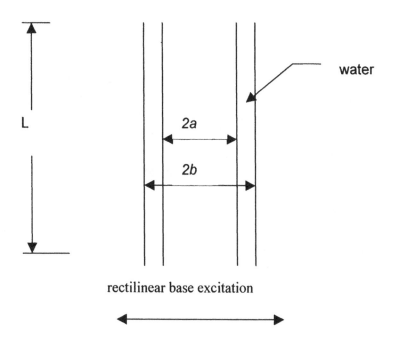

Figure 5.3   Coaxial cylinders excited by base motion

Here $\{\Psi\} = \begin{Bmatrix} 1 \\ 1 \end{Bmatrix}$ is the column vector representing the mode shape of the two cylinders

and $[M_H]$ is the 2 x 2 effective generalized hydrodynamic mass discussed in Subsection 4.4. Using the slender cylinder approximation with $n=1$, we get

$$\hat{M}_H^a = \pi\rho a^2 L\left(\frac{b^2 + a^2}{b^2 - a^2}\right), \quad \hat{M}_H^{ab} = -2\pi\rho abL\left(\frac{ab}{b^2 - a^2}\right)$$

$$\hat{M}_H^{ab} = -2\pi\rho baL\left(\frac{ba}{b^2 - a^2}\right), \quad \hat{M}_H^b = \pi\rho b^2 L\left(\frac{b^2 + a^2}{b^2 - a^2}\right)$$

Carrying out the matrix multiplication we get,

$$\hat{M}_{Hsystem} = \rho\pi(b^2 - a^2)L$$

This is exactly the same as the inertia loading on the cylinders due to the mass of fluid contained in the annular gap, as it should be if the hydrodynamic mass formulation is consistent.

The Enclosed Cavity Paradox

As discussed in Section 4.4, the Fourier coefficient is a measure of the compatibility between the structural mode shape and the acoustic mode shape, and plays an essential role in the calculation of the hydrodynamic masses. Coupling to a certain acoustic mode takes place only if that particular Fourier coefficient is non-zero. Failure to observe this, as is often done in the literature, would lead to paradoxical results, as is illustrated in the following example. Consider a slender cylindrical shell a with both ends simply-supported and coaxially placed inside a cylindrical cavity of slightly larger radius b, and with the cavity filled with water (Figure 5.4). In case I the end condition of the water annulus is pressure-released ($p'=0$ at $x=0$ and $x=L$), while in case II the cavity is closed so that the acoustic pressure p' is maximum at $x=0$ and at $x=L$. Using the "slender cylinder" approximation, the hydrodynamic mass component for the beam mode vibration of this shell is, from Equation (5.2), equal to

$$h_\alpha^a = \frac{\rho a}{1} \frac{b^2 + a^2}{b^2 - a^2}$$

It is often stated in the literature that the effective hydrodynamic mass of the cylinder is

$$\hat{M}_H^a = \rho \pi a^2 \frac{b^2 + a^2}{b^2 - a^2}$$

Based on this, the in-water frequencies of the shell would be the same in both cases. Our intuition, however, tells us that they should be different. The truth is that while the hydrodynamic mass *components* h are the same in both cases, the compatibility of the structural mode shape to the acoustic mode shapes are not the same. In both cases, the structural mode shape is:

$$\psi_1(x) = \sqrt{\frac{2}{L}} \sin \frac{\pi x}{L}$$

In case I, the acoustic mode shapes are, from Equation (4.18),

$$\phi_\alpha(x) = \sqrt{\frac{2}{L}} \sin \frac{\alpha \pi x}{L}$$

The Fourier coefficients are from Equation (4.35), $C_{11} = 1.0$ and $C_{1\alpha} = 0$ for $\alpha \neq 1$. Coupling is perfect with the first acoustic mode and the effective hydrodynamic mass is indeed equal to:

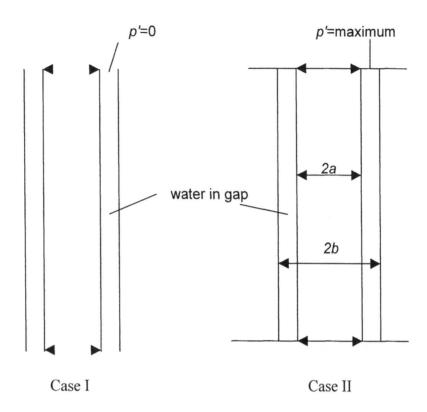

$p'=0$ $p'=\text{maximum}$

water in gap

2a

2b

Case I   Case II

Figure 5.4 The enclosed cavity paradox (The inner shells in both cases are identical; the end conditions of the fluid annuli are different in these two cases. A correct formulation of the hydrodynamic mass approach to the coupled fluid-shell problem should predict that the beam mode frequency in Case I is lower than that in Case II.)

$$\hat{M}_H^a = \rho\pi a^2 \frac{b^2 + a^2}{b^2 - a^2}$$

In Case II, the acoustic mode shapes are, from Equation (4.19),

$$\phi_0 = \frac{1}{\sqrt{L}} ; \qquad \phi_\alpha(x) = \sqrt{\frac{2}{L}} \cos\frac{\alpha\pi x}{L} \qquad \alpha=1, 2, 3 \ldots$$

The structural mode shape is the same as in Case I. The Fourier coefficients are, from Equation (4.35),

$$C_{10} = \int_0^L \frac{1}{\sqrt{L}} \sqrt{\frac{2}{L}} \sin\frac{\pi x}{L} dx = \frac{2}{\pi} = 0.64$$

$$C_{1\alpha} = 0 \quad \alpha \neq 0$$

Coupling is less than 100% in Case II. The hydrodynamic mass is only

$$\hat{M}_H^a = 0.41 \rho \pi a^2 \frac{b^2 + a^2}{b^2 - a^2}$$

The in-water frequencies of the cylinder in Case II will be higher than those in Case I, as we may expect.

5.10   Hydrodynamic Masses for Other Geometry

In this and the last chapters, the dynamics of a coupled fluid-circular cylindrical shell system is studied in great detail. Although this is one of the most common geometry in the power and process industries, it is by no means the only geometry in which fluid-structure interaction plays an essential role in the response of the system to external forces. For geometry of arbitrary shape, some commercial finite-element structural analysis computer programs offer "fluid elements" to account for the effect of fluid-structure interaction. Special techniques have also been developed to solve the general fluid-structure interaction problem with general finite-element computer programs that do not have special "fluid elements" to address this problem. The readers are referred to a review article by Brown (1982). Specific techniques have also been developed to address fluid-structure interaction problems of specific geometry. It is not the intention of this book to be a treatise on the subject of fluid-structure interaction. The following quotes some of the more common results. The readers are referred to the literature for the details.

Square Cylinders

Nuclear fuel assemblies and spent fuel storage racks often contain co-axial square cylinders with fluid gaps between them. The readers are referred to a paper by Scavuzzo (1979) for a detailed discussion of the dynamics such systems. When the fluid gaps are extremely narrow, the ripple approximation, discussed in Section 5.6, can be used. As discussed before, the hydrodynamic mass M_H is effective only in the direction normal to the surface of the square. Thus, if the motion is in a direction normal to two of the faces, hydrodynamic masses should be applied only to these faces, as shown in Figure 5.5.

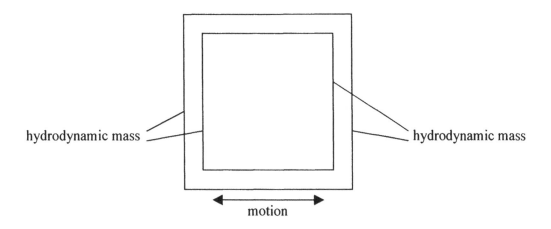

Figure 5.5 Hydrodynamic mass for fluid-coupled square cylinders

Sloshing Problems

The dynamics of large storage tanks containing fluid is important in seismic engineering and is a subject of intensive study by the American Society of Civil Engineers. The readers are referred to a survey report by Kana et al (1984) for these kinds of problems.

Hydrodynamic Mass for Tube Bundles

Fluid coupling is of paramount importance in the dynamics of heat exchanger tube and nuclear fuel bundles. This subject has been extensively studied, notably by Chen (1985), and is well beyond the scope of this book. In this case, the diagonal elements of the hydrodynamic mass matrix is the added mass of a tube due to its own motion, while the off-diagonal elements couple the neighboring tubes in organized, orbital motions. Very often, only the diagonal elements (generally just called added mass in tube bundle dynamics) is needed to estimate tube bundle stability or turbulence-induced vibration (see Chapters 7 and 9). Figures 5.6 and 5.7 show the "added mass coefficients"—defined as:

$$Added\ Mass\ per\ Unit\ Length = Added\ mass\ cofficient\ \times\ \frac{\rho \pi D^2}{4} \qquad (5.10)$$

computed by Chen (1985)—and their comparison with the experimental data. A much simpler, though much less vigorous, expression for the added mass coefficient for tube bundles can be found in Blevins (1979):

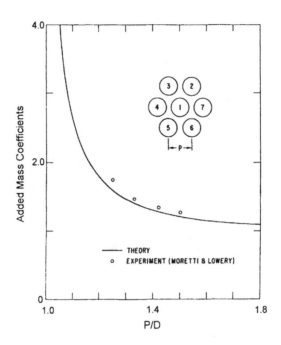

Figure 5.6 Added mass coefficients for a tube in a tube bundle: triangular array
(Chen, 1985)

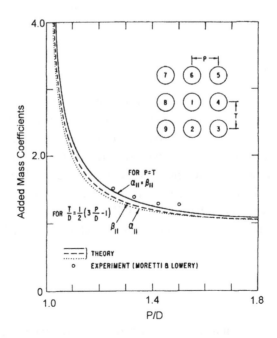

Figure 5.7 Added mass coefficients for a tube in a tube bundle: square array
(Chen, 1985)

$$Added\ mass\ coefficient = \left[\frac{(D_e/D)^2 + 1}{(D_e/D)^2 - 1} \right]$$ (5.11)

where

$$D_e/D = (1 + \frac{P}{2D})(\frac{P}{D})$$

Equation (5.11) applies only to a vibrating tube surrounded by stationary tubes and thus has, strictly speaking, very limited application. However, it has been used extensively to estimate the added mass in tube bundle dynamic analysis. In general, it gives a value lower than what was measured experimentally when the surrounding tubes are also flexible.

5.11   Hydrodynamic Damping

Just as the fluid inertia adds to the effective mass of cylindrical shells, the viscosity of the entrained fluid adds to the effective damping of the coupled fluid-shell system. This hydrodynamic damping, or added damping, due to coupling of a cylindrical structure and a viscous fluid, was studied by Chen and Rosenberg (1975) and Yeh and Chen (1977). In the general case, the expression for the added damping can be very complicated and is beyond the scope of this book. The readers are referred to the cited references for the details. In the special case of beam mode ($n=1$) vibration of a slender cylinder of radius a (such as tubes) inside a rigid outer cylinder of radius b, the following simplified expressions can be derived from Chen (1985, Chapter 2). Chen showed that the added modal damping ratio is given by:

$$\zeta_1 = -[0.5\rho\pi a^2 /(M_H + \mu\pi a^2)]\mathrm{Im}(H)$$ (5.12)

Here the subscript 1 denotes the $n=1$ (beam) mode. $\mathrm{Im}(H)$ is a fairly complicated function of modified Bessel functions and a parameter S, defined as:

$$S = \omega a^2 /v$$ (5.13)

where v is the kinematic viscosity of the entrapped fluid. Figure 5.8, reproduced from Chen (1985), gives theoretical values of $-\mathrm{Im}(H)$ as a function of the radius ratio b/a, for different values of S. It shows that as $S \rightarrow \infty$, $-\mathrm{Im}(H)$ becomes infinitesimal even for very narrow gaps ($b/a \rightarrow 1$), indicating that hydrodynamic damping is negligible for large values of S. Examples given at the end of this subsection seem to confirm this. For most practical problems, $S \rightarrow 10^8$, well out of the range of S given in Figure 5.8.

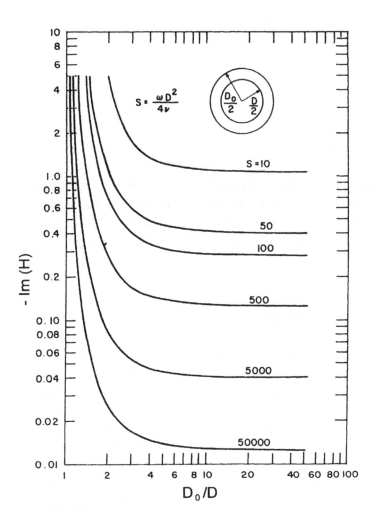

Figure 5.8 Imaginary part of H (Chen, 1985)

For most applications, the following simplified equation for the damping ratio can be derived. From Chen (1985), for the beam ($n=1$) mode, the damping coefficient (see Chapter 2) is given by:

$$C = \frac{4\pi\mu a}{\sqrt{\dfrac{2v}{\omega}}}\left[\frac{b^4 + ba^3}{(b^2 - a^2)^2}\right] \qquad (5.14)$$

where $\mu = v\rho$ is the dynamic viscosity. The damping ratio is related to the damping coefficient by (see Chapter 2),

$$C = 2\omega_1(\hat{M} + m)\zeta_1$$

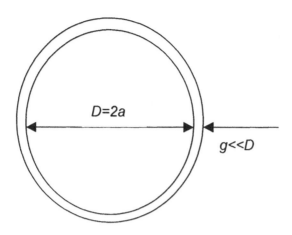

Figure 5.9 Vibration of a slender cylinder in a thin annular fluid gap

where m is the physical mass *per unit length* and $\hat{M}_H$ is the effective mass *per unit length* of the cylinder. From the above two equations we get:

$$\zeta_1 = \frac{\rho a}{\hat{M}_H + m} \sqrt{\frac{\pi v}{f_1} \left[\frac{b(b^3 - a^3)}{(b^2 - a^2)^2} \right]}$$ (5.15)

The reader should check that the right-hand side of Equation (5.15) is indeed dimensionless, as it should be. When the annular gap width is very small compared with the radius of the cylinder, (Figure 5.9), the above equation can be simplified to

$$\zeta_1 = \frac{3}{16} \frac{\rho D^2}{(\hat{M}_H + m)} \sqrt{\frac{\pi v}{g^2 f_1}} \qquad \text{for } g << D$$ (5.16)

where D is the diameter of the tube, and $\hat{M}_H + m$ is the physical and hydrodynamic *linear* mass density.

Example 5.4:   Hydrodynamic Damping in Nuclear Reactor Internal Component

Return to the nuclear reactor case in Example 4.2; calculate the hydrodynamic damping of the inner cylinder due to the annular water gap.

Solution

Since the equation for hydrodynamic damping (5.16) is based on the "slender cylinder" approximation, it is appropriate to estimate the hydrodynamic damping using the hydrodynamic mass computed with the same approximation. From the calculations in Example 4.4,

	US Unit	SI Unit
Total (physical+added) linear mass density $\hat{M}_H + m$	10.16 (lb-s²/in)/in	70105 Kg/m
ρD^2	1.665 (lb-s²/in)/in	11499 Kg/m
Kinematic viscosity ν (ASME Steam Table)	1.876E-4 in² /s	1.214E-7 m² /s

Using f= 13.5 Hz as computed in Example 4.2, and Equation (5.13), we get

S=2.8E+9

This is completely out of the range in Figure 5.8. We use instead the approximate Equation (5.16), even though g/a =0.13 is not quite $\ll 1.0$, and get:

$$\zeta_1 \approx 2.0E - 5 \quad \text{(US or SI unit)}$$

Added damping is utterly negligible in this case. Although the above example is for the beam mode vibration of an infinitely long cylinder, the same conclusion holds true qualitatively for the beam mode or shell mode vibration of finite cylindrical shells. This analytical result is also supported by experimental observations. Damping ratios of thermal shields measured during pre-operational tests of pressurized water reactors showed little difference from their corresponding values measured in air. Thus, hydrodynamic damping in large cylindrical structures can be ignored, especially at elevated temperatures.

Example 5.5: Spent Nuclear Fuel Racks

Figure 5.9 shows a typical arrangement of spent nuclear fuel racks in a pool. The dimensions and masses are given in Table 5.3. The racks are closely packed with extremely narrow water gaps separating them. The widths of these water gaps are important factors in preventing the spent fuel from collectively reaching the critical mass and thus causing a nuclear chain reaction. The concern is that during an earthquake, these water gaps might close in, thus decreasing their neutron absorption capacity. The racks are designed with short aspect ratios so that they would not tilt over in an earthquake. However, they may slide along the bottom of the pool. Estimate the

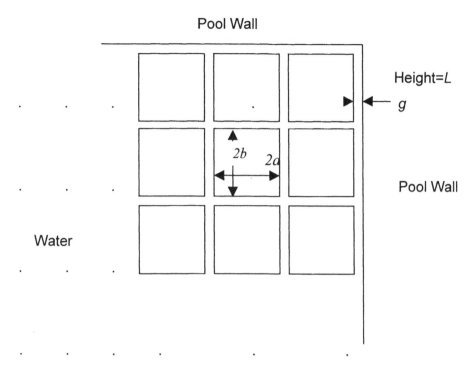

Figure 5.9 Spent nuclear fuel rack

hydrodynamic mass and damping of one of these racks sliding toward the others, assuming the dominant frequency in the seismic spectrum is 10 Hz.

Table 5.3: Data for Spent Fuel Rack

	US Unit	SI Unit
No. of fuel assembly per rack	12 x 12	12 x 12
Dimension of each rack ($2a$ x $2b$ x L)	11' x 11' x 16'	3.353 x 3.353 x 4.877 m
Gap between rack and pool wall, g	0.5 in	0.0127 m
Mass per rack	15000 lb	6804 Kg
	(38.82 (lb-s^2/in))	
Temperature of pool water	70 deg F	21 deg C
Pressure	14.5 psi	101325 Pa
Seismic frequency	10 - 25 Hz	10 - 25 Hz

Solution

Continuing with our practice, we immediately convert the US mass unit into (lb-s^2/in). From the ASME Steam Table (1979), we find at the given temperature and pressure,

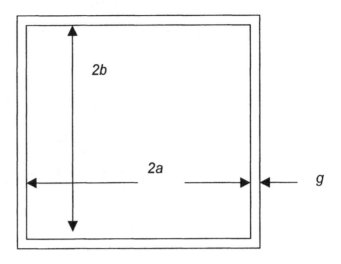

Figure 5.10 Approximate model of each rack

	US Unit	SI Unit
Density of water	62.4 lb/ft^3	998.1 Kg/m^3
	(1.938 slug/ft^3 , 9.3455 (lb-s^2/in)/in^3)	
Dynamic viscosity, μ	2.038E-5 lbf-s/ft^2	9.7856E-4 Pa-s
Kinematic viscosity, $\nu=\mu/\rho$	(1.052E-5 ft^2/s=1.514E-3 in^2/s)	9.804E-7 m^2/s
Velocity of sound	4852 ft/s	1479 m/s
	(58224 in/s)	

The numbers in the parentheses are converted immediately from those given in the ASME Steam Table to be consistent with the rest of the calculations. For calculating the kinematic viscosity in the US unit system, the mass density must be in slug/cu.ft. The resulting kinematic viscosity has unit ft^2 /s. This must be multiplied by 144 to convert it to in^2 /s for later use. Since the gaps between the racks and between the racks and the pool walls are obviously much smaller than the cross-sectional dimensions of the racks, we can use the ripple approximation Equation (5.6) to estimate the hydrodynamic mass, even though that equation is originally derived for circular cross-sections. We model one rack as two concentric square cylinders separated by a uniform water gap, as shown in Figure 5.10, and assume the pool is pressure-released at the top ($p'=0$) and closed at the bottom (p'=maximum). First, we see if we can use the graph in Figure 5.8. At a frequency of 10 Hz,

$$S = \omega^2 a / \nu = 1.8E + 8$$

Again, the parameter S is completely out of the range in Figure 5.8, indicating that the hydrodynamic damping is probably small. We use instead the approximate Equation (5.16), which requires that we have to find the hydrodynamic mass first. From Equation (4.20), the acoustic mode shapes are,

$$\phi_1(x) = \sqrt{\frac{2}{L}} \sin\frac{\pi x}{2L}, \quad \phi_2(x) = \sqrt{\frac{2}{L}} \sin\frac{3\pi x}{2L}, \quad \phi_3(x) = \sqrt{\frac{2}{L}} \sin\frac{5\pi x}{2L}, \quad \cdot$$

The structural mode shape is, for lateral rigid body motion, given by

$$\psi_1(x) = \frac{1}{\sqrt{L}}$$

The Fourier coefficients are computed from Equation (4.35). This has been done in Section 4.6. The results are:

$$C_{11} = \sqrt{8}/\pi, \quad C_{12} = \sqrt{8}/3\pi, \quad C_{13} = \sqrt{8}/5\pi, \quad \dots\dots$$

Using Equation (5.6), we find λ and h for each acoustic mode. These values are shown in Table 5.4.

Table 5.4: Fourier Coefficients (squared), λ and h

US units

Mode	1	2	3	4	5	6	7	8
C^2=	0.810571	0.090063	0.032423	0.016542	0.010007	0.006699	0.004796	0.003603
λ=	0.02224	0.01517	0.01507	0.01506	0.01506	0.01506	0.01506	0.01506
h=	0.378067	0.812739	0.823035	0.824184	0.824441	0.824523	0.824556	0.824571
C^2*h=	0.306450	0.073198	0.026685	0.013634	0.008250	0.005523	0.003955	0.002971

SI units

C^2=	0.810571	0.090063	0.032423	0.016542	0.010007	0.006699	0.004796	0.003603
λ=	0.87548	0.59719	0.59345	0.59303	0.59294	0.59291	0.59290	0.59289
h	1.0254E+5	2.2037E+5	2.2316E+5	2.2347E+5	2.2354E+5	2.2356E+5	2.2357E+5	2.2358E+5
C^2*h=	8.312E+04	1.985E+04	7.235E+03	3.697E+03	2.237E+03	1.498E+03	1.072E+03	8.054E+02

Summing over the acoustic modes, we find the hydrodynamic mass density of each rack:

$$\hat{M}_H = 0.541 \text{ (lb-s}^2/\text{in)/in}^2 = 1.467\text{E+5 Kg/m}^2$$

As discussed in Section 5.10, this should be acting on only two of the four surfaces of the rack. Thus the effective *linear* hydrodynamic and physical mass density is

$$\hat{M}_H = 142.8 \ \text{(lb-s}^2\text{/in)/in} = 9.835\text{E+5 Kg/m}$$

and the total effective hydrodynamic mass is,

$$2.74\text{E+4 (lb-s}^2\text{/in)} \quad \text{or} \quad 4.80\text{E+6 Kg}$$

This is more than 700 times the physical mass of each rack. Finally, using Equation (5.16) and the *linear* effective hydrodynamic mass density, we get

$$\zeta_H = 9.3\text{E-5}$$

Even in this case of a very tiny water gap and with the water at room temperature, the hydrodynamic damping ratio is still negligibly small. The reason is that, as discussed in Chapter 3, the damping ratio is not a direct measure of energy dissipation. When the water gaps are narrow, energy dissipation due to squeezed film indeed increases, however, so does the hydrodynamic mass. The two effects counteract each other as can be seen from Equation (5.16). The damping ratio due to the squeezed film may not be significant.

References

ASME Steam Tables, 1979, Fourth Edition.

Au-Yang, M. K., 1986, "Dynamics of Coupled Fluid-Shells," ASME Transaction, Journal of Vibrations, Acoustics, Stress, Reliability in Design, Vol.108, pp. 339-347.

Blevins, R. D. 1979, Formulas for Natural Frequencies and Mode Shapes, Van Nostand, New York.

Brown, S. J., 1982, "A Survey of Studies into the Hydrodynamic Response of Fluid-Coupled Cylinders," ASME Transaction, Journal of Pressure Vessel Technology, Vol. 104, pp. 2-19.

Chen, S. S. and Rosenberg, G. S., 1975, "Dynamics of a Coupled Fluid-Shell System," Journal Nuclear Engineering and Design, Vol. 32, pp. 302-310.

Chen, S. S., 1985, Flow-Induced Vibration of Circular Cylindrical Structures, Argonne National Laboratory Report ANL-85-51, Chapter 3.

Fritz, R. J. and Kiss, E., 1966, The Vibration Response of a Cantilever Cylinder Surrounded by an Annular Fluid, Knolls Atomic Power Laboratory Report KAPL M-6539.

Kana, D. D. et al, 1984, Fluid-Structure Interaction During Seismic Excitation, American Society of Civil Engineers Report.

Scavuzzo, R. J., Stokey, W. F. and Radke, E. F., 1979, "Fluid-Structure Interaction of Rectangular Modules in Rectangular Pools" in Dynamics of Fluid-Structure Systems in the Energy Industry, ASME Special Publication PVP-39, edited by M. K. Au-Yang and S. J. Brown, , pp. 77-86, ASME Press, New York.

Yeh, T. T. and Chen, S. S., 1977, "Dynamics of a Cylindrical Shell System Coupled by a Viscous Fluid," Journal of Acoustical Society of America, Vol. 62(2), pp. 262-270.

CHAPTER 6

VORTEX-INDUCED VIBRATION

Summary

A stationary cylinder subject to cross flow sheds vortices alternately from one side and then the other side of the cylinder. The vortex-shedding frequency is given by,

$$f_s = S \frac{V}{D} \tag{6.1}$$

where S is the Strouhal number. Over a range of Reynolds number

$$\mathrm{Re} = \frac{VD}{v} \tag{6.2}$$

the value of the S can be taken as

$$S = 0.2 \quad for \quad 1.0E+3 \le \mathrm{Re} \le 1.0E+5$$
$$0.2 \le S \le 0.47 \quad for \quad 1.0E+5 < S < 2.0E+6$$
$$0.2 \le S \le 0.3 \quad for \quad 2.0E+6 \le \mathrm{Re} \le 1.0E+7$$

The vortices exert a fluctuating reaction force on the cylinder. The component of this force in the lift direction (perpendicular to the flow direction) has a frequency equal to vortex-shedding frequency f_s, while that in the drag direction (direction of flow) has a frequency equal to $2f_s$. In general, the force component in the drag direction is much smaller than the force component in the lift direction.

When the cylinder is flexible with characteristic natural frequencies, a phenomenon called lock-in may happen. When one of the structural modal frequencies are close to f_s or $2 f_s$, the vortex-shedding frequency f_s (or $2 f_s$) may actually shift from its value for a stationary cylinder to the nearest natural frequency of the cylinder, resulting in large amplitude, resonant vibration. Lock-in can occur in either the lift or the drag direction and both have been observed in power plant components, resulting in substantial financial losses.

The most important rule in designing components against vortex-induced vibration damage is to avoid lock-in. Although there are several alternate rules, the most common one is the "separation rule." If the natural frequency f_n is more than $1.3f_s$, or less than $0.7 f_s$, then lock-in in the lift direction will be avoided. Likewise, if f_n is more than $1.3*2f_s$, or less than $0.7 *2f_s$, then lock-in in the drag direction will be avoided.

If lock-in is avoided, the forced response of the structure can be computed by standard methods of structural dynamic analysis, using either a finite-element computer program or hand calculation. The forcing function in this case is given by:

$$F_L = (\rho V^2 / 2) D C_L \sin(2\pi f_s t) \tag{6.9}$$

in the lift direction.

A conservative value for the lift coefficient is C_L=0.6. The vibration amplitude in the drag direction is generally much smaller and can be ignored. If lock-in occurs, the vibration amplitude of the structure can be estimated by the following three semi-empirical equations recommended by the ASME Boiler and Pressure Vessel Code (1998):

Griffin-Ramberg: $$\frac{y_n^*}{D} = \frac{1.29\gamma}{[1 + 0.43(2\pi S^2 C_n)]^{3.35}} \tag{6.11}$$

Ivan-Blevins: $$\frac{y_n^*}{D} = \frac{0.07\gamma}{(C_n + 1.9)S^2}[0.3 + \frac{0.72}{(C_n + 1.9)S}]^{1/2} \tag{6.12}$$

Sarpkaya: $$\frac{y_n^*}{D} = \frac{0.32\gamma}{[0.06 + (2\pi S^2 C_n)^2]^{1/2}} \tag{6.13}$$

where γ is the mode shape factor that varies between 1.0 for a spring-supported rigid cylinder and 1.305 for the first modes of a cantilevered cylinder. Depending on the value of S and the reduced damping ratio C_n, these three equations give results that vary from reasonable to poor agreement. Except for the most flexible structure, most power and process plant components cannot tolerate lock-in vortex-induced vibration.

Nomenclature

C_D Drag coefficient

C_L Lift coefficient

$C_n = \dfrac{4\pi\zeta_n m_n}{\rho D^2 \int_L \psi_n^2(x)dx}$ = Reduced damping parameter

D Diameter of cylinder, or width of structure perpendicular to flow

f_S Vortex-shedding frequency

f_D Frequency of drag-direction forcing function due to vortex-shedding

f_L Frequency of lift-direction forcing function due to vortex-shedding

f_n Natural frequency of structure

F_D Drag-direction fluctuating force due to vortex-shedding

F_L Lift-direction fluctuating force due to vortex-shedding

L Length of structure

m Total (physical + hydrodynamic) mass per unit length of structure

$$m_n = \int_0^L m\ (x)\psi_n^2(x)dx = \text{Generalized mass}$$

Re Reynolds number

S Strouhal number

V Cross-low velocity

y Vibration amplitude

y^* Peak vibration amplitude

ψ Mode shape function

μ Dynamic viscosity

ν Kinematic viscosity

ρ Mass density of fluid

ζ_n Modal damping ratio

6.1   Introduction

Isolated structures (e.g., cylinders) subject to cross-flow shed vortices. As the vortices are shed alternately from one side and then the other side of the structure (Figure 6.1), the pressure distribution on the structure varies, causing a net fluctuating lift force acting on the structure. The frequency of this fluctuating force is equal to the vortex-shedding frequency f_s. In addition, these vortices also impose a net force in the drag direction. The frequency of this fluctuating drag force, based on the geometry of the vortex street (see Figure 6.1), is equal to twice the shedding frequency. Both components of this fluctuating force cause the structure to vibrate. As long as the natural frequencies of the structure are "well separated" from both the lift- and the drag-direction shedding frequencies, structural responses are of the forced vibration type and seldom cause any problems to the integrity of the structure. However, if one of the structural frequencies is "close" to either the lift- or the drag-direction shedding frequency, then a phenomenon called lock-in will occur. The shedding frequency may actually shift from its nominal value for a stationary structure, to coincide with the nearest natural frequency of the flexible structure. The resulting structural response will then change from forced vibration to resonant vibration (see Chapter 3), with much larger vibration amplitudes and possible subsequent structural damage.

Vortex-induced vibration is one of the earliest known flow-induced vibration phenomena. Extensive literature exists in this area, which is still very much a subject of active research. The objective of this chapter is to outline the steps to avoid lock-in with vortex-shedding, and the methods to estimate the structural responses. For a more in-depth treatment of vortex-induced vibration, the reader is referred to Chapter 3 of Blevins (1990) and the references cited there.

One cycle, induced drag force

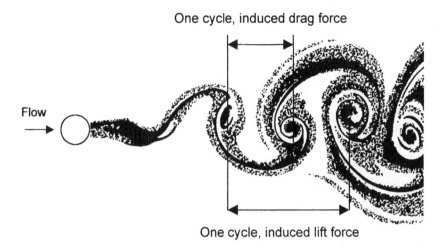

Flow

One cycle, induced lift force

Figure 6.1 Vortex street off a circular cylinder subject to cross-flow (Griffin and Ramberg, 1974, reproduced with permission from Cambridge University Press)

6.2   Vortex-Shedding Frequency and the Strouhal Number

Based on the above introduction, one of the most important parameters to assess the concern of vortex-induced vibration is obviously the vortex-shedding frequency f_s , in both the lift and drag directions.   This has been determined experimentally with a stationary circular cylinder in uniform cross-flow.   It was found that the shedding frequency f_s was proportional to the velocity V and inversely proportional to the diameter of the cylinder D.   The constant of proportionality is subsequently call the Strouhal number S:

$$f_s = SV / D \tag{6.1}$$

The value of S had been measured experimentally and was found to be dependent on the Reynolds number:

$$\mathrm{Re} = VD / v \tag{6.2}$$

where v is the kinematic viscosity; v is related to the dynamic viscosity μ by

$$v = \mu / \rho \tag{6.3}$$

where ρ is the fluid density. In the ASME Steam Table (1979), μ is given in lbf-s/ft^2 (Pa-s in SI units). To be consistent with these units, the fluid density ρ must be in slug/ft^3 and V, D must be in ft/s and ft (Kg/m^3, m/s and m in SI units; see Chapter 1 for a brief but very important explanation of units) when we calculate the kinematic viscosity v and the

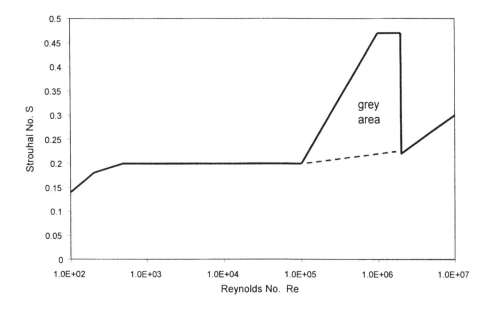

Figure 6.2 Strouhal number as a function of Reynolds number
for a stationary circular cylinder

the dimensionless Reynolds number Re. The example given in this chapter should make it clear.

Figure 6.2, based on data from Lienhard (1966) and Achenbach and Heinecke (1981), gives an approximate Strouhal number as a function of the Reynolds number. Even though the data were obtained based on measurements with a circular cylinder, the relationship is generally used for structures of arbitrary cross-sectional shapes. In the case of non-circular cross-sections, D should be taken as the width of the structure perpendicular to the flow. Figure 6.2 shows that for the purpose of engineering assessment of vortex-induced vibration concerns in power and process plant components, one can use the following Strouhal numbers:

$$S = 0.2 \quad for \quad 1.0E+3 \leq \mathrm{Re} \leq 1.0E+5$$
$$0.2 \leq S \leq 0.47 \quad for \quad 1.0E+5 < S < 2.0E+6 \qquad\qquad (6.4)$$
$$0.2 \leq S \leq 0.3 \quad for \quad 2.0E+6 \leq \mathrm{Re} \leq 1.0E+7$$

Equation (6.1) gives the alternate vortex-shedding frequency in the *lift* (perpendicular to the flow) direction. Based on the geometry of the vortex street (Figure 6.1), it can be deduced that there is also a fluctuating force component in the drag (in the direction of flow) direction, with a frequency equal to *twice* the shedding frequency,

$$f_L = f_s, \quad f_D = 2f_s \qquad\qquad (6.5)$$

where f_L, f_D are the frequencies of the vortex-induced forces on the structure in the lift and drag directions.

6.3   Lock-in

In the case of a flexible structure subject to cross-flow, the vortex-shedding frequency is not as definite as that in a stationary cylinder. If the shedding frequency, as defined in Equation (6.1), is *closed* to one of the structure's flexural frequencies, a phenomenon called lock-in occurs. The vortex-shedding frequency can actually shift from its value for a stationary cylinder to the natural frequency of the structure, resulting in much larger amplitude, resonant vibration of the structure. From Equation (6.5), lock-in can occur either in the lift direction or in the drag direction at twice the cross-flow velocity for lift-direction lock-in.

Experimental data showed that the fluctuating force in the drag direction is generally almost one order of magnitude smaller than that in the lift direction. This, together with the fact that lock-in in the drag direction occurs at *twice* the flow velocity for lock-in in the lift direction, indicates that lock-in in the lift direction is a much more common and serious concern. However, structural failure due to lock-in vortex-induced vibration in both the lift and drag directions had been observed (see Case Studies at the end of this chapter).

Rules to Avoid Lock-In

Based on experimental data from many researchers, ASME International recommends the following rules to avoid lock-in synchronization (ASME Boiler and Pressure Vessel Code, Sec III Appendix N-1324, 1998):

(1)   Low Velocity Rule

If the velocity for the fundamental mode ($n=1$) of vibration of the structure satisfies:

$$V / f_1 D < 1.0 \tag{6.6}$$

then lock-in in both the lift and the drag directions are avoided.

(2)   Large Damping Rule

If the reduced damping C_n, defined as

$$C_n = \frac{4\pi \zeta_n m_n}{\rho D^2 \int_L \psi_n^2(x)dx} \tag{6.7}$$

$$m_n = \int_0^L m\ (x)\psi_n^2(x)dx \qquad \text{(See Equation (3.32))} \tag{6.8}$$

is larger than 64, then lock-in will be suppressed in the nth mode of vibration.

(3)   Combined Low Velocity and Large Damping Rule

If $V / f_n D < 3.3$ *and* $C_n > 1.2$ are *both* satisfied, then lock-in in the lift direction is avoided and that in the drag direction is suppressed for mode n.

(4)   Separation Rule

If the structural modal frequency is at least 30% below *or* 30% above the vortex-shedding frequency, i.e., if

$$f_n < 0.7 f_s \quad or \quad f_n > 1.3 f_s$$

then lock-in in the lift direction is avoided in the nth mode. Likewise, if

$$f_n < 0.7*(2f_s) \quad or \quad f_n > 1.3*(2f_s)$$

then lock-in in the drag direction is avoided in the nth mode.

6.4   Vortex-Induced Vibration Amplitudes

Off lock-in, vortex-induced vibration is of the forced vibration type, with the forcing function given by:

$$F_L = (\rho V^2 / 2)DC_L \sin(2\pi f_s t) \tag{6.9}$$

in the lift direction and

$$F_D = (\rho V^2 / 2)DC_D \sin(4\pi f_s t) \tag{6.10}$$

in the drag direction. A conservative value for the lift coefficient is C_L=0.6 (Blevins, 1990, Chapter 3). The drag coefficient C_D is generally one order of magnitude smaller than the lift coefficient. Using Equation (6.9) or (6.10), the vibration amplitudes in the lift and drag directions can be computed either by using a standard commercial finite-element computer program, or with the equation of structural dynamics in the case of simple structures (see Chapter 3). Even with a conservative value for C_L, vortex-induced vibration is seldom a problem in the lift direction if lock-in is avoided. The drag coefficient C_D is generally one order of magnitude smaller than the lift coefficient, so vortex-induced vibration in the drag direction is of even less concern if lock-in in the drag direction is also avoided.

The same cannot be said about lock-in vortex-induced vibration. When lock-in occurs, not only do the fluid-dynamic coefficients become much larger, but non-linear effects due to strong fluid-structure coupling (see Chapter 4) invalidate the linear structural dynamics equation in Chapter 3. The responses in this case are computed by empirical equations. The following three equations are recommended by the ASME Boiler and Pressure Vessel Code Sec III Appendix N1324 (1998; see also references cited there):

Griffin-Ramberg:
$$\frac{y_n^*}{D} = \frac{1.29\gamma}{[1 + 0.43(2\pi S^2 C_n)]^{3.35}}$$
(6.11)

Ivan-Blevins:
$$\frac{y_n^*}{D} = \frac{0.07\gamma}{(C_n + 1.9)S^2}[0.3 + \frac{0.72}{(C_n + 1.9)S}]^{1/2}$$
(6.12)

Sarpkaya:
$$\frac{y_n^*}{D} = \frac{0.32\gamma}{[0.06 + (2\pi S^2 C_n)^2]^{1/2}}$$
(6.13)

Here y_n^* is the maximum response amplitude for the mode (nth) that locks in with the vortex-shedding frequency, S is the Strouhal number and γ is a mode shape factor. Table 6.1 gives the mode shape factors for several common structural components (Blevins, 1990).

Table 6.1: Mode Shape Factors

Structure	γ
Spring-supported rigid cylinder	1.0
Pivoted rod	1.91
Taut string	1.155, all modes
Simply-supported beam	1.155, all modes
Cantilever	1.305, $n=1$, 1.499, $n=2$, 1.537, $n=3$

Figures 6.3 and 6.4 show comparisons of the computed normalized responses $y*/(\gamma D)$ as a function of the reduced damping parameter C_n for Strouhal numbers $S=0.2$ and 0.47, using Equations (6.11), (6.12) and (6.13). It can be seen that the agreements among the three equations are good for $S=0.2$, at which most of the test data were acquired. At $S=0.47$ (for Reynolds number 1.0E+5 and 2.0E+6), agreement is poor. This is because little test data is available at this Reynolds number. These figures show that estimates of

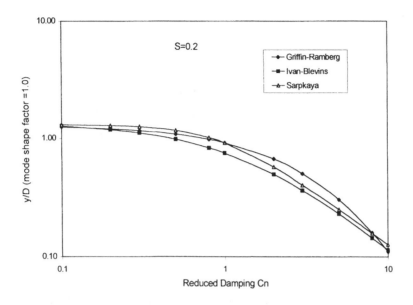

Figure 6.3 Comparison of normalized responses computed with
the three equations for S=0.2

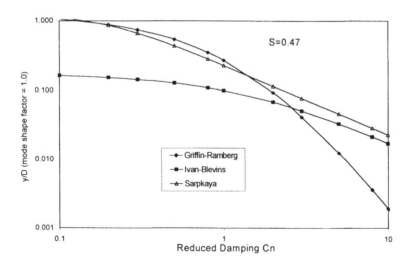

Figure 6.4 Comparison of normalized responses computed with
the three equations for S=0.47

lock-in vortex-induced vibration are at the best semi-quantitative. Considerable
engineering judgment should be exercised when using the computed results. With few
exceptions, most structural components in the power and process industries cannot
survive lock-in vortex-induced vibration. On the other hand, vortex-induced vibration
seldom poses any danger to such components if lock-in is avoided.

Example 6.1

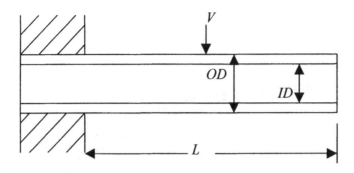

Figure 6.5 Instrument nozzle subject to cross flow

Figure 6.5 shows an instrument nozzle extending inside a pipe carrying superheated steam. The known parameters of this system are given in Table 6.2. One end of the nozzle can be assumed clamped at the pipe wall while the other end is free. Assess the vortex-induced vibration concern of this instrument nozzle for this application.

Table 6.2: Know Parameters for Instrument Nozzle

	US Units	SI Units
Steam pressure	1075 psi	7.42E+06 Pa
Steam temperature	700 F	371 C
Cross flow velocity V	70 ft/s	21.336 m/s
OD	1.495 in	0.03797 m
ID	0.896 in	0.02276 m
L	14.5 in	0.368 m
Density of material	481 lb/ft^3	7706 Kg/m^3
Young's modulus	25E+6 psi	1.72E+11 Pa
Damping ratio	0.005	0.005

Solution

The first concern is whether there is any danger of the vortex-shedding frequency locking into the fundamental modal frequency of the cantilever. For this we first calculate the shedding frequency. From the ASME Steam Table (1979), we find that at the given temperature and pressure, the steam is superheated with properties given in Table 6.3:

Table 6.3: Properties of Steam at the Given Ambient Conditions

Steam Properties	US Units	SI Units
Density	1.79 lb/ft^3 $=5.559$E-2 slug/ft^3	28.68 Kg/m^3
Dynamic viscosity	4.82E-7 lb-s/ft^2	2.31E-5 Pa-s

Therefore,

kinematic viscosity $v = \mu / \rho$ = 4.82E-7/5.559E-2 .... 8.05E-7 m-s

$=8.68$E-6 ft-s

$Re = \dfrac{VD}{v} =$ 70*(1.495/12)/8.68E-6   21.336*0.03797/8.05E-7

$=1.0$E+6 $=1.0$E+6

From Figure 6.2, at a Reynolds number of 1.0E+6, the Strouhal number can be anywhere between 0.2 and 0.47. The lower and upper limits for the shedding frequencies are therefore:

S	US Unit	SI Unit
0.2	f_s=0.2*(70*12)/1.475=112.4	0.2*21.336/0.03797=112.4
0.47	f_s=0.47*(70*12)/1.475=264.1	0.47*21.336/0.03797=264.1

The fundamental modal frequency can be computed assuming the instrument nozzle is a cantilever beam (Blevins, 1979):

$$f_1 = \frac{1.875^2}{2\pi L^2} \sqrt{\frac{EI}{m}}$$

Here the total linear mass density m must include the mass of steam its displaces (hydrodynamic mass, see Chapter 4). The mass of air contained inside the instrument nozzle, however, can be ignored. The following structural parameters are computed:

Structural Parameters	US Unit	SI Unit
$I = (\pi/64)(OD^4 - ID^4)$	0.2136 in^4	8.889E-8 m^4
Inside area $A_i = \pi ID^2/4$	0.6305 in^2	4.068E-4 m^2
Outside area $A_o = \pi OD^2/4$	1.7554 in^2	1.133E-3 m^2
Material density =481/(32.2*12*1728)= 7.204E-4 (lb-s^2/in)/in^3		7.706E+3 Kg/m^3

Mat. linear den.=7.204E-4*(1.7554-0.6305)=

	8.103E-4 (lb-s^2/in)/in	5.592 Kg/m
Steam density=1.79/(32.2*12*1728)=	2.681E-6 (lb-s^2/in)/in^3	28.68 Kg/m^3
Hydrodyn. mass=2.681E-6*1.7554=	4.706E-6 (lb-s^2/in)/in	3.248E-2 Kg/m
Total linear mass density m	8.150E-4 (lb-s^2/in)/in	5.625 Kg/m
f_l=	215 Hz	215 Hz

The fundamental modal frequency of the nozzle is within the possible range of vortex-shedding frequencies. We must assume that the possibility of lock-in vortex-induced vibration exists. The vibration amplitudes as a result of this will be estimated by Equations (6.11), (6.12) and (6.13). For that, we must first calculate the reduced damping, Equation (6.7), and the generalized mass, Equation (6.8). From Equation (6.8), since the total linear mass density m is constant,

$$m_1 = m \int_0^L \psi^2 dx$$

The integrals in Equation (6.7) cancel out, the reduced damping for the fundamental mode can be calculated in a straightforward manner, using Equation (6.7):

	US Unit	SI Unit
C_l	$\dfrac{4\pi*0.005*8.150E-4}{2.681E-6*1.495^{\wedge}2} = 8.547$	$\dfrac{4\pi*0.005*5.625}{28.68*0.03797^{\wedge}2} = 8.547$

At the lock-in frequency of 215 Hz,

$S = \dfrac{f_l D}{V} = \dfrac{215*1.495}{840} = 0.383$		$\dfrac{215*0.03797}{21.336} = 0.383$

From Equation (6.11), $y_1^* = 0.018$ in   4.45E-4 m

From Equation (6.12), $y_1^* = 0.061$ in   1.56E-3 m

From Equation (6.13) $y_1^* = 0.079$ in   2.00E-3 m

The stress induced at the clamped end of the instrument nozzle can be estimated from an equivalent static model (Chapter 3, Section 3.8). From Roark (1989), for a uniformly loaded cantilever, the maximum bending moment (at the clamped end) and the maximum deflection (at the free end) are given by:

$$M_{max} = \frac{wL^2}{2} \quad and \quad y_{max} = \frac{wL^4}{8EI}$$

where w is the static load per unit length of the cantilever. Using the middle number (y^*=0.061 in) as computed by Equation (6.12), It follows that the maximum moment at the clamped end of the instrument nozzle is,

US Unit   SI Unit

$$M_{max} = \frac{4EI}{L^2} y_{max} = \frac{4*25E+6*0.2136*0.061}{14.5^2} \qquad \frac{4*1.72E+11*8.889E-8}{0.368^2}$$

$$=6243 \text{ lb-in} \qquad\qquad =704.3 \text{ n-m}$$

The maximum induced stress is,

$$\sigma = M(D/2)/I = \frac{6243*1.495}{2*0.2136} = 21839 \ psi \qquad =1.504E+8 \text{ Pa}$$

Both are 0-peak values at 215 Hz. This is a very high stress level. Even if the nozzle is made of high tensile steel, it probably will fail by fatigue in less than 1.0 million cycles, or about 1.3 hours (ASME Boiler and Pressure Vessel Code, Sec III Appendix I, 1998). Note that the Griffin-Ramberg Equation (6.11) gives an answer that does not agree with those from the other two equations. The responses computed from Equation (6.12) were used to calculate the final stress.

Example 6.2

Suppose the instrument nozzle in Example 6.1 is shortened from 14.5 in. (0.368 m) to 9.875 in. (0.251 m) while all the other parameters remain the same. Estimate the vortex-induced vibration amplitude and stress in the nozzle.

Solution

Following the same calculation as in Example 6.1 but with L=9.875 in. (0.251 m), we find the first cantilever modal frequency f_l=464 Hz. Since the vortex-shedding frequencies are, from the results of **Example 6.1**, between 112 and 264 Hz, the fundamental modal frequency is more than 130% of the highest possible vortex-shedding frequency. Therefore, lock-in is avoided. The forced vibration amplitude of this re-designed instrument nozzle can be estimated following the standard technique of structural dynamic analysis, with a forcing function in the lift direction equal to:

$$F_L = (C_L D_0 \rho V^2 / 2)(\sin 2\pi f_s t) \quad \text{per unit length}$$

which is a harmonic forcing function with a frequency equal to the vortex-shedding frequency. Using a conservative equivalent lift coefficient from Blevins (1990), C_L=0.6, we get,

$$| F_L | = 0.6*1.495*2.681E-6*840^2/2 = 0.848 \text{ lb/in} \quad \text{in US units,}$$
$$| F_L | = 0.6*0.03797*28.68*21.336^2/2 = 148.7 \text{ N/m} \quad \text{in SI units,}$$

The vibration amplitudes can be calculated with standard finite-element computer programs with the above forcing function as input. In the present simple case of uniform cross-flow over a cylinder of uniform cross-sectional properties, we can calculate the response at the tip of the nozzle using Equations (3.16) and (3.28). The rest of the solution is given in Examples 3.1 and 3.2 following Sections 3.7 and 3.8 in Chapter 3, where it is found that the maximum stress is,

$$\sigma_{max} = \frac{M_{max}(OD/2)}{I} = 195.5 \text{ psi} \qquad (1.35E+6 \text{ Pa})$$

As we see, if lock-in is avoided, the vibration amplitude and stress are negligible.

6.5   Vortex-shedding inside a Tube Bundle

Unlike vortex-shedding off an isolated cylinder, vortex-shedding in a tightly packed tube bundle is a much more controversial subject. Even the very basic question of whether or not vortex-shedding exists inside a tube bundle is a subject of debate among the researchers. Over the last 40 years, the opinions among the experts flip-flopped a few times. At one time, it was generally accepted that vortex-shedding in a tube bundle could excite acoustic modes in a heat exchanger, until Owen (1965) argued that vortex-shedding could not exist inside a tightly packed tube bundle. From the late 1960s to the early 1970s, theories about flow-induced vibration in heat exchangers was very much influenced by Chen (1968), who tried to explain virtually every flow-induced vibration phenomenon in terms of Karman vortex streets. His model was widely accepted at that time. Since the discovery of fluid-elastic instability phenomenon by Connors in the early 1970s (see Chapter 7), the question of vortex-shedding in heat exchangers suddenly was largely ignored, at least by the designers. Instead, practicing engineers tended to focus on fluid-elastic instability and turbulence-induced vibration as the criteria for designing heat exchangers. The very existence of vortex streets inside a tube bundle was once again in doubt until the early 1990s, when photographs from experiments conducted by Weaver et al (1988) clearly showed the formation of vortices behind the first few rows of tubes in a tube bundle (see Chapter 12). The question of vortex-induced vibration in heat exchanger tube bundles once again was revived. The readers are referred to an excellent review paper by Weaver (1993) for the historical background and the latest developments in vortex-induced vibration in tube bundles.   This paper also contains extensive bibliography on this topic, including the papers cited earlier in this paragraph.

Based on all the available experimental data, there is no doubt that even if vortex-shedding does occur in a tube bundle, it is much less clearly defined compared with vortex-shedding from an isolated cylinder.   When the shedding frequency is well separated from the tubes' natural frequencies, based, for example, on the 30% rule, the magnitudes of forced tube vibration are usually small and are bounded by turbulence-induced vibration (see Chapter 9).   The main concern of vortex-shedding in a tube bundle is its potential to excite the acoustic modes in the heat exchanger. This will be addressed in Chapter 12 under acoustically induced vibration.

6.6   Strouhal Numbers for Tube Arrays

Unlike the case of an isolated cylinder, there is a bewildering quantity of experimental data on the Strouhal numbers for tube arrays.   Apart from the complication that the Strouhal number for tube bundles apparently is dependent on the array geometry, there is the further confusion that different investigators expressed their data in terms of different parameters. Even worse, different investigators interpreted "vortex-shedding" differently and might have intermixed their interpretation with other phenomena such as jet-switching, acoustic resonance or even amplitude-limited fluid-elastic instability.   There are considerable discrepancies amongst various sets of data.   The following empirical equation from Weaver and Frizpatrick (1988) seems to be the simplest and easiest to apply (see Chapter 7, Figure 7.6 for the definitions of tube bundle geometry):

In-line square array: $\qquad S = \dfrac{1}{2(P/D-1)}$ $\qquad\qquad\qquad\qquad$ (6.14)

Rotated square array: $\qquad S = \dfrac{1}{2(P/D-1)}$ $\qquad\qquad\qquad\qquad$ (6.15)

Normal triangular array: $\quad S = \dfrac{1}{1.73(P/D-1)}$ $\qquad\qquad\qquad$ (6.16)

Rotated triangular array: $\quad S = \dfrac{1}{1.16(P/D-1)}$ $\qquad\qquad\qquad$ (6.17)

The above equations are based on the approach velocity, rather than the pitch velocity commonly used in tube bundle dynamic analyses (see Chapters 7 and 9). As in the case of isolated cylinders, the vortex-induced vibration of tube bundles can be computed by standard methods of structural dynamic analysis, with the forcing function given by

$$F_L = (C_L D_0 \rho V^2 /2)(\sin 2\pi f_s t) \quad \text{per unit length} \qquad\qquad (6.18)$$

Pettigrew et al (1978) found that the upper bound for C_L for a tube bundle is 0.07, which is almost one order of magnitude smaller than that in an isolated cylinder. Thus, vortex-induced vibration in a tube bundle is of much smaller concern than that in an isolated cylinder, and is best handled together with turbulence-induced vibration (see Chapter 8). The concern here is the excitation of acoustic modes inside the heat exchanger by vortex-shedding.

6.7   Departure from the Ideal Situation

In the real world, one seldom encounters the ideal situation of uniform flow over the entire span of a uniform circular cylinder with the flow perpendicular to the axis of the cylinder. Very often the cylinder is tapered, or the flow is not uniform or over the entire length of the cylinder, or the flow is not perpendicular to the cylinder axis. Under these conditions, engineering judgment will be necessary. The following gives some common practices to handle such situations. Some are backed up by test data, while others are purely engineering judgments that seem to work but are awaiting rigorous experimental verifications.

Bluff Bodies of Non-Circular Cross-Sections

In this case the diameter D is taken as the width of the structure perpendicular to the flow. The Strouhal number will be different from that for a circular cross-section but is surprisingly close to it. Data compiled by several researchers showed that S generally lies between 0.1 and 0.2. See Blevins (1990; Figure 3-6).

Tapered Cross-Sections

In this case the Renolds numbers and the shedding frequencies should be computed with the smallest and largest diameters. Lock-in must be assumed possible for any mode with a natural frequency less than 130% of the upper shedding frequency limit, and more than 70% of the lower shedding frequency limit.

Non-Uniform Structural Properties

With the wide availability of finite-element structural analysis computer programs, the natural frequencies, mode shapes and vibration responses of structures of non-uniform cross-sectional properties can easily be computed, provided that the forcing function is known. After the natural frequencies are computed, one should check for the possibility of lock-in for all the lower modes. If lock-in is avoided, usually vortex-induced vibration is not a concern. However, one can compute the forced vibration responses by inputting the forcing function from Equations (6.9) and (6.10), with conservative values for the lift and drag coefficients.

Non-Uniform Flow Over Structure

In this case the bounding vortex-shedding frequencies should be computed with the cross-flow velocity $V(x)$ over the segment of structure with diameter $D(x)$. If any of the natural frequencies of the structure lies within the range of vortex-shedding frequencies, lock-in in that particular mode should be assumed.

Oblique Flow

When the flow is not exactly perpendicular to the axis of the structure, it is customary to resolve the velocity into two components—one parallel and the other perpendicular to the axis of the cylinder—and just use the normal velocity component to calculate the vortex-shedding frequency and the harmonic forcing function.

Structures Subject to Two-Phase Cross-Flows

Probably because of the extra energy dissipation capability in a two-phase mixture, vortex-shedding from a structure subject to two-phase cross-flow has so far not been observed. The entire area of two-phase flow-induced vibration is still very much un-explored.

6.8   Case Studies

Although it is one of the easiest flow-induced vibration problems to avoid, vortex-induced vibration has caused large financial losses in the power industry. The following are two of the well publicized and documented (at least within the nuclear industry) examples and show how these financial losses can be avoided by straightforward flow-induced vibration analysis.

Case Study 6.1: Nuclear Reactor In-core Instrument Guide Tube Broken by Vortex-Induced Vibration (Paidoussis, 1980)

In-core instrument guide tubes, also called thimble tubes or nozzles, are sheaths that allow interments to be moved inside the core of a nuclear reactor. These are pressure boundaries designed to withstand the high pressure inside the nuclear reactor (Figure 6.6, see also Figure 2.20). In one original design, the in-core guide tubes were too flexible. During the pre-operational test, which lasted only a few days, these guide tubes experienced lock-in vortex-shedding. Post-test inspection revealed that most of them broke, with pieces lying on the bottom of the reactor vessel. Since the fuel assemblies were not loaded into the core (see Figure 2.20) during the pre-operational test, some of these broken loose parts were carried by the coolant flow into the upper plenum of the

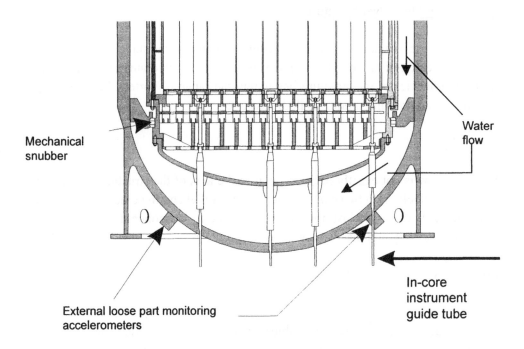

Mechanical
snubber

Water
flow

External loose part monitoring
accelerometers

In-core
instrument
guide tube

Figure 6.6 Lower part of a pressurized water nuclear reactor showing the
in-core instrument guide tubes

steam generators. Trapped between the tubesheet and the upper plenum head (see Figure 9.2 in Chapter 9 for the schematics of this steam generator), these loose parts were whirled around by the coolant flow and repeatedly impacted the tube ends, causing extensive damage to the steam generators. These guide tubes were subsequently re-designed to be stiffer. The re-designed in-core guide tubes did not encounter any flow-induced vibration problems. However, thimble tubes and guide tubes in other nuclear plants have experienced wear due to confined axial flow or cross-flow turbulence-induced vibration (see Chapters 9 and 10).

Case study 6.2:   Thermocouple Well Failure (Yamaguchi et al, 1997)

In 1995, a thermocouple well in Japan's Monju fast breeder reactor failed. This thermocouple well served as a pressure boundary between the liquid sodium flowing in a pipe in the intermediate heat transport system, and the pipe's exterior insulation (Figure 6.7). The well was a fairly slender, stepped tube design with the outer diameter changing quite abruptly from 10 mm to 25 mm at the neck. Post-incident inspection showed that the well broke at this diameter transition, causing liquid sodium to leak from the pipe. Operation history showed that the thermocouple well had operated for a period at 100% flow without any problem. Failure seemed to occur when the flow was decreased to 40%. Laboratory tests using full-scale models of the thermocouple well showed that at

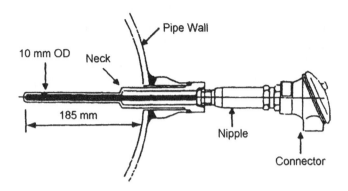

Figure 6.7 Thermocouple well. The component broke at the neck at 40% full flow.
(Yamaguchi A. et al, 1997)

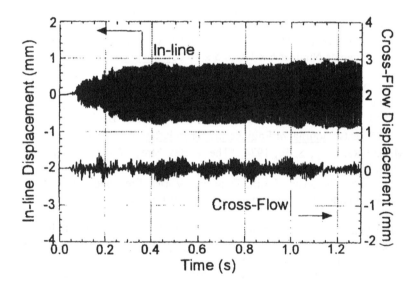

Figure 6.8 Vibration amplitudes of the thermocouple well in the drag and
lift directions at 40% of the full flow velocity. (Ogura et al, 1999)

the flow velocity corresponding to either 40% or 100% power, vibration amplitudes in
both the lift and drag directions were small as long as the thermocouple well was in an
un-degraded condition. However, at a flow velocity corresponding to 40% power and
when a crack was artificially etched in at the neck of the thermocouple, large amplitude
vibration in the drag direction was observed (Figure 6.8). This amplitude of vibration
was much larger than that in the lift direction and was enough to cause rapid failure of the
thermocouple.

Root cause analysis of the problem: The thermocouple was fairly slender and had a
very small damping ratio. On top of this, there was no attempt to "round off" the corner

at the step transition, which therefore had a high stress concentration and was a breeding ground for fatigue crack growth. It was likely that some minute fatigue cracks were initiated by turbulence-induced vibration, which in turn reduced the overall stiffness of the thermocouple well.   However, being off resonance with the vortex-shedding frequency (both in the lift and drag directions), this crack growth rate was slow (but probably would still lead to failure ultimately). When the system was operated at 40% of the full flow velocity, the drag-direction vortex-shedding frequency, which is twice the lift-direction vortex-shedding frequency, synchronized with the fundamental mode of the tip of the thermocouple well clamped at the transition point of the diameters.   The resulting large alternating stress rapidly caused the thermocouple well to break at this transition point.

References:

ASME Boiler and Pressure Vessel Code, 1998, Section III, Appendix N1324.

ASME Steam Tables, 1979, Fourth Edition.

Achenbach, E and Heinecke, E, 1981, "On Vortex Shedding from Smooth and Rough Cylinders in the Range of Reynolds Number 6 x 10^3 to 5 x 10^6, Journal of Fluid Mechanics, Vol. 109, pp. 239-251.

Blevins, R. D., 1990, Flow-Induced Vibration, Second Edition, Chapter 3, Van Nostrand Reinhold, New York.

Chen, Y. N., 1968, "Flow-Induced Vibration and Noise in Tube Bank Heat Exchangers due to Von Karman Streets," Journal of Engineering for Industry, Vol. 90. pp. 134-146.

Griffin, O. M. and Ramberg, S. E., 1974, "The Vortex Street Wakes of Vibrating Cylinders," Journal of Fluid Mechanics, Vol. 66, pp. 553-576.

Lienhard, J. H., 1966, "Synopsis of Lift, Drag and Vortex Frequency Data for Rigid Circular Cylinders," Washington State University, College of Engineering, Research Division Bulletin 300.

Ogura, K., Morishita, M. and Yamaguchi, A., 1998, "Cause of Flow-Induced vibration of Thermocouple well" in Flow-Induced Vibration and Transient Thermal-Hydraulics, ASME Special Publication PVP-363, edited by M. K. Au-Yang, pp. 109-117.

Owen, P. R., 1965, "Buffeting Excitation of Boiler Tube Vibration," Journal of Mechanical Engineering Science, Vol. 7, pp. 431-439.

Paidoussis, M. P., 1980, "Flow-Induced Vibration in Nuclear Reactors and Heat Exchangers," in Practical Experience with Flow-Induced Vibrations, edited by E. Naudascher and D. Rockwell, Springer-Verlag, Heidelberg, pp. 1-81.

Pettigrew, M. J. et al 1978, "Vibration Analysis of Heat Exchanger and Steam Generator Designs," Journal Nuclear Engineering and Design, Vol. 48, pp. 97-115.

Roark, 1989, Roark's Formula for Stress and Strain, 6th Edition revised by W.C. Young, McGraw Hill, New York.

Weaver, D.S., 1993, "Vortex Shedding and Acoustic Resonance in Heat Exchanger Tube Arrays," in Technology for the 90s, edited by M.K. Au-Yang, ASME Press, pp. 775-810.

Weaver, D. S.; Fitzpatrick, J. A., 1988, "A Review of Cross-Flow-Induced Vibration in Heat Exchanger Tube Arrays," Journal of Fluids and Structures, Vol. 2, pp. 73-93.

Yamaguchi, A. et al, 1997, "Failure Mechanism of a Thermocouple Well Caused by Flow-Induced Vibration" in Proceeding 4th International Symposium on Flow-Induced Vibration, Vol I, edited by M. P. Paidoussis, ASME Special Publication AD-Vol. 53-1, pp. 139-148.

CHAPTER 7

FLUID-ELASTIC INSTABILITY OF TUBE BUNDLES

Summary

When a tube bundle is subject to cross-flow with increasing velocity, it will come to a point at which the responses of the tubes suddenly rapidly increase without bound, until tube-to-tube impacting or other non-linear effects limit the tube motions. This phenomenon is known as fluid-elastic instability. The velocity at which the vibration amplitudes of the tubes suddenly increase is called the critical velocity. Unlike vortex shedding, the vibration amplitudes of a fluid-elastically unstable tube bundle will continue to increase even when the critical velocity is exceeded. The motions of the tubes in the bundle become correlated and bear definite phase relationship to one another. Based on dimensional analysis, Connors, who first discovered the phenomenon of fluid-elastic instability in 1970, derived the following simple equation to predict the critical velocity of a tube bundle:

$$V_c = \beta f_n \left(\frac{2\pi \zeta_n m_t}{\rho} \right)^{1/2}$$
(7.4)

The constant β is subsequently known as the Connors' constant, or constant of fluid-elastic instability. Later theoretical formulation of tube bundle dynamics by Chen (1983) showed that β is not a true constant, especially for tube bundles vibrating in heavy fluids. However, when the fluid density is small, such as air or super-heated steam, β is approximately constant. There are several other theoretical formulations of the theory of fluid-elastic instability. However, all of them lead to equations very similar to the original Connors' Equation (7.4), and none of them offer distinct advantages over the Connors' equation in practical industrial applications. The following values for the constant of fluid-elastic instability and damping ratios are recommended for use with Connors' equation:

Table 7.4: Recommended Connors' Constant and Damping Ratios

	Connors' constant β	ζ_n, water or wet steam, tightly supported tube	ζ_n, air or gas, tightly supported tube	ζ_n, loosely supported tube
Mean Value	4.0	0.015	0.005	0.05
Conservative Value	2.4	0.01	0.001	0.03

One can define the fluid-elastic stability margin as

$$FSM = V_c / V_p \tag{7.5}$$

$FSM > 1.0$ implies that the tube bundle is stable while $FSM < 1.0$ implies that the tube bundle is unstable. When the velocity and density of cross-flow over the tube are not constant, an equivalent mode shape-weighted pitch velocity is defined:

$$\overline{V}_{pn}^2 = \frac{\dfrac{1}{\rho_0} \displaystyle\int_0^L \rho(x) V_p^2(x) \psi_n^2(x) dx}{\dfrac{1}{m_0} \displaystyle\int_0^L m_t(x) \psi_n^2(x) dx} \tag{7.6}$$

$$FSM = V_c / \overline{V}_{pn} \tag{7.7}$$

In applications to industrial heat exchangers with multiple spans and tube-support clearances, Connors' equation often encounters problems because of closely-spaced normal modes and because tube-support plate interactions often require non-linear structural dynamic analysis, which is usually carried out in the time domain. It can be shown that in the time domain, Connors' equation is equivalent to

$$M\ddot{y} + (C_{sys} + C_{fsi})\dot{y} + ky = F_{inc} \tag{7.14}$$

with

$$C_{fsi} = -4\pi M f_c \frac{\rho V_p^2 / 2}{\pi f_c^2 m \beta^2} = -\frac{4qL}{f_c \beta^2} \tag{7.19}$$

where

$$q = \rho V_p^2 / 2 \tag{7.20}$$

When the pitch velocity and the linear mass density of the tube is non-uniform, Equation (7.19) is replaced with the following more general equation:

$$\sum C_{fsi}^{(i)} = -\frac{4\sum q_i dx_i}{f_c \beta^2} \tag{7.21}$$

$$q_i = \rho V_{pi}^2 / 2 \tag{7.22}$$

where the summation is over the nodes i. The fluid-elastic stability margin is given by

$$FSM = \sqrt{\frac{q_c L}{\sum_i q_i dx_i}} \qquad (7.23)$$

$$q_c L = -f_{cc}\beta^2 (C_{fsi})_c / 4 \qquad (7.24)$$

where $(C_{fsi})_c$ is the negative damping coefficient and f_{cc} is the crossing frequency at the threshold of instability.

The time domain Equation (7.14) can be used to predict the critical velocity in an industrial heat exchanger tube with multiple spans, closely-spaced modal frequencies and tube-to-support plate interactions, by gradually increasing the magnitude of C_{fsi} until the solution starts to diverge. In practice, non-linear finite-element structural dynamic analysis computer programs with gap and damping elements are used to perform this time domain analysis.

Acronyms

FSM   Fluid-elastic Stability Margin
OTSG   Once Through Steam Generator
RSG   Re-circulating Steam Generator

Nomenclature

C_M Added mass coefficient of tube
C_{sys} System damping coefficient (excluding that due to fluid-structure interaction)
C_{fsi} Damping coefficient due to fluid-structure interaction
D Outside Diameter of tube
E Young's modulus
F_{fsi} Force due to fluid-structure interaction
F_{inc} Incident (external) force
f Frequency, in Hz
f_c Crossing frequency, in Hz
f_{cc} Critical crossing frequency, in Hz
$G_d(f)$ Power spectral density of tube response (see Chapters 8 and 9)
k Tube stiffness
L Length of tube or beam
ℓ_i Length of one span
m_0 Mean total mass per unit length of tube
m_t Total mass per unit length of tube

m_n	Generalized mass
M	Total mass, $= m_l L$
P	Pitch of a tube bundle
q	Generalized coordinate, or dynamic pressure ($= \rho V_p^2 / 2$)
V	Free stream, or approach velocity
V_c	Critical velocity
V_p	Pitch velocity
$\bar{V}_{pn}$	nth mode shape-weighted mean pitch velocity
x	Coordinate
Δx	Finite-element length
y	Vibration amplitude
β	Constant of fluid-elastic instability (Connors' constant)
δ	Reduced damping ($2\pi m_v \zeta_v / \rho D$) or logarithmic decrement ($= 2\pi\zeta$)
ρ	Shell-side (outside of tube) fluid density
ρ_0	Mean shell-side fluid density
ρ_i	Tube-side (inside tube) fluid density
ρ_m	Mass density of tube material
ψ	Mode shape function
ζ	Damping ratio
ω	Frequency, in radian/s

Subscripts and Indices

fsi	Force, damping coefficient or damping ratio due to fluid-structure interaction
sys	System properties excluding that due to fluid-structure interaction
.	Time derivative
c	Critical values (values at the threshold of instability)
n	Modal index
v	In-vacuum values

7.1   Introduction

Connors (1970) reported an experiment on cross-flow-induced vibration of a row of tubes suspended by piano wires in a wind tunnel. He observed that as the mean flow velocity gradually increased the vibration amplitudes of the tubes first gradually increased with the flow velocity typical of turbulence-induced vibration. However as the cross-flow velocity continued to increase, at one point the vibration amplitudes of the tubes suddenly increased rapidly, until tube-tube impacting occurred. The tube row basically became unstable. The vibration amplitude increased without bound, except by non-linear effects and tube-to-tube impacting. Connors had discover one very basic excitation mechanism

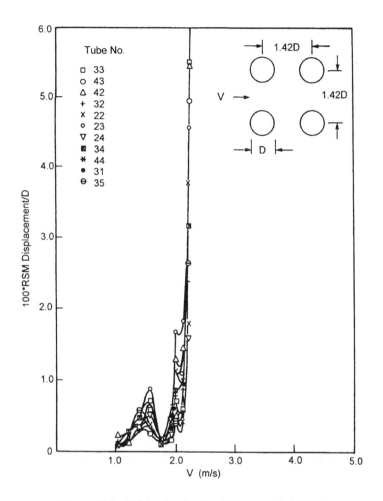

Figure 7.1 Critical velocity (Chen et al, 1978)

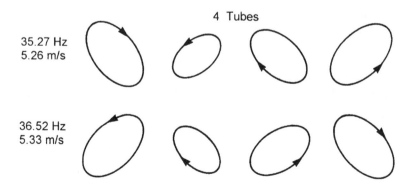

Figure 7.2 Post-instability motion of tubes (Southworth et al, 1975; Reproduced
with permission from Journal of Mechanical Engineering Science)

in tube bundles that in the subsequent 20 years, became the mostly intensely researched subject in flow-induced vibration.

Fluid-elastic instability (of tube bundles) differs from vortex shedding in that the vibration of the tubes is not caused by unsteady fluid forces. Rather, it is caused by the interaction of the flow field and the adjacent tubes. Unlike vortex-induced vibration in which the tube vibration amplitude will decrease once the cross-flow velocity exceeds the lock-in Strouhal number range (see Chapter 6), the vibration amplitudes of a tube bundle will continue to increase without bound once the critical velocity is exceeded (Figure 7.1). In practice, non-linear effect and tube-to-tube impacting will limit the vibration amplitudes. Also, unlike vortex shedding, which can occur in an isolated tube, fluid-elastic instability can occur only in a row or array of tubes, even if just one tube in the array is flexible. The latter, of course, is of academic interest, as such a situation almost never occurs in industrial applications. The resulting motions are highly organized orbital motions, with definite phase relationship between neighboring tubes, as shown in Figure 7.2.

Fluid-elastic instability received wide attention because it can cause rapid structural damage to the tubes, through one of the following mechanisms:

(1)   Tube-to-tube impacting occurs when the 0-peak vibration amplitudes of the neighboring tubes exceed the half-gap clearance between the tubes. Neighboring tubes start to impact one another, usually at the mid-span of the tubes. This causes impact wear, which ultimately will remove enough material from the tube wall to enable internal pressure to burst the tube. Tube failure of this type usually has long, longitudinal, fish-mouth type of cracks (Chapter 11, Figure 11.12). Failure can occur within days or even hours.

(2)   Fatigue failure occurs when the vibration amplitudes of the tubes become so large that the resulting stress in the tubes exceeds the endurance limit. The tube is then suddenly severed at the section where the stress is highest, which is usually at the tube supports, especially if the gap between the tube support plate and the tube is "packed" (i.e., filled) by oxide buildups.

(3)   Severe fretting wear occurs when the tube bundle is near the critical velocity. The tube bundle would have been unstable except for the interaction between the tubes and support plates, which helps dissipate energy and increase the apparent damping ratio of the tubes. The tubes then become stable again momentarily, resulting in less tube-to-tube support interaction and hence smaller damping. The vibration amplitude then starts to grow again. Taylor et al (1995) called this "amplitude limited " fluid-elastic instability. A tube bundle vibrating in the amplitude-limited fluid-elastic instability mode will experience much higher fretting wear rate at the supports. The tubes may have to be removed from service by plugging and stabilization (Au-Yang, 1987).

In the 1970s and 1980s, for a period of about 20 years, a huge number of technical papers and reports on fluid-elastic instability was published in the literature and many alternate forms of Equation (7.4) were proposed for predicting the critical velocity. Most of them offered little advantage over the original simple equation proposed by Connors (discussed later). Some are so complex and require so many input parameters, which usually have to be measured experimentally, that they are not practical to use under the industrial environment, although they may give some insight on the origin of the instability mechanism. These large but uncoordinated efforts among so many researchers, who invariably presented their data in different forms and very often used the same terminology to mean different variables—or use different terminologies to mean the same variable—has caused a lot of confusion among the industrial engineers who must deal with systems that do not resemble the highly idealized, uniform-flow and single-span tube bundle configurations that are so frequently reported in the literature. In the following sections, basic equations and input parameters for estimating the critical velocity are discussed. It is not the objective of this chapter to give an exhaustive treatise on fluid-elastic instability. Readers who are interested in an in-depth understanding of this subject should refer to several excellent review papers by Price (1993), Paidoussis (1987) and Chen (1987), as well as several books and reports by Paidoussis (1998), Chen (1985) and Blevins (1990). Each of these publications in turn contain extensive bibliography on the subject of fluid-elastic instability.

7.2   Geometry of Heat Exchanger Tube Arrays

Modern heat exchangers are complex and expensive components. Those used in nuclear plants for converting water into steam are called steam generators. These are high-performance heat exchangers with high flow rates and thin-walled tubes to maximize the heat transfer rates. Figure 7.3 is a cutaway view of a typical horizontal U-tube heat exchanger. Figure 7.4 shows the U-tube arrangement in a typical recirculating nuclear steam generator. In this design, hot water from the nuclear reactor enters from one side (called the hot leg) of the inverted U and flows up, around the U bend and then down the other side (cold leg), thereby heating up the water (shell-side water) surrounding the tubes. More schematics of nuclear steam generators can be found in Chapter 9. Figure 9.13, for example, shows a steam generator with straight tubes. In this design, the shell-side water enters from the bottom of the steam generator, up through the tubes and then exits from the top of the steam generator. The tube-side water enters from the top of the steam generator and then exits from the bottom. This design is called a once-through steam generator (OTSG).

Figure 7.6 shows the common arrangements of tube arrays in heat exchangers and nuclear steam generators. The distance between the centers of the adjacent tubes is the pitch, P, of the tube array. The parameter P/D, commonly referred to as the pitch-to-diameter ratio, is an important parameter in tube bundle dynamic analysis. The cross-flow velocity through the gap clearance between the tubes, commonly referred to as the

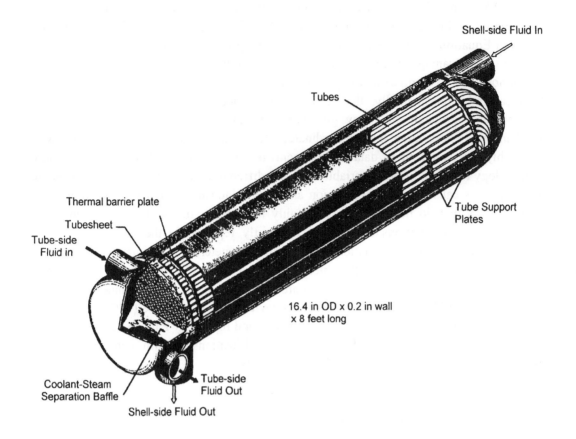

Figure 7.3 Cutaway drawing of a heat exchanger (Nelms, 1970)

gap, or pitch velocity, is related to the free stream, or approach velocity V, by the equation,

$$V_p = \frac{P}{P-D}V = \frac{P/D}{(P/D)-1}V \qquad (7.1)$$

Most heat exchangers tube bundles have P/D ratios between 1.2 and 1.5. Many researchers believe the critical velocity is dependent on the geometry of the tube array. As a result, publications in the literature often link the data to specific array geometry and P/D ratios. In actual applications, the distinction is less important because one seldom can determine, with certainty, the precise direction of the cross-flow. In addition, the scatter of the data within each test often is as high as that between data sets from tests involving different tube array geometry and P/D ratios.

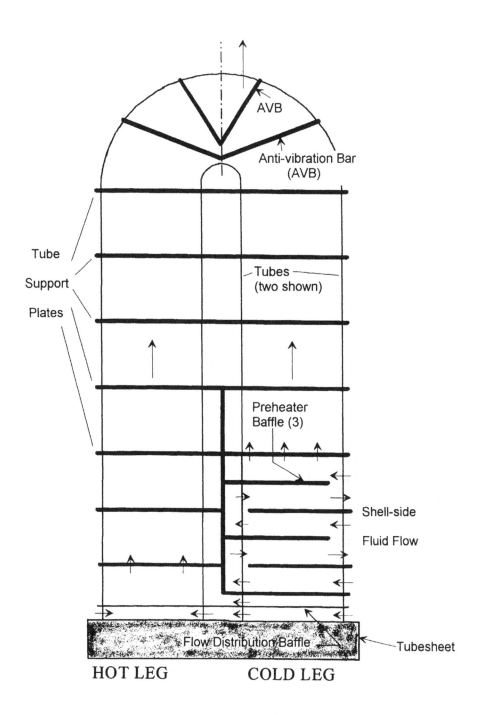

Figure 7.4 Schematic of a typical recirculating steam generator (RSG)

1. Upper dome, Steam Outlet Nozzle and Flow Restrictor (one piece forging)

2. Dryers (star arrangement)

3. Secondary manway

4. Separators (cyclones)

5. Main feedwater distribution system

6. Main feedwater nozzle

7. Auxiliary feedwater ring

8. Tube bundle

9. Divider plate (economizer)

10. Tube bundle wrapper

11. Double wrapper (economizer)

12. Pressure vessel

13. Tube support plates

14. Flow distribution baffle

15. Tubesheet

16. Partition plate

17. Primary coolant outlet nozzle

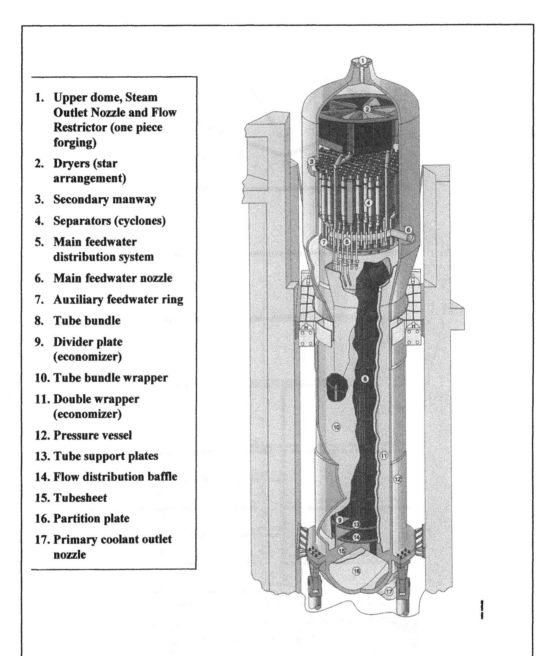

Figure 7.5 Schematic of a high-performance nuclear steam generator
(Courtesy: Framatome)

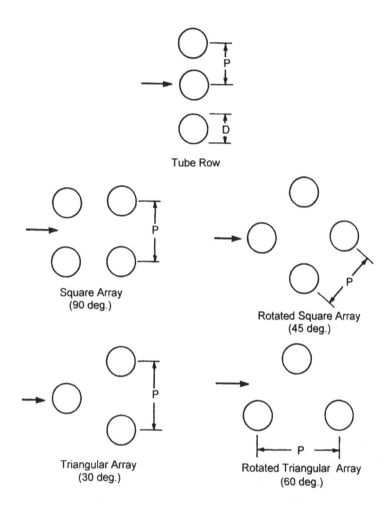

Figure 7.6 Geometry of Heat Exchanger Tube Arrays
(ASME Boiler Code, 1998)

7.3   Connors' Equation

In 1970, after discovering the fluid-elastic instability phenomenon in an experiment consisting of one row of rigid tubes, with a P/D ratio of 1.42 and supported by piano wires in a wind tunnel, Connors derived an empirical equation for predicting the critical velocity based on simple dimensional analysis. Obviously, the critical velocity V_c, that is, the velocity through the gaps between the tubes at which the tube bundle will become unstable, will depend on the fluid density ρ, the total mass per unit length of the tube m_t, the tube outside diameter D, the modal frequency of the tube f_n and the modal damping ratio of the tube ζ_n. Being an experimentalist, however, Connors chose to use the logarithmic decrement $\delta_n = 2\pi\zeta_n$ instead of the damping ratio ζ_n that is more commonly used among the analytical engineers. We shall use ζ_n in our discussions. From these variables, one can form three dimensionless parameters: $(V_c / f_n D)$, $(m_t / \rho D^2)$ and, finally, $\delta_n = 2\pi\zeta_n$. Therefore, it is reasonable to assume that,

$$\frac{V_c}{f_n D} = \beta \left(\frac{m_t}{\rho D^2} \right)^a (2\pi\zeta_n)^b \tag{7.2}$$

where a, b and β are dimensionless constants. β has since been commonly called the Connors' constant or constant of fluid-elastic instability. Connors then found that,

$a=b=0.5$
$\beta=9.9$

seemed to fit his data best. With these, Equation (7.2) simplifies to:

$$\frac{V_c}{f_n D} = 9.9 \left(\frac{2\pi\zeta_n m_t}{\rho D^2} \right)^{1/2} \tag{7.3}$$

Connors' discovery stirred up intensive research effort on flow-induced vibration of tube bundles throughout the 1970s and 1980s. Literally hundreds of technical papers were published in this period, many of them duplications of Connors' tests with different tube array geometry, fluid media and flow regimes, in attempts to more precisely determine the constants a, b and β. These later tests found that the constant β depends on the tube bundle geometry. In particular, for cross-flow over an array, instead of a row of tubes, the critical velocity is significantly lower than what is predicted by Equation (7.2), which was then put into a more general form,

$$V_c = \beta f_n \left(\frac{2\pi\zeta_n m_t}{\rho} \right)^{1/2} \tag{7.4}$$

with the constant β to be determined from tests based on the specific tube array geometry. Equation (7.4) is widely known as the Connors' equation.

Fluid-Elastic Stability Margin (FSM)

For application to heat exchanger tube bundle design, it is convenient to define a fluid-elastic stability margin (FSM) as the ratio of actual operating gap (pitch) velocity over the critical velocity; it has the same meaning as the factor of safety:

$$FSM = V_c / V_p \qquad\qquad (7.5)$$

With this definition, $FSM > 1.0$ means tubes are stable, whereas $FSM < 1.0$ means the tubes are unstable. Some authors use the term "margin" to mean V_p / V_c. In that case the margin should be less than 1.0 to ensure stability. The first definition is more inline with the usual engineering meaning for "margin." The second definition is more like a "usage."

<u>Non Uniform Cross-Flow Velocities</u>

When the cross-flow over the tube is not uniform, an effective pitch velocity is obtained by mode shape weighting (Connors, 1980):

$$\overline{V}_{pn}^2 = \frac{\dfrac{1}{\rho_0} \displaystyle\int_0^L \rho(x) V_p^2(x) \psi_n^2(x)\,dx}{\dfrac{1}{m_0} \displaystyle\int_0^L m_t(x) \psi_n^2(x)\,dx} \qquad\qquad (7.6)$$

with the corresponding fluid-elastic stability margin given by

$$FSM = V_c / \overline{V}_{pn} \qquad\qquad (7.7)$$

where ψ_n is the nth mode shape function, $\rho(x)$, $m_t(x)$ are the fluid mass and total tube linear mass densities along the tube and ρ_0, m_0 are the corresponding average densities. The integration is over the entire single or multi-span tube. Thus, there is an effective pitch velocity and a fluid-elastic margin (FSM) for each mode. Note that,

- As defined in Equation (7.6), $\overline{V}_{pn}$ is independent of mode shape normalization.
- To calculate the stability margin of a multi-span tube, it is not enough to consider only the spans with cross-flow over them, as can be seen from Example 7.2.
- It is not necessarily true that the fundamental mode has the lowest FSM, as is shown by the following example.

Figure 7.7 shows the mode shapes and frequencies of a heat exchanger tube with seventeen spans. Only the bottom and the sixteenth (second from the top) spans are subject to cross- flows, with the flow over the sixteenth span being the highest. Because of this, the mode with the lowest FSM is not the first or the second mode with natural frequencies of 33.1 and 36 Hz, but the third mode with a natural frequency of 37.4 Hz. In another similar design, the mode with the lowest FSM is the twenty-fifth.

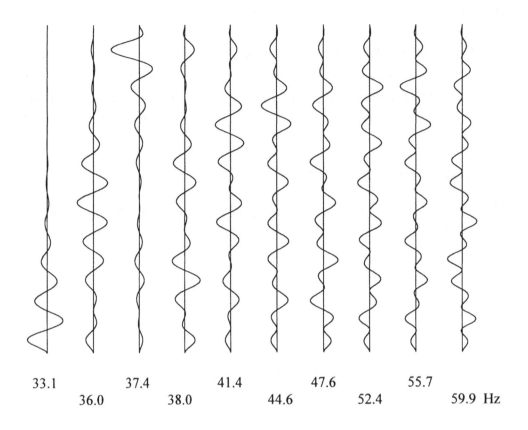

33.1   37.4   41.4   47.6   55.7

36.0   38.0   44.6   52.4   59.9 Hz

Figure 7.7 Frequencies and mode shapes of a typical multi-span heat exchanger tube

Other Forms of Connors' Equation

Because of its simplicity, Connors' Equation (7.4) was widely accepted by the industry as the standard way of assessing fluid-elastic instability concerns in heat exchanger designs. It was adopted by the ASME Boiler Code (1998), Standard of the Tubular Exchanger Manufacturers Association Design Guide (THEMA Standard, 1988), as well as many other individual technical papers on flow-induced-vibration design guides. Being an empirical, highly simplified equation on a very complex phenomenon, Equation (7.4) is not without its flaws both conceptually and in applications to industrial heat exchanger tubes with multiple spans and with non-uniform cross-flow velocities. In addition, it often cannot predict with reasonable accuracy the critical velocity even in simple laboratory tube bundle models. Part of the problem may be due to the inherent difficulties in determining the other parameters, such as natural frequencies and damping ratios. Other problems arise because of misunderstandings of Equation (7.4). Published data using Equation (7.4) often do not explicitly mention whether the frequency f_n and the damping ratio ζ_n were measured in quiescent air or water or at flow, and whether m_t includes the added or hydrodynamic mass (see Chapter 4).

In attempts to "improve" Connors' Equation (7.4), many authors had proposed alternate forms to the basic equation by introducing other dimensionless parameters, such as Reynolds' number, Strouhal number and P/D ratio to Equation (7.4). Others proposed to keep the more general Equation (7.1), and assign different values to the parameters a, b and β in accordance with the tube bundle geometry. At the end, after 20 years of intensive research effort by the industry, as well as by those in the academia, none of these alternate empirical equations offer significant advantage over the original Connors' equation, other than that β is typically in the range of 2.5 to 6.0 in tube arrays, instead of 9.9 as measured by Connors in his tube row experiment. While there is evidence that the parameters a and b may differ from 0.5 for different array geometry and orientations to the cross-flow, data scatter in any specific test seemed to mask this weak dependence. In addition, under actual industrial environment, the direction of flow itself is subject to large margins of uncertainties.

Example 7.1

A heat exchanger tube bundle consists of single-span, square-pitch tubes (Figure 7.8). The tubes are pin-supported at the two ends. The dimensions of the tubes and the velocity, densities of the fluids are given in Table 7.1. Estimate the critical velocity and the fluid-elastic stability margin (FSM) of this tube bundle.

Table 7.1: Tube Bundle Parameters for Example 7.1

	US Unit	SI Unit
Length of tube, L	24 in	0.6096 m
OD of tube, D	0.75 in	0.01095 m
ID of tube	0.66 in	0.01676 m
Tube pitch, P	1.00 in	0.0254 m
Density of tube material	511.5 lb/ft^3	
	(7.661E-4 (lb-s^2/in)/in^3)	8193 Kg/m^3
Young's modulus, E	2.81E+7 psi	1.937E+11 Pa
Damping ratio, ζ	0.01	0.01
Tube-side fluid density, ρ_i	37.55 lb/ft^3	
	(5.624E-5 (lb-s^2/in) /in^3)	601.5 Kg/m^3
Shell-side fluid density, ρ	47.55 lb/ft^3	
	(7.11E-5 (lb-s^2/in) /in^3)	760.9 Kg/m^3
Cross-flow velocity, V	2 ft/s (24 in/s)	3.658 m/s
Connors' constant, β	3.3	3.3

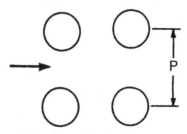

Figure 7.8   Geometry of the tube bundle for Example 7.1

Table 7.2: Computed Parameters for Example 7.1

	US Unit	SI Unit
Inside x'section area of tube, A_i	0.3421 in^2	2.207E-4 m^2
Outside x'section area of tube, Ao	0.4418	2.850E-4 m^2
Metal x'section area of tube, A_m	0.09967	6.430E-5 m^2
Area moment of inertia, I	6.217E-3 in^4	2.588E-9
D_e/D	2.2222	2.2222
Added mass coeff., C_M	1.51	1.51
Total mass/unit length, m_t	2.377E-4 (lb-s^2/in)/in	0.9866 Kg/m^3
Fundamental frequency, f_1	95.3 Hz	95.3 Hz
Pitch velocity, V_p	96 in/s	2.44 m/s
Critical velocity, V_c	111.8 in/s	2.838 m/s
$FSM = V_c/V_p$	1.164	1.164

Solution

Following our standard practice, we immediately convert all the units in the US unit system into a consistent unit set (see Chapter 1). The mass densities are converted into (lb-s^2/in)/in^3, while the velocity is converted into in/s. These values are given in parenthesis in Table 7.1 and are used in the calculation. Readers who are still not familiar with the importance of using consistent unit sets should study Chapter 1 thoroughly before proceeding. The next step is to calculate the natural frequencies of the tube. For this, we must first calculate the inside and outside cross-sectional area of the tube, the mass per unit length of fluid carried inside the tube, the added mass due to the shell-side fluid (see Chapters 4 and 5), the area moment of inertia of the tube cross-section, and so on. These values are given in rows 1 - 4 in Table 7.2. The following outlines some of the steps without going into details.

From Equation (5.11),

$$Added\ Mass\ Coefficient\ C_M = \left[\frac{(D_e/D)^2 + 1}{(D_e/D)^2 - 1} \right]$$

where

$$D_e/D = (1 + \frac{P}{2D})(\frac{P}{D})$$

The values of D_e/D and the added mass coefficient are computed from the above equations and shown in rows 5 and 6 in Table 7.2. Note that both are dimensionless and are the same in both unit systems. The total mass per unit length of the tube is equal to the sum of the material mass, the mass of water contained in the tube, and the added mass due to the shell-side water (see Chapter 4). Using the notations in Tables 7.1 and 7.2, this is equal to

$$m_t = \rho_m A_m + \rho_i A_i + C_M \rho A_o$$

Using the values of ρ_m, A_m, ρ_i, A_i, C_M, ρ, A_o in Tables 7.1 and 7.2, m_t is computed in both US and SI units and is given in Table 7.2. From Table 3.1, the natural frequency of the fundamental mode (which will have the lowest FSM from Equation (7.4)) is given by:

$$f_1 = \frac{\pi}{2L^2} \sqrt{\frac{EI}{m_t}}$$

Using the given values of L, E from Table 7.1 and the computed values m_t and I in Table 7.2, the fundamental modal frequency is computed and given in row 7 of Table 7.2. Next, we find the pitch velocity V_p. From Equation (7.1),

$$V_p = \frac{P}{P - D} V$$

Using the values of P, D and V given in Table 7.1, we calculate the pitch velocities in both US and SI units. These are given in row 8 in Table 7.2. The critical velocity is given by Connor's Equation (7.4),

$$V_c = \beta f_n \left(\frac{2\pi \zeta_n m_t}{\rho} \right)^{1/2}$$

Using the given values of β, ζ and ρ in Table 7.1 and the computed values of f_1 and m_t in Table 7.2, the critical velocity can be computed from the above equation. Its values in both US and SI units are given in Table 7.2. Finally, from Equation (7.5),

$$FSM = V_c / V_p$$

Using the values of the just-computed critical and pitch velocities, the fluid-elastic stability margin is found to be

FSM=1.16

in both unit systems. Thus, the tube bundle has 16% margin of safety against fluid-elastic instability.

Example 7.2

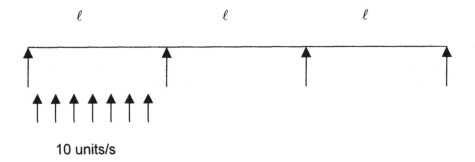

10 units/s

Figure 7.9 A tube bundle with three equal spans

Figure 7.9 shows a tube bundle with three equal spans, each of length ℓ and total length $L=3\,\ell$. The tube is simply-supported at the two ends and at each of the intermediate supports. There is a cross-flow with uniform pitch velocity of 10 units/s over only the first span. Assume:

$f_1 = 10$ Hz
$\beta = 3.4$
$\zeta = 0.01$
$m_t = 1$ unit
$\rho = 1$ unit

What is the fluid-elastic stability margin (FSM) for the first mode?

Solution

The critical velocity is, from Equation (7.4),

$$V_c = 3.4 * 10 * \sqrt{2\pi * 0.01} = 8.5$$

If we consider only the first span, the mode shape function is [1]

$$\psi = \sqrt{\frac{2}{\ell}} \sin \frac{\pi x}{\ell}$$

The effective pitch velocity is

$$\overline{V}_p^2 = \frac{\int_0^\ell (10)^2 (\frac{2}{\ell}) \sin^2 \frac{\pi x}{\ell} dx}{\int_0^\ell (\frac{2}{\ell}) \sin^2 \frac{\pi x}{\ell} dx} = 100$$

or

$$\overline{V}_p = 10$$

$$FSM = 8.52/10 = 0.85$$

We would have concluded that the tube is unstable. But if we consider the entire tube, as we should, we would get,

$$\overline{V}_p^2 = \frac{\int_0^\ell (10)^2 (\frac{2}{L}) \sin^2 \frac{3\pi x}{L} dx}{\int_0^L (\frac{2}{L}) \sin^2 \frac{3\pi x}{L} dx} = 100/3$$

---

[1] Since Equation (7.6) is independent of mode shape normalization, we can calculate the mode shape-weighted pitch velocity without properly normalizing the modes shape functions (see Chapter 3, Equation (3.30)). Since this will generally not be the case, especially when we come to turbulence-induced vibration in the following chapters, it is advisable to form the habit of properly normalizing the mode shape functions.

$\overline{V}_p = 5.77$

and

$FSM = 8.5/5.77 = 1.48$

The tube bundle is actually stable. This example shows that in calculating the mode shape-weighted velocity with Equation (7.6), we must integrate over the entire tube, not just over the spans over which there is cross-flow.

Example 7.3

By this time it is assumed that we have stressed the importance of the proper handling of units enough that this example is given without specifically mentioning the units, other than that they are given in *consistent* units. Readers who are familiar with the dimensions of heat exchanger tubes will recognize from the given data, that the units used are those of the US system. However, in going through the sample, one does not have to pay attention to which unit system is used.

Figure 7.10 shows a heat exchanger tube with three spans. The tube is simply-supported at one end ($x=0$) and at two intermediate tube support plates, and is clamped at the other end ($x=L$). The tube is subject to cross-flow over its spans, with the cross-flow velocities given in the third column in Table 7.3 and plotted in Figure 7.11. The density of the fluid in each span can be assumed to be approximately constant. Other relevant data are given below, all in a consistent unit set.

OD of tube, D_o=0.625 length units
ID of tube, D_i=0.555 length units
Young's modulus, E=2.9228E+7 units
Shell-side fluid densities, ρ

Span 1: 2.816E-6, Span 2: 2.711E-6, Span 3: 2.726E-6 mass unit/unit vol.

The total mass densities per unit length, including mass of water contained and the added mass due to the shell-side fluid (see Chapter 4), m_t are:

Span 1: 6.9811E-5; Span 2: 6.9770E-5; Span 3: 6.9728 E-5 mass unit/unit length

Damping ratio ζ=0.03 for all modes
Connors' constant, β=3.3

Compute the mode shape-weighted pitch velocity, the critical velocity and the fluid-elastic stability margin of this tube bundle for the fundamental mode.

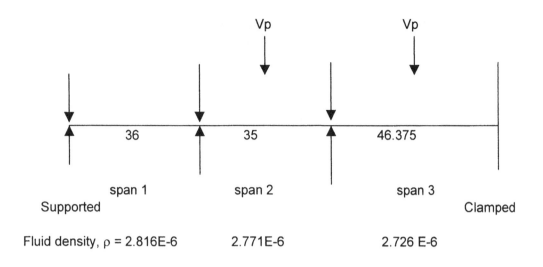

Figure 7.10 A three-span heat exchanger tube

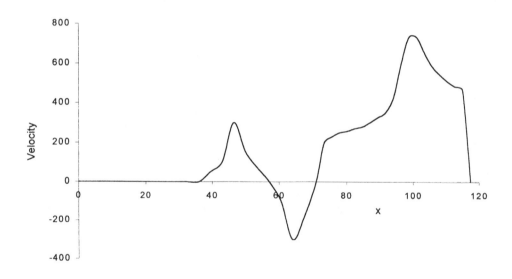

Figure 7.11 Cross-flow pitch velocity distribution over length of tube

Solution

As in Examples 7.1 and 7.2, the first step is to calculate the natural frequencies and mode shapes. We shall use a finite-element structural analysis computer program, NASTRAN, to solve this problem. Since the objective is to calculate the frequencies and mode shapes, other much less complex finite-element modal analysis program can be used. A finite-element model is set up, with 10 elements each in the first two spans and 20 elements in the third span.   The node numbers and their x-coordinates are shown in the first two columns of Table 7.3.   Using this model, it is found that

f_I=41.37 Hz

Table 7.3: Nodal Properties for Example 7.3

Node	x	V_p	ρ	ψ	$m_i\psi_i^2\Delta x_i$	$\rho_i V_p^2\psi_i^2\Delta x_i$
30020	117.38	0	2.726E-06	0	0.000E+00	0.000E+00
30019	115.061	465	2.726E-06	0.4754	3.655E-05	3.089E-01
30018	112.742	482	2.726E-06	1.765	5.037E-04	4.575E+00
30017	110.423	508	2.726E-06	3.666	2.173E-03	2.192E+01
30016	108.104	543	2.726E-06	5.98	5.782E-03	6.665E+01
30015	105.785	586	2.726E-06	8.51	1.171E-02	1.572E+02
30014	103.466	655	2.726E-06	11.07	1.982E-02	3.324E+02
30013	101.147	732	2.726E-06	13.5	2.947E-02	6.173E+02
30012	98.828	732	2.726E-06	15.64	3.955E-02	8.286E+02
30011	96.509	603	2.726E-06	17.36	4.873E-02	6.927E+02
30010	94.19	431	2.726E-06	18.56	5.570E-02	4.045E+02
30009	91.871	353	2.726E-06	19.17	5.942E-02	2.895E+02
30008	89.552	327	2.726E-06	19.12	5.911E-02	2.471E+02
30007	87.233	302	2.726E-06	18.43	5.492E-02	1.958E+02
30006	84.914	280	2.726E-06	17.09	4.723E-02	1.448E+02
30005	82.595	271	2.726E-06	15.15	3.711E-02	1.066E+02
30004	80.276	258	2.726E-06	12.7	2.608E-02	6.787E+01
30003	77.957	250	2.726E-06	9.835	1.564E-02	3.822E+01
30002	75.638	228	2.726E-06	6.668	7.190E-03	1.461E+01
30001	73.319	194	2.726E-06	3.441	1.915E-03	2.817E+00
30000	71	0	2.726E-06	0	0.000E+00	0.000E+00
20009	67.5	-181	2.771E-06	-4.758	5.528E-03	7.193E+00
20008	64	-300	2.771E-06	-8.846	1.789E-02	6.830E+01
20007	60.5	-100	2.771E-06	-11.95	3.487E-02	1.385E+01
20006	57	0	2.771E-06	-13.84	4.677E-02	0.000E+00
20005	53.5	70	2.771E-06	-14.38	5.050E-02	9.827E+00
20004	50	150	2.771E-06	-13.54	4.477E-02	4.001E+01
20003	46.5	300	2.771E-06	-11.42	3.185E-02	1.138E+02
20002	43	100	2.771E-06	-8.241	1.658E-02	6.587E+00
20001	39.5	50	2.771E-06	-4.308	4.532E-03	4.500E-01
20000	36	0	2.771E-06	0	0.000E+00	0.000E+00
10009	32.4	0	2.816E-06	4.391	4.846E-03	0.000E+00
10008	28.8	0	2.816E-06	8.324	1.741E-02	0.000E+00
10007	25.2	0	2.816E-06	11.43	3.283E-02	0.000E+00
10006	21.6	0	2.816E-06	13.41	4.519E-02	0.000E+00
10005	18	0	2.816E-06	14.08	4.982E-02	0.000E+00
10004	14.4	0	2.816E-06	13.37	4.493E-02	0.000E+00
10003	10.8	0	2.816E-06	11.36	3.243E-02	0.000E+00
10002	7.2	0	2.816E-06	8.252	1.711E-02	0.000E+00
10001	3.6	0	2.816E-06	4.337	4.727E-03	0.000E+00
10000	0	0	2.816E-06	0	0.000E+00	0.000E+00

Span 1: Δx=3.6   sum = 1.02   sum = 4.493E+03
Span 2: Δx=3.5
Span 3: Δx=2.319

The corresponding mode shape is given in column 5 of Table 7.3. To find the critical velocity, we first compute the mean total mass per unit length of the tube and the mean shell-side fluid density. This can be done either by averaging over the three spans, or, more likely in practice, using the same spreadsheet that is used in the subsequent calculations. It can be readily shown that

m_0=6.9766E-5 mass unit per unit length and ρ_0=2.7670E-6 mass unit per unit volume

Using Connors' Equation (7.4):

$$V_c = \beta f_1 \left(\frac{2\pi \zeta_1 m_0}{\rho_0} \right)^{1/2}$$

and substituting the values of $\beta, f_1, \zeta, m_0, \rho_0$ from above, we get

V_c=297.6 units

The mode shape-weighted mean velocity is best obtained from a spreadsheet. Table 7.3 shows an example of spreadsheet setup used in the calculations. First, we observe the way the mode shape is normalized in this model (see Chapter 3 on mode shape normalization). It can be readily found from the spreadsheet that:

$$\sum m_i \psi_i^2 \Delta x_i \approx 1.0$$

Thus, Equation (7.6) can be simplified to

$$\overline{V}_{pn}^2 = \frac{m_0}{\rho_0} \int_0^L \rho(x) V_p^2(x) \psi_n^2(x) dx = \frac{m_0}{\rho_0} \sum \rho_i V_i^2 \psi_i^2 \Delta x_i$$

Using the mean linear mass density and the mean shell-side fluid density computed above together with the sum computed with the spreadsheet, we can readily find,

$\overline{V}_p = 336.6$ units and

FSM =0.884

The tube bundle is therefore fluid-elastically unstable.

7.4   Theoretical Approaches to Fluid-Elastic Instability

As expected, since Connors discovered the tube bundle fluid-elastic instability mechanism, numerous authors tried to derive an equation to predict the onset of instability based on vigorous equations of fluid and structural dynamics. It is not the objective of this book to get deep into the various theories of fluid-elastic instability. Interested readers are referred to the treatise by Chen (1985), or a review paper by Price (1993), and the extremely extensive bibliography contained in both. The following brief review, however, shows why even after 30 years, Connors original simple semi-empirical Equations (7.4) and (7.6) have survived numerous criticisms and remained as the most widely accepted way to estimate the critical velocity of tube bundles.

<u>Displacement Controlled Instability Theory</u>

In the early days of fluid-elastic instability theory development, it was thought that instability was caused by a fluid displacement mechanism, and that coupling of a flexible tube to its flexible neighboring tubes through this displacement-induced fluid stiffness force was a requirement leading to tube bundle whirling (instability). Based on this approach, Blevins (1974) derived analytically the equation for the critical velocity:

$$\frac{V_c}{f_n D} = \beta \left(\frac{2\pi \zeta_n m}{\rho D^2} \right)^{1/2}$$

which of course, is the same as Connors' Equation (7.4). Based essentially on the fluid displacement approach, Paidoussis et al (1984) derived a similar but slightly more complicated equation,

$$\frac{V_c}{f_n D} = \alpha_1 [1 + \left(1 + \alpha_2 \frac{2\pi \zeta_n m}{\rho D^2} \right)^{1/2}] \qquad (7.7)$$

Neither equation offers significant advantage over the original Connors' equation because, even though the constants α and β can be expressed in terms of fluid stiffness forces, these forces need to be measured experimentally just like the Connors' constants need to be measured experimentally. In addition, the displacement-controlled instability theory has one serious flaw: it predicts that since fluid-elastic coupling is a requirement for instability, a flexible tube surrounded by rigid tubes cannot be unstable—a prediction that was spectacularly shattered by the experiment of Lever and Weaver (1982). In a videotape shown in flow-induced vibration symposia all over the world, a flexible tube surrounded by rigid tubes clearly became unstable as the cross-flow velocity increased beyond a certain critical value.

<u>Velocity Controlled Instability Theory</u>

Lever and Weaver (1982) suggested that a negative flow velocity dependent damping force controlled fluid-elastic instability. When this negative damping force numerically exceeds the system damping force of the tube bundle, the latter will be unstable. Based on this theory, Lever and Weaver arrived at an equation that took exactly the same form as Connors' equation:

$$\frac{V_c}{f_n D} = \beta \left(\frac{2\pi \zeta_n m}{\rho D^2} \right)^{1/2}$$

Leaver and Weaver's theory correctly predicts that a flexible tube surrounded by rigid tubes can be unstable. Other than this, it again offers little advantage over the original Connors' equation for practicing engineers to estimate the critical velocity of a tube bundle.

<u>Chen's Theory</u>

Using a highly mathematical model taking into account the complete sets of equations of fluid dynamics and structural mechanics, Chen (1983) formulated the equations of motion of a group of cylinders subject to cross-flow, and showed that the instability of a tube bundle depends on both fluid displacement (stiffness-controlled mechanism) and fluid velocity (fluid-damping-controlled mechanism). With either mechanism, the equation for the critical velocity again takes the form of the original Connors' equation,

$$\frac{V_c}{f_v D} = \beta \left(\frac{2\pi \zeta_v m_v}{\rho D^2} \right)^{1/2}$$

except that the modal frequency, modal damping and linear mass density are all "in-vacuum" values and the constant β is expressed in terms of fluid force coefficients that must be measured experimentally. Thus again, Chen's theory offers little relief to the practicing engineers, most of whom do not have the time, the budget or the facility to measure the many fluid force coefficients to calculate the single Connors' constant. Many of them would rather design an experiment to measure the single Connors' constant.

Chen's theory is generally regarded as the most comprehensive theory to explain the phenomenon of tube bundle instability. Researchers such as Tanaka and Takahar (1980) and Hara (1987) did carry out experiments to measure the fluid force coefficients. In addition, Chen's theory can be used to deduce many effects that have important consequences in tube bundle designs. Among these are:

Constancy of β

Based on Chen's theory, Connors' "constant" β should be more or less constant for a tube bundle vibrating in a light fluid such as air or superheated steam. However, it will be a function of the dimensionless velocity V / fD for tube bundles vibrating in a heavy fluid such as water, and is thus not a true constant.

Fluid-Structure Coupling

Chen predicts that at low reduced damping $\delta = 2\pi m_v \zeta_v / \rho D^2$, which prevails in heavy fluids such as water, the instability mechanism is controlled by the fluid velocity mechanism (fluid-damping-controlled). In this case, fluid coupling is not necessary to cause instability. A flexible tube surrounded by rigid tubes can be unstable in a heavy fluid such as water. For the same reason, the effect of de-tuning tubes (making the natural frequencies of the tubes different) on the critical velocity is not significant for a tube bundle vibrating in a heavy fluid such as water. On the other hand, at high reduced damping $\delta = 2\pi m_v \zeta_v / \rho D^2$, which prevails in light fluids, the instability mechanism is controlled by fluid stiffness that requires fluid-structure coupling. Thus, detuning the tube bundle can increase the critical velocity of a tube bundle vibrating in a light fluid such as superheated steam or gas.

7.5   ASME Guide

From the above discussions, it can be appreciated that as of the end of the millennium, and after 30 years of intensive effort, no one theory of fluid-elastic instability emerged as distinctly better suited for design engineers than the original Connors equation. In the 1998 version of the ASME Boiler Code, the basic Connors' Equation (7.4) and its generalization to non-uniform cross-flow, Equation (7.6) are the only recommended equations for predicting the critical velocity in a tube bundle, with the provision that m_t must be the total mass (including added mass: see Chapter 4) per unit length and ζ_n includes any damping caused by the fluid.

Values for Connors' Constant and Damping Ratios

Equation (7.4) and its generalization Equation (7.6) require as input the values of two parameters, the Connors' constant β and the modal damping ratio ζ_n, both of which must be measured experimentally. Considerable confusion exists in the literature on measured values of these two parameters. Theoretically, the damping ratio and the modal frequency are inherent properties of the structure and can therefore be measured separately. Knowing the velocity at which the vibration amplitude suddenly increases, the Connors' constant can then be deduced from Equation (7.4). In practice, both the

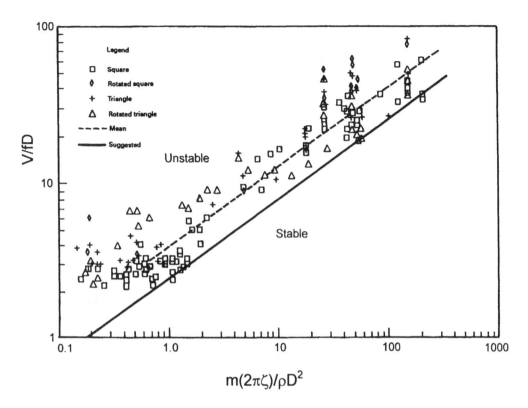

Figure 7.12 Stability diagram (ASME Boiler Code, 1998)

Table 7.4:   Recommended Connors' Constant and Damping Ratios

	β	ζ_n, water or wet steam, tightly supported tube	ζ_n, air or gas, tightly supported tube	ζ_n, loosely supported tube
Mean Values	4.0	0.015	0.005	0.05[1]
Conservative Values	2.4	0.01	0.001	0.03[1]

[1] This value is not from ASME, but is based on the author's own experience.

damping ratio and the modal frequency of a tube bundle subject to cross-flow will be different from those measured in air, especially if the fluid is a liquid or wet steam. It is this "in-flow" damping ratio and modal frequency that should be used in Equation (7.4). To deduce the Connors' constant, one must measure the critical velocity, the modal frequency and the modal damping ratio "at flow" separately. Unfortunately, there is no method of determining the last parameter, the damping ratio at flow, reliably. It is relatively meaningless to report, as is often done in the literature, the measured Connors' constant without stating the damping ratio used in deducing its value. In short, the experimentalist can only measure $\beta\sqrt{\zeta_n}$ together, and deduce β by assuming, based on separate in-air measurement coupled with theoretical analysis, a certain value for ζ_n.

From the above discussion, values for the Connors' constant and the damping ratio must be considered together. Based on 170 data points for the onset of instability in tube bundles of different array geometry, ASME (1998) recommended the values given in Table 7.4 for the Connors' constant as well as the damping ratios for the typical heat exchanger tube bundles with *P/D* ratio in the range between 1.25 and 1.50. Figure 7.12, reproduced from the ASME Boiler Code (1998), shows the mean and lower bound fit to the 170 data points. From this figure, it is obvious that the data scatter in the tests outweighs any merit of trying to use different values of β for different array geometry.

7.6   Time Domain Formulation of Fluid-Elastic Instability

From an analytical viewpoint, Connors' Equation (7.4), or its generalization, Equation (7.6) works well in the case of a single-span tube with well-separated normal modes. Unfortunately, industrial heat exchangers almost always contain tubes with multiple supports. As shown in Figure 7.7, this results in many modes with closely-spaced natural frequencies. To make the situation even worse, industrial heat exchanger tubes are often supported by oversized holes in the tube support plates. The gaps between the support plates and the tubes give rise to tube-support interaction, which can only be analyzed with non-linear structural dynamics in which the very existence of normal modes becomes an issue. The validity of the Connors' equations, which is basically a frequency domain equation, in such real-life industrial systems has been a question frequently asked, far more so than the exact values of the Connors' constant or the exact form of Connors' equation.

In applications to recirculating heat exchangers with U-shaped tubes (Figure 7.4), Connors' equation poses another problem. It often predicts the lower in-plane modes (Figure 7.13), including the pogo mode, to be unstable. Yet, in practice, these modes have never been observed to be unstable even when the computed theoretical stability margin is well below 1.0. All these indicate that being such a complex system; a heat exchanger tube may not be amenable to the simplified, modal analysis treatment.

Equation of Fluid-Elastic Stability in the Time Domain

It was shown by Chen (1983), Axisa, et al (1988) and Sauve' (1996) that Connors' Equation (7.4) is equivalent to a coupled fluid-structural dynamic equation in the time domain. The following simplified derivation follows that of Sauve'. The more mathematically inclined readers may want to follow Chen's more rigorous derivation.

The equation of motion of a tube vibrating in a quiescent fluid is addressed in Chapter 4, where it is shown that the equation of motion can be written in the form,

$$M\ddot{y} + C_{sys}\dot{y} + ky = F_{fsi} \tag{7.8}$$

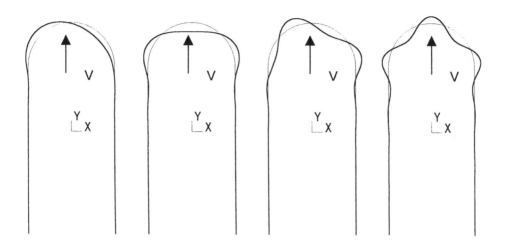

Figure 7.13 In-plane modes that have never been observed to be unstable even though the computed fluid-elastic stability margins are well below 1.0

where F_{fsi} is the force due to fluid-structure interaction. In Chapters 4 and 5, the interaction is assumed to be weak, so that in the presence of an external force (incident force), the total force on the right hand side of the equation of motion is just the sum of the incident force and the force due to fluid-structure interaction,

$$M\ddot{y} + C_{sys}\dot{y} + ky = F_{inc} + F_{fsi}$$

This "linear superposition" assumption subsequently leads to the added mass, or hydrodynamic mass, approach to the solution of weakly coupled fluid-structure systems that were treated extensively in Chapters 4 and 5. A fluid-elastically unstable tube bundle, on the other hand, is a strong-interacting fluid-structure system, as evidenced by the large displacement of the tubes compared with the flow channels (gaps between tubes). Therefore, the fluid-structure coupling force, F_{fsi}, will not be linearly superimposed on the incident force, which is proportional to the dynamic pressure,

$$F_{inc} = \frac{C}{2}\rho V_p^2 \tag{7.9}$$

where C is a constant. Sauve' (1996) assumed that the fluid-structure interaction force is proportional to a dimensionless velocity of tube response, $\dot{y}/\omega D$, and that instead of linearly superimposing on the incident fluid force F_{inc}, it multiplies the fluid force to form the resultant force acting on the tube. The equation of motion becomes

$$M\ddot{y} + C_{sys}\dot{y} + ky = \frac{C}{2\omega D}\rho V_p^2 DL\dot{y}$$

where $M=mL$. Rearranging,

$$M\ddot{y} + (C_{sys} - \frac{C}{2\omega}\rho V_p^2 L)\dot{y} + ky = 0 \qquad (7.10)$$

Following the modal decomposition technique described in Chapter 3, Equation (7.10) can be written in modal form (see notations defined in Chapter 3):

$$m_n\ddot{q} + (2m_n\omega_n\zeta_n - \frac{C}{2\omega_n}\rho V_p^2)\dot{q} + m_n\omega_n^2 q = 0 \qquad (7.11)$$

Instability occurs when the quantity in parentheses is equal to zero,

$$(2m_n\omega_n\zeta_n - \frac{C}{2\omega_n}\rho V_p^2) = 0$$

or

$$V_c = f_n\sqrt{\frac{2\pi m_n\zeta_n}{\rho}\left(\frac{8\pi}{C}\right)} \qquad (7.12)$$

If we put

$$\beta = \left(\frac{8\pi}{C}\right)^2 \qquad (7.13)$$

Equation (7.12) then becomes

$$V_c = \beta f_n\sqrt{\frac{2\pi m_n\zeta_n}{\rho}}$$

which is the Connors' equation, Equation (7.4). This shows that in the time domain, the equation of motion for tube bundle dynamics can be written as

$$M\ddot{y} + (C_{sys} + C_{fsi})\dot{y} + ky = F_{inc} \qquad (7.14)$$

where

$$C_{fsi} = 4\pi M f_o \zeta_{fsi} \quad and \quad \zeta_{fsi} = -\frac{\rho V_p^2 / 2}{\pi f_o^2 m \beta^2} \tag{7.15}$$

The question then arises: What is f_0? Axisa et al (1988) used the modal frequency f_n for f_0, defeating the purpose of writing the stability equation in the time domain, with the objective of non-linear analysis in which the very concept of normal modes can no longer be used. Sauve' (1996) called f_0 a participation frequency and computed and updated it at every time step in the solution. As Sauve' put it, the major task of the entire procedure of solving the time domain stability equation was to compute this "participation frequency." The need to continuously update this frequency resulted in a very time consuming algorithm. Au-Yang (2000) suggested that in the presence of closely spaced modal frequencies and non-linear effects, f_0 is the crossing frequency. This concept is discussed in the following paragraph.

The Crossing Frequency of a Vibrating Structure

The crossing frequency, or more accurately the positive crossing frequency, of a vibrating structure is defined as the number of times per second the response (displacement, stress, strain, etc.) at a point of a structure crosses the zero (or mean, if there is a static mean) response line from the negative amplitude to the positive amplitude. In a linear structure vibrating in one mode, the crossing frequency is equal to the modal frequency of the structure, as shown in Figure 7.14. In a linear structure vibrating with more than one mode, it can be shown that (Rice, 1945; Crandall and Mark, 1973) the crossing frequency is an effective modal participation-weighted mean frequency of the structure:

$$f_c^2 = \frac{\int_0^\infty f^2 G_d(f) df}{\int_0^\infty G_d(f) df} \tag{7.16}$$

As shown in Figures 7.14 and 7.15, if the structure vibrates predominantly in the first mode, its crossing frequency will be almost equal to its first modal frequency; while if it vibrates predominantly in the second mode, its crossing frequency will be almost equal to its second modal frequency. If both modes contribute equally to the response, its crossing frequency will be a weighted average of these two frequencies. Figure 7.16 shows the crossing frequency of a heat exchanger tube with oversized tube support plate holes. The motion in this case is chaotic.

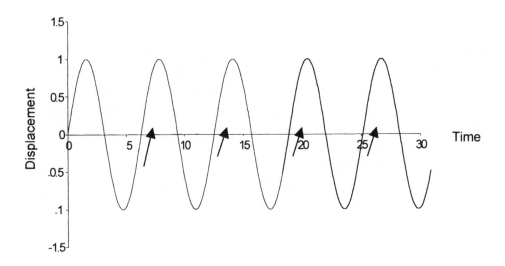

Figure 7.14 In a structure vibrating in one mode, the positive crossing (arrows) frequency is equal to its modal frequency

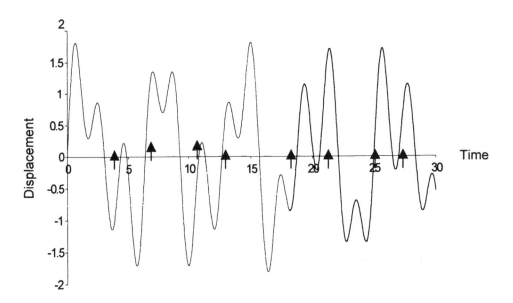

Figure 7.15 In a structure vibrating in two modes, the crossing frequency is a modal participation-weighted average of the two modal frequencies.

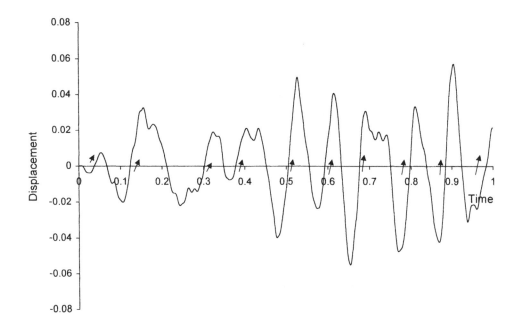

Figure 7.16 Crossing frequency of a system undergoing chaotic motion

The crossing frequency concept has been used by engineers to predict the fatigue usage of structural components vibrating with a combination of normal modes, based on fatigue curves derived from tests using cyclic loads, each at one specific frequency. Since the modal frequency is a measure of the structure's stiffness vibrating in that particular mode, and the crossing frequency is a modal participation-weighted average of the modal frequencies, one can deduce that the crossing frequency is a modal participation-weighted measure of the system stiffness of the structure, vibrating in a combination of normal modes. Since the modal critical velocity is completely governed by the tube's modal stiffness, it is reasonable to assume that when more than one mode becomes unstable at the same time, the system critical velocity (instead of the modal critical velocity) would depend on the system stiffness (or crossing frequency) of the tube.

<u>Critical Velocity for a Tube Bundle with Non-Linear Supports</u>

From Equation (7.15), with f_0 replaced by f_c,

$$C_{fsi} = -4\pi M f_c \frac{\rho V_p^2 / 2}{\pi f_c^2 m \beta^2} = -\frac{4qL}{f_c \beta^2} \qquad (7.17)$$

where

$$q = \rho V_p^2 / 2 \tag{7.18}$$

When the pitch velocity and the linear mass density of the tube is non-uniform, Equation (7.19) is replaced with the following more general equation, just as Connors' Equation (7.4) is replaced with the more general Equation (7.6):

$$\sum C_{fsi}^{(i)} = -\frac{4\sum q_i dx_i}{f_c \beta^2} \tag{7.19}$$

$$q_i = \rho V_{pi}^2 / 2 \tag{7.20}$$

where the summation is over the nodes i. The fluid-elastic stability margin, FSM, in this case can be defined in an analogous way to Equation (7.5),

$$FSM = \sqrt{\frac{q_c L}{\sum_i q_i dx_i}} \tag{7.21}$$

$$q_c L = -f_{cc} \beta^2 (C_{fsi})_c / 4 \tag{7.22}$$

$(C_{fsi})_c$ is the negative damping coefficient and f_{cc} is the crossing frequency at the threshold of instability.

The time domain Equation (7.14) can be used to predict the critical velocity in an industrial heat exchanger tube with multiple spans, closely-spaced modal frequencies and with both tube-to-support plate and fluid-structure interactions, by gradually increasing the magnitude of C_{fsi} until the solution starts to diverge. In practice, a non-linear finite-element structural dynamic analysis computer program with gap and damping elements will be used to perform this time domain analysis. Following this method, Au-Yang (2000) found that (see Figures 7.17, 7.19 and 7.20):

- Below the instability threshold, the crossing frequency depends on the tube-to-support plate clearance.
- After the initial transient, the crossing frequency is fairly time-independent.
- At and beyond the instability threshold, the crossing frequency is independent of the tube-support plate clearance.

Based on the second conclusion, in a time domain numerical analysis of tube dynamics with tube-to-support clearance, one does not have to update the crossing frequency at every time step except for the short initial transient. This is what makes the crossing frequency concept particularly useful.

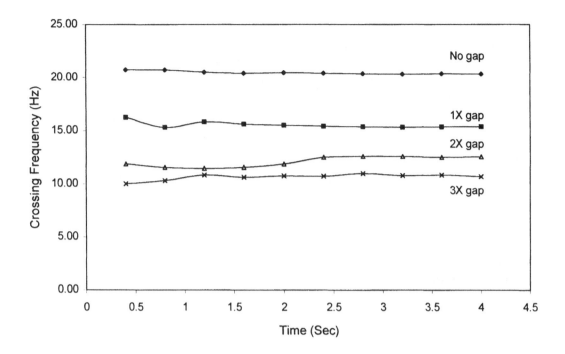

Figure 7.17 Crossing frequencies as a function of time for different tube-support
plate clearances (below critical velocity, Au-Yang, 2000)

Example 7.4

In this example, a finite-element model of an inner-row tube in a re-circulating nuclear
steam generator (see Figure 7.4) is used. Since the purpose of this example is to show the
steps to estimate the fluid-elastic stability margin (FSM) of a multi-span tube with tube-
support clearances, the units are not explicitly mentioned except that they form a
consistent unit set (see Chapter 1). The readers who are familiar with the dimensions of
nuclear steam generators will probably guess correctly that the US unit set is used.

Figure 7.18 shows the finite-element model of a tube in an inner row of a nuclear
steam generator. This tube is clamped at the lower ends marked HL (hot leg) and CL
(cold leg), and is supported at the intermediate points with oversized holes. The main
excitation source is the cross-flow in the U-bend. It is assumed that the velocity vector is
radial in the U-bend, so that it is always perpendicular to the tube axis. However, neither
the velocity nor the density is constant along the U-bend. The following values can be
assumed:

Connors' constant, $\beta=5.2$

$$\sum q_i \Delta x_i = 7.77$$

In the model, the element lengths associated with each node in the U-bend is Δx_i=1.3086 units. The radius of the semi-circle is 15.0 units. Two (negative) damping elements are put into nodes 80 and 100, each in the out-of-plane (z) direction. The values of these damping elements are gradually changed until the response at the apex of the U-bend node 90) starts to diverge. When the gap clearances between the tube and the support plate are equal to the normal, as-built, tube-to-support plate clearance, the time history responses of node 90 in the out-of-plane (z) at C_{fsi} =-0.0592   and C_{fsi} =-0.0595 are shown in Figures 7.19 and 7.20, while Figures 7.21 shows the crossing frequencies as a function of time at C_{fsi} =-0.0595. From Figures 7.19 and 7.20, it is obvious that the critical value of C_{fsi} , that is, the value at which the tube just becomes unstable, is between C_{fsi} =-0.0592 and C_{fsi} =-0.0595. From Figure 7.21, we deduce that the critical crossing frequency, i.e., the crossing frequency after the tube becomes unstable, is f_{cc} =20.1 Hz. To find the fluid-elastic stability margin, we have from Equation (7.19), since the damping element is applied at *two* nodes 80 and 100,

$$2*(C_{fsi})_c = \frac{4*q_c*\pi*15}{20.1*5.2^2} = 2*0.0595$$

$$q_c = \frac{2*0.0595*20.1*5.2^2}{4*\pi*15} = 0.343 = \frac{\rho V_c^2}{2}$$

Thus from Equation (7.23),

$$FSM = \sqrt{\frac{0.343*15\pi}{7.77}} = 1.44$$

The fluid-elastic stability margins at zero, two times (2X) and three times (3X) the normal tube-support clearances can be obtained similarly. Figure 7.21 shows the crossing frequencies as a function of time in these four cases. Note that as the tube becomes unstable, the crossing frequencies all approach 20.1 Hz, which is the constant crossing frequency at zero tube-support clearance. Table 7.5 shows the fluid-elastic stability margins at 0, 1X, 2X and 3X tube-support clearances, together with the values of C_{fsi} at the threshold of instability. Note that in each case, this critical damping value is applied at two nodes so that the total critical fluid-structure interaction damping coefficient is $2C_{fsi}$. As a comparison, the fluid-elastic stability margin for the zero-gap case was also computed with the frequency-domain Connors' Equation (7.6), using the same input parameters and the same finite-element model. It was found that FSM=0.89 for the U-bend out-of-plane mode, compared with FSM=1.32 based on time domain analysis. The reason is that in the frequency domain analysis, we assume each mode is isolated from the others. The computed critical velocity is that for the least stable mode. In the time domain analysis, more than one mode participates in the response at any time. The critical velocity computed is a system, rather than a modal, parameter.

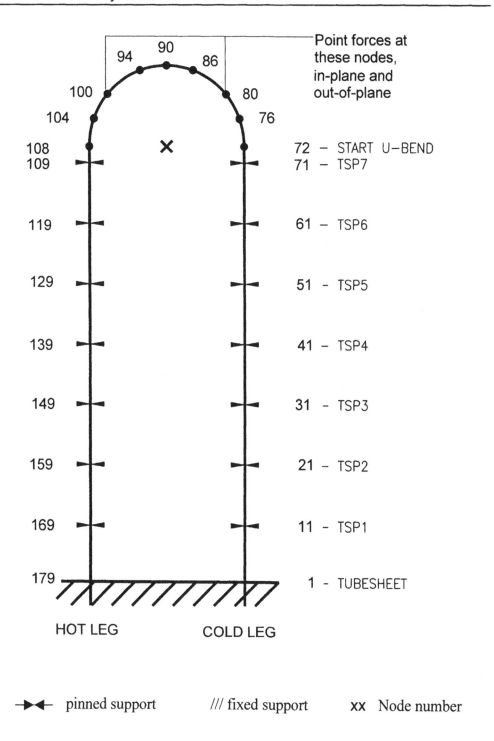

Figure 7.18 Finite-element model of a tube in a re-circulating steam generator

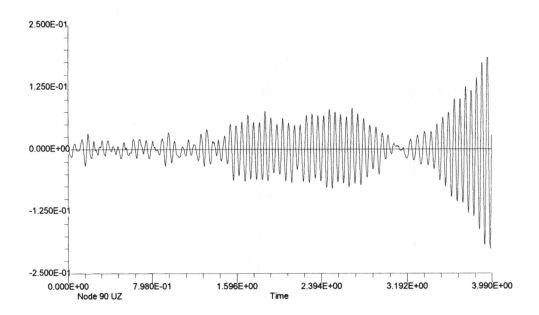

Figure 7.19 Time history response of a point at the apex of the
U-bend in the out-of-plane (z) direction when C_{fsi} =-0.0592

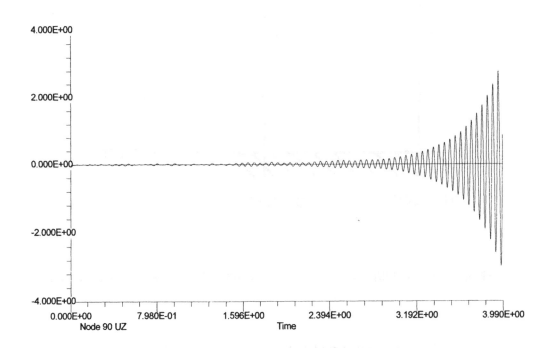

Figure 7.20 Time history response of the same point the
out-of-plane (z) direction when C_{fsi} =-0.0595

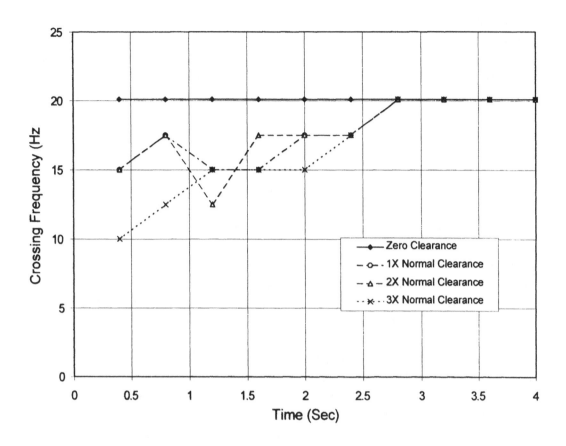

Figure 7.21 Crossing frequency as a function of time
with different tube-support clearances

Table 7.5: Fluid-elastic stability margins (FSM) as a function of tube-support clearance

	0 Clearance	1X As-Built Clearance	2X As-Built Clearance	3X As-Built Clearance
C_{fsi} (each node)	-0.05	-0.0595	-0.0775	-0.090
q_c	0.2883	0.3431	0.4469	0.5190
FSM	1.32[1]	1.44	1.65	1.77

[1] *FSM* computed from Equations (7.6) and (7.7) is 0.89.

Case Study 7.1: Sudden Tube Crack in Nuclear Steam Generators

In 1988, a tube in the inner row of a recirculating nuclear steam generator in the United States cracked, leaking radioactive coolant into the environment. This steam generator had been in service for a number of years and the tube in question was inspected just prior to the incident. No indication of degradation was found during this inspection. Post-incident inspection of the steam generator using video equipment found that the anti-vibration bars (AVB; see Figure 7.4), which restraint the out-of-plane motion of the U-bend, were not inserted far enough into the tube bundle. As a result, this particular tube was not supported by any AVB, contrary to the design specifications. Metallurgical examination showed that the tube failed because of high cycle fatigue. Failure occurred at the top tube support plate. The cause was determined to be fluid-elastic instability, resulting in excessive vibration amplitudes in the out-of-plane direction, with resulting high bending stress at the top support plate (US Nuclear Regulatory Commission, Docket 1988).

In 1991, a tube in another, almost identical steam generator in Japan experienced a similar failure, after 19 years of operation. Post-incident examination again showed that the AVBs did not reach the inner row, as they should. Metallurgical examination again showed that the tube failed because of high cycle fatigue and that failure occurred at the top tube-support plate, similar to the one that failed in the U.S. (Ministry of International Trade and Industry of Japan, 1991).

Under normal conditions, a fluid-elastically-unstable tube will vibrate with such large amplitudes that either tube-tube impacting will occur or large bending stress will be induced at the support locations. In either case, failure usually occurs within months, or even days, of operation. Yet both of these two steam generators had been operating for many years before these tube failures occurred.

Post-incident inspection also found out that in both steam generators, the carbon steel tube-support plates had chemically reacted with the shell-side water, forming oxides call magnetite. Since the volume of the magnetite is larger than the volume of the base metal, this magnetite started to fill up the crevice between the tube and the support plates. When the steam generators were new, because of the design margins of safety, these tubes were fluid-elastically stable even without the support of the AVBs. However, as the tube-support plate clearances decreased, the margins against fluid-elastic instability decreased, as is shown in Example 7.4, until at some point the tubes suddenly became unstable. The resulting large induced bending stresses at the top support plate locations rapidly caused fatigue failure of the tubes.

References

ASME, 1998, Boiler Code Sec III Appendix N-1300 Series.

Au-Yang, M. K., 1987, "Development of Stabilizers for Steam Generator Tube Repair," Journal of Nuclear Engineering and Design, Vol. 107, pp. 189-197.

Au-Yang, M. K., 2000, "The Crossing Frequency as a Measure of Heat Exchanger Tube Support Plate Effectiveness," in Flow-Induced Vibration, edited by S. Ziada and T. Staubli, Balema, Rotterdam, pp. 497-504.

Axisa, F., Antune, J. and Villard, B., 1988, "Overview of Numerical Methods for Predicting Flow-Induced Vibration," ASME Transaction, Journal of Pressure Vessel Technology, Vol. 110, pp. 6-14.

Blevins, R. D., 1974, "Fluid-Elastic Whirling of a Tube Row," ASME Transaction, Journal of Pressure Vessel Technology, Vol. 96, pp. 263-267.

Blevins, R. D., 1990, Flow-Induced Vibration, Second Edition, Van Nostrand Reinhold, New York.

Chen, S. S., Jendrzejczyk, J. A. and Lin, W.H., 1978, Experiments on Fluid-Elastic Instability in Tube Banks Subject to Liquid Cross-Flow, Argonne National Laboratory Report ANL-CT-78-44.

Chen, S. S., 1983, "Instability Mechanisms and Stability Criteria for a Group of Circular Cylinders Subject to Cross-Flow, Part I and Part II," ASME Transaction, Journal of Vibration, Acoustics and Reliability in Design, Vol. 105, pp. 51-58 and pp. 243-260.

Chen, S. S., 1985, Flow-Induced Vibration of Circular Cylindrical Structures, Argonne National Laboratory Report No. ANL-85-51.

Chen, S. S., 1987, "A General Theory for the Dynamic Instability of Tube Arrays in Cross-Flow," Journal of Fluids and Structures, Vol. 1, pp. 35-53.

Connors, H. J., 1970, "Fluid-Elastic Vibration of Tube Arrays Excited by Cross-Flow," in Flow-Induced Vibration of Heat Exchangers, ASME Special Publication, edited by D. D. Reiff, pp. 42-56.

Connors, H. J., 1980, "Fluid-Elastic Vibration of Tube Arrays Excited by Non-uniform Cross-Flow," in Flow-Induced Vibration of Power Plant Components, ASME Century 2 Special Publication PVP-41, edited by M. K. Au-Yang, pp. 93-107.

Crandall, S. H. and Mark, W. D., 1973, Random Vibrations in Mechanical Systems, Academic Press, New York.

Hara, F., 1987, "Unsteady Fluid Dynamic Forces Acting on a Single Row of Cylinders," in Flow-Induced Vibration, ASME Special Publication PVP Vol. 122, edited by M. K. Au-Yang and S. S. Chen, pp. 51-58.

Lever, J. H. and Weaver, D. S., 1982, "A Theoretical Model for Fluid-Elastic Instability in Heat Exchanger Tube Bundles," ASME Transaction, Journal of Pressure Vessel Technology, Vol. 104, pp. 147-158.

Ministry of International Trade and Industry of Japan, 1991, Final Report on Steam Generator Tube Break at Mihama Unit 2 on February 9, 1991.

Nelms, H. A., 1970, "Flow-Induced Vibrations: A Problem in the Design of Heat Exchangers for Nuclear Services," Paper presented at the ASME Winter Annual Meeting, Dec. 1970.

Price, S. J., 1993, "Theoretical Model of Fluid-Elastic Instability for Cylinder Arrays Subject to Cross-Flow," in Technology for the 90s, ASME Special Publication, edited by M. K. Au-Yang, pp. 711-774.

Paidoussis, M. P., Mavriplis, D. and Price, S. J., 1984, "A Potential Theory for the Dynamics of Cylinder Arrays in Cross-Flow," Journal of Fluid Mechanics, Vol. 146, pp. 227-252.

Paidoussis, M. P., 1987, "Flow Induced Instabilities of Cylindrical Structures," Applied Mechanics Reviews, Vol. 40, pp. 163-175.

Paidoussis, M. P., 1998, Fluid-Structure Interaction: Vol. 1, Academic Press, New York.

Rice, S. O., 1954, "Mathematical Analysis of Random Noise," Bell System Technical Journal Vol. 23, pp. 282-332, and Vol. 24, pp. 46-156.

Sauve', R. G., 1996, "A Computational Time Domain Approach to Solution of Fluid-Elastic Instability for Non-linear Tube Dynamics" in Flow-Induced Vibration-1966, ASME Special Publication, PVP Vol. 328, edited by M. J. Pettigrew, pp. 327-336.

Southworth, D. J. and Zdravkovich, M. M., 1975, "Cross Flow Induced Vibration of Finite Tube Banks with In-Line Arrangements, Journal of Mechanical Engineering Sciences, Vol. 17, pp. 190-198.

THEMA Standard, 1988, Sec. V, "Flow-Induced Vibration," Tubular Heat Exchanger Manufacturer Association.

Tanaka, H. and Takahara, S., 1980, "Unsteady Fluid Dynamic Force on Tube Bundle and Its Dynamics Effect on Vibration," in Flow-Induced Vibration of Power Plant Components, ASME Century 2 Special Publication PVP-41, edited by M. K. Au-Yang, pp. 93-107.

Taylor, C. E., Bourcher, K. M. and Yetisir, M., 1995, "Vibration and Impact Forces Due to Two-Phase Cross-Flow in U-Bend Region of Heat Exchangers," Proceeding of Sixth International Conference on Flow-Induced Vibration, London, UK, pp. 404-411.

US Nuclear Regulatory Commission Bulletin No. 88-02, 1988, "Rapidly Propagating Fatigue Cracks in Steam Generator Tube."

CHAPTER 8

TURBULENCE-INDUCED VIBRATION IN PARALLEL FLOW

Summary

In spite of recent advances in computational fluid dynamics, today the most practical method of turbulence-induced vibration analysis follows a hybrid experimental/analytical approach. The forcing function is determined by model testing, dimensional analysis and scaling, while the response is computed by finite-element probabilistic structural dynamic analysis using the acceptance integral approach formulated by Powell in the 1950s. The following equation is often used to estimate the root-mean-square (rms) response, or to back out the forcing function from the rms response, of structures excited by flow turbulence:

$$< y^2(\vec{x}) >= \sum_\alpha \frac{A G_p(f_\alpha) \psi_\alpha^2(\vec{x}) \vec{J}_{\alpha\alpha}(f_\alpha)}{64\pi^3 m_\alpha^2 f_\alpha^3 \zeta_\alpha} \tag{8.50}$$

where $\vec{J}_{\alpha\alpha}$ is the familiar joint acceptance. As it stands Equation (8.50) is general and applicable to one-dimensional as well as two-dimensional structures in either parallel flow or cross-flow; the latter will be covered in the following chapter. It is also independent of mode shape normalization as long as the same normalization convention is used throughout the equation. However, Equation (8.50) is derived under many simplifying assumptions, of which the most important ones are that cross-modal contribution to the response is negligible, and the turbulence is homogeneous, isotropic and stationary. In addition, if one assumes that the coherence function can be factorized into a streamwise component, assumed to be in the x_1, or longitudinal direction, and a cross-stream x_2, or lateral component, *and* that each factor can be completely represented by two parameters—the convective velocity and the correlation length—then the acceptance integral in the two directions can be expressed in the form:

$$\mathrm{Re}\, J_{mr} = \frac{1}{L_1} \int_0^{L_1} dx'' \psi_m(x'') \int_0^{x''} \psi_r(x') e^{-(x'-x'')/\lambda} \cos\frac{2\pi f(x'-x'')}{U_c} dx' +$$

$$\frac{1}{L_1} \int_0^{L_1} dx'' \psi_m(x'') \int_{x''}^{L_1} \psi_r(x') e^{-(x''-x')/\lambda} \cos\frac{2\pi f(x''-x')}{U_c} dx' \tag{8.42}$$

$$\mathrm{Im}\, J_{mr} = \frac{1}{L_1} \int_0^{L_1} dx'' \psi_m(x'') \int_0^{x''} \psi_r(x') e^{-(x'-x'')/\lambda} \sin\frac{2\pi f(x'-x'')}{U_c} dx' +$$

$$\frac{1}{L_1} \int_0^{L_1} dx'' \psi_m(x'') \int_{x''}^{L_1} \psi_r(x') e^{-(x''-x')/\lambda} \sin\frac{2\pi f(x''-x')}{U_c} dx'$$
(8.43)

Likewise from Equations (8.34) and (8.40), the lateral acceptance is always real and is given by:

$$J'_{ns} = \frac{1}{L_2} \iint_{L2} \phi_n(x_2') e^{-|x_2'-x_2''|/\lambda} \phi_s(x_2'') dx_2' dx_2''$$
(8.44)

where ψ, ϕ are the mode shape functions in the streamwise and cross-stream directions, and $\alpha=(m, n)$ and $\beta=(r, s)$ are the mode numbers, with m, r in streamwise and n, s in the cross-stream directions. Equations (8.42) to (8.44) are the simplest forms of the acceptance integrals in the general case of parallel flow-induced vibration. The more complicated Bull's (1967) representation is used to calculate the acceptance integrals given in the charts contained in Appendix 8A.

The fluctuating pressure power spectral density (PSD), G_p due to boundary layer type of turbulence is given in Figures 8.15 to 8.17. For low frequencies, to which Figure 8.16 applies, it is also given by the empirical equation:

$$\frac{G_p(f)}{\rho^2 V^3 D_H} = 0.272E - 5/S^{0.25}, \quad S < 5;$$
$$= 22.75E - 5/S^3, \quad S > 5$$
(8.68)

where

$$S = 2\pi f D_H / V$$
(8.69)

The pressure PSD in extremely turbulent flows in confined channels, such as those encountered in industrial piping systems with elbows, valves and changes in cross-section in which there are no well-defined boundary layers, are given in Figures 8.17 and 8.18, or by the following empirical equations.

For turbulent flow without cavitation,

$$\frac{G_p(f)}{\rho^2 V^3 R_H} = 0.155e^{-3.0F}, \quad 0 < F < 1.0$$
$$= 0.027e^{-1.26F} \quad 1.0 \le F \le 5.0$$
(8.70)

where

$$F = f R_H / V$$
(8.71)

is the dimensionless frequency and R_H is the hydraulic radius.

For turbulent flow with light cavitation,

$$\frac{G_p(f)}{\rho^2 V^3 R_H} = \min\left\{20F^{-2}(-|x|/R_H)^{-4}, 1.0\right\}$$
(8.72)

or values from Equation (8.70), whichever is larger, where $|x|$ is the absolute value of the distance from the cavitating source, such as an elbow or a valve.

For boundary layer type of turbulence, the convective velocity is given by the empirical equation:

$$U_c/V = 0.6 + 0.4e^{-2.2(\omega\delta^*/V)}$$ (Chen and Wambsganss, 1970)   (8.56)

or

$$U_c/V = 0.59 + 0.30e^{-0.89(\omega\delta^*/V)}$$ (Bull, 1967)   (8.57)

For extremely turbulent flow in confined channels without a well-defined boundary layer,

$$U_c \approx \overline{V}$$ (Au-Yang, 1980, 1995)   (8.58)

For boundary layer type of turbulence, the correlation length is approximately equal to the displacement boundary layer thickness,

$$\lambda \approx \delta *$$

For extremely turbulent flow in confined channels, the correlation length is approximately equal to 0.4 times the hydraulic radius of the channel,

$$\lambda \approx 0.4R_H$$
(8.65)

Because of the presence of the convective factor, which alters the phase of the forcing function as the wave front propagates along the surface of the structure, the acceptance integral in parallel flow usually has to be evaluated by numerical integration. Charts of the acceptance integrals for boundary layer type turbulent flow over rectangular panels are given in Appendix 8A. These acceptance integrals were obtained by numerical integration based on more complicated Bull's representation (1967) of the coherence function. Since in the cross-stream direction, the forcing function is always in phase over the surface of the structure, the cross-stream (or lateral) acceptances in parallel flow are the same as the longitudinal acceptances in cross-flow covered in the following chapter.

The value of the acceptance integral depends on mode-shape normalization. In the case of structures with uniform surface mass density, as is assumed in this chapter, the mode-shape function can be normalized to unity,

$$\int_A \phi_\alpha(\vec{x})\phi_\beta(\vec{x})d\vec{x} = \delta_{\alpha\beta}$$

In that case it can be shown that

$$\sum_\alpha \vec{J}_{\alpha\alpha} = 1.0$$

and the acceptance integral has the physical meaning of the transition probability amplitude between two modes. This is a measure of the compatibility of the coherence of the forcing function and the two structural mode shapes involved. For a point force acting on a point mass-spring system, the acceptance integral is always unity.

In the general case when the modal frequencies are closely-spaced, cross-modal contribution to the response will no longer be negligible. Under this condition the mean square response can be obtained by numerical integration under the response PSD curve:

$$< y^2(\vec{x}) > = \int_0^{f_{max}} G_y(\vec{x}, f)df \tag{8.45}$$

$$G_y(\vec{x}, f) = AG_p(f)\sum_\alpha \psi_\alpha(\vec{x})H_\alpha(f)H_\alpha{}^*(f)\psi_\alpha(\vec{x})\vec{J}_{\alpha\alpha}(f)$$
$$+ 2AG_p(f)\sum_{\alpha \neq \beta} \psi_\alpha(\vec{x})H_\alpha(f)H_\beta{}^*(f)\psi_\beta(\vec{x})\vec{J}_{\alpha\beta}(f) \tag{8.46}$$

$$H_\alpha(f) = \frac{1}{(2\pi)^2 m_\alpha[(f_\alpha{}^2 - f^2) + i2\zeta_\alpha f_\alpha f]} \tag{8.47}$$

with $\vec{J}_{\alpha\beta} = J_{mr}J'_{ns}$, where the two factors are the longitudinal and lateral acceptances given in Equations (8.45) and (8.46) above. This direct approach is well within the capability of today's personal computers.

Acronym

PSD   Power Spectral Density

Nomenclature

A Total surface area of structure
D_H Hydraulic diameter=4*cross-section area/wetted perimeter
f Frequency, in Hz

F Dimensionless frequency

g Gap width in annular flow channel, $=R_H$

$G_p(f)$ Single-sided fluctuating pressure power spectral density (PSD), in (force/area)^2/Hz, due to turbulence

$G_y(f)$ Single-sided response power spectral density

H Transfer function

j $=\sqrt{-1}$

$\vec{J}_{\alpha\beta}$ $=J_{mr}J'_{ns}$, acceptance integral (note: some authors write it as J^2)

J_{mr} Longitudinal acceptance (note: some authors write it as J^2)

J'_{ns} Lateral acceptance (note: some authors write it as J'^2)

L Length of structure

ℓ_c $=2\lambda$, correlation length as defined by some authors

m Surface mass density (mass per unit area)

m_α Generalized mass

$p(t)$ Fluctuating pressure due to turbulence

$P_\alpha(t)$ Generalized pressure

$P_\alpha(\omega)$ Fourier transform of generalized pressure

R Correlation function

R_H Hydraulic radius, $=D_H/2$

$S_p(\vec{x}',\vec{x}'',\omega)$ Cross-spectrum of fluctuating pressure

$S_y(\omega)$ Power spectral density of y, as a function of ω and defined from $-\infty$ to $+\infty$

T Half of the total time interval of integration

U_c Convective velocity

U_{phase} Phase velocity

V Velocity

$\bar{V}$ Mean velocity

x_1 Position in the "1" direction (normally the longitudinal or the streamwise direction)

x_2 Position in the "2" direction (normally the lateral or the cross-stream direction)

$\vec{x}$ Position vector on surface of structure

$d\vec{x}$ Integration area, same as dA or dS

$y(t)$ Response

$Y(\omega)$ Fourier transform of $y(t)$

Γ Coherence function

δ Characteristic length of structure or flow channel

Δ_1, Δ_2 $=\delta^*/L_1, \delta^*/L_2$

δ^* Displacement boundary layer thickness

ζ Damping ratio

λ Correlation length $= \ell_c / 2$

$\psi_\alpha(\vec{x}) = \psi_m(x_1)\phi_n(x_2)$, mode shape function

σ Poison ratio

σ_y^2 Variance of the collection of records y_i

Φ Normalized pressure PSD

ω Frequency in rad/s

Indices

1,2   Longitudinal (normally also the stream), and lateral (normally cross-stream) directions

α $=(m,n)$, modal index, two directions

β $=(r,s)$, modal index, two directions

< >   Mean value

8.1   Introduction

Turbulence-induced vibration distinguishes itself from all other flow-induced vibration phenomena in that, like dust or noise or weed is to our everyday lives, it is a necessary "evil" in the power and process industries. Unlike vortex-induced vibration, fluid-elastic instability or acoustic resonance, which can be eliminated or their effects minimized by design, turbulence-induced vibration cannot be avoided in virtually all industrial applications. When the fluid velocity exceeds any but the smallest values characteristic of "seepage" flows, eddies will form even if the surface of the flow channel is perfectly smooth. The flow is said to be turbulent (versus "laminar" when there are no eddies in the flow). The Reynolds number (see Chapter 6) at which the flow regime changes from laminar to turbulent is treated in any textbook on fluid dynamics. In industrial applications, very often the flow channels are designed to generate turbulence in order to improve the heat or mass transfer efficiencies or to suppress vortex shedding or other instability phenomena. As a result, turbulence-induced vibration cannot be avoided in industrial applications, and in most cases it cannot even be reduced.

Figure 8.1 is a time history of the fluctuating pressure in a turbulent flow measured with a dynamic pressure transducer. It is obviously a random force. Like the force that induces the responses, turbulence-induced vibration is also a random process that can be dealt with only by probabilistic methods. In using this approach, we sacrifice any attempt to calculate the detailed time histories of the responses. Instead, we are satisfied with calculating only the *root mean square* (rms) values of the responses. Based on this rms response (displacement amplitudes and stresses), one then proceeds to assess the potential of damage (impact wear, fatigue and fretting wear) on the structure. This damage assessment based on the rms responses involves yet other major steps of analysis, which will be treated in a Chapter 11. This chapter outlines the steps leading to an estimate

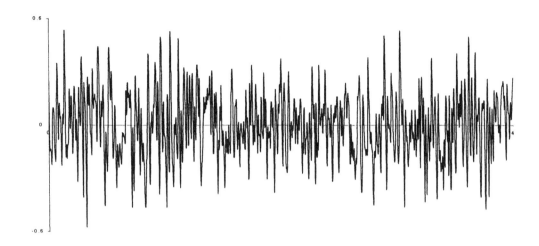

Figure 8.1 Time history of fluctuating pressure measured in a turbulent flow

of the rms responses due to turbulent parallel flow over the surfaces of structures. Elements of probabilistic methods will be stated without proof in this and the following Chapters. Readers should refer to Chapter 13 on digital signal processing for more information on the power spectral density function and the Fourier transform; and to books on statistics for probability density distribution functions. Condensed statistical theories that are absolutely necessary for more in-depth understandings of the subject can also be found in books on structural dynamics, such as Hurty and Rubinstein (1964) or books on signal analysis (Bendat and Piersol, 1971).

In spite of the latest development in computational fluid mechanics, it is still not feasible to determine the turbulent forcing function by numerical techniques. Today most of the turbulence-induced vibration analysis is based on a combination of experimental and analytical techniques. The random pressure as well as other parameters that characterizes the turbulent flow field are measured either in field tests, scale model tests or in the case of heat exchange tube bundles, by measurement based on models with limited numbers of tubes and spans. Data from model tests are then scaled up to prototype dimensions by dimensional analysis. Using this forcing function as input, the responses are then estimated by probabilistic structural dynamics, usually with the help of specialized finite-element computer programs. The determination of the forcing function involves only fluid mechanical parameters and the technique of model testing, scaling and dimensional analysis are well established in fluid mechanics. On the other hand, the response analysis involves only structural parameters with the forcing function as input. The finite-element method, which is well established in structural dynamic analysis, is frequently used. Thus this approach combines two well-established techniques in two difference engineering disciplines. It is hoped that one day, computational fluid dynamics will be developed to the point that the forcing function can also be determined analytically. Until then, we have to follow this hybrid approach.

8.2   Random[1] Processes and Probability Density Functions

A collection of records, such as the time histories of the fluctuating pressure $p(t)$ in a turbulent flow, or the structural response $y(t)$ to it,

$$y_i = y(t_i), \quad i=1,2,3,4, \ldots \ldots n \tag{8.1}$$

where t_i denotes a point in time between $-T$ and $+T$ as T becomes very large, is called a random process if the records can be characterized by statistical properties. That is, if it's mean value can be estimated by:

$$<y> = \lim(n \to \infty) \frac{1}{n} \sum_{i=1}^{n} y_i \tag{8.2}$$

and its variance σ_y^2 is given by

$$\sigma_y^2 \equiv \lim(n \to \infty) \sum_{i=1}^{n} (y_i - <y>)^2 = \lim(n \to \infty) \sum_{i=1}^{n} (y_i^2 - 2y_i <y> + <y>^2)$$
$$= <y^2> - <y>^2 \tag{8.3}$$

where $<y^2>$ is the mean square, and $<y>^2$ is the square of the mean value. The variance is important in random vibration analysis for several reasons, one of which is that in most random vibrations, including turbulence-induced vibration, only the variation about the mean value is of interest. If the mean value is not zero, it can be estimated separately. Assuming from now on that the mean is zero, the variance is, from Equation (8.3), equal to the mean square value:

$$<y^2> = \sigma_y^2 = \lim(n \to \infty) \sum_{i=1}^{n} (y_i - <y>)^2 \tag{8.4}$$

Equation (8.4) shows that to find the mean value, one must average over a large number of data points. Likewise, to find the mean square value, one must subtract the mean from each data point, square the resultant and then sum and average the resulting values. In general, the mean and mean square values may be different dependent on the time at which we select these "large" numbers of data points to average. In this case, the random

---

[1] The term stochastic is also commonly used in the literature. Without getting into the vigorous statistical definitions, in this book these two terms will be used interchangeably based on whichever term is more commonly used for a specific process. When both terms are commonly used, "random" will be used.

process is called non-stationary random. However, if the resulting mean and mean square values are independent of the time we select the data points, then the random process is *stationary*. If, in addition, each record is statistically equivalent to any other record—that is, independent of where, and how many time points we use to calculate the sum, as long as the number of time points are large enough—the process is *ergodic*. If the mean and the variance vary slowly with time, we call the process quasi-stationary or quasi-ergodic. In all of our calculations, we always assume the process is at least quasi-stationary, *even if it is not*. This is the reason why in many power spectral density plots of structural responses, we occasionally see anomalies such as double resonance peaks.

From Figure 8.1, it is apparent that the time histories of random processes, either the forcing function or the responses, do not convey a lot of meaning to the design engineers whose ultimate goal is to assess the vibration amplitudes and stresses of the structural components. In addition, measuring and recording the time history of the forcing function such as that given in Figure 8.1, and using it in a time history response analysis, is not only prohibitively time-consuming, but the resulting time history of the responses, which also resemble what is shown in Figure 8.1, cannot be easily related to failure mechanisms like fatigue. This is the reason why until very recently when both the memory and speed in electronic computers became less expensive, responses to turbulence-induced vibration were almost exclusively carried out in the frequency domain using the method of spectral analysis. By following this approach, we sacrifice the details of the time history response, and settle for an overall picture by computing the response power spectrum and the rms response. Time-domain analysis of turbulence-induced vibration will be briefly covered in the chapter on fatigue and wear analysis.

Returning to Equation (8.1), in the limit when the time steps are infinitesimal, y becomes a continuous function of the time variable t. That is, $y=y(t)$. The mean and variance of y are given by:

$$< y >= \lim(T \to \infty)\frac{1}{2T} \int_{-T}^{T} y(t)dt = 0$$

$$\sigma_y^2 = \lim(T \to \infty)\frac{1}{2T} \int_{-T}^{T} (y- < y >)^2 dt = < y^2 >$$

assuming again that we are concerned only with the variation about the mean, so that $y(t)$ has zero mean.

Over a long time record, the probability of finding $y(t)$ in a certain bandwidth of values follows a certain function $f(y)$, so that the probability of finding y between y_0 and $y_0+\delta y$ is given by

$$\text{Prob } (y_0 < y < y_0 + \delta y) = \int_{y0}^{y0+\delta y} f(y)dy \qquad (8.5)$$

The function $f(y)$ is called the probability density function of the random variable y. In terms of the probability density function, the mean and the variance of the random variable y is given by

$$< y >= \int_{-\infty}^{\infty} yf(y)dy \tag{8.6}$$

$$\sigma_y^2 = \int_{-\infty}^{\infty} (y- < y >)^2 f(y)dy \tag{8.7}$$

If, as in an ensemble of random variables, the mean is zero or is estimated separately, the variance is the mean square value of the random process. The probability density function has to satisfy two more properties: First, since probabilities cannot be less than zero, $f(y)$ must be a positive function of y:

$$f(y) \geq 0 \tag{8.8}$$

for all y. Second, since the total probability of finding y to have a value between $-\infty$ *and* $+\infty$ must be unity, we have,

$$\int_{-\infty}^{\infty} f(y)dy = 1 \tag{8.9}$$

The following function:

$$f(y) = \frac{1}{\sqrt{2\pi}\sigma_y} e^{-(y-<y>)^2 / 2\sigma_y^2} \tag{8.10}$$

satisfies both Equations (8.6) and (8.7) and is a well-behaved function of y. Using Equations (8.6) and (8.7), one can readily show that in a random process with probability distribution function, Equation (8.10), the mean and variance are given by $< y >$ and σ_y^2. This function, called the Gaussian distribution function, is therefore a potential candidate for a probability density function. What makes the Gaussian distribution function particularly important is the central limit theorem, which states that if the random processes y_i are statistically independent of each other, then the random variable

$$y = \sum_i y_i$$

follows the Gaussian distribution function. If we consider only the variation about the mean value by subtracting the mean values first, the mean square values would be equal to the variance. If, in addition, we choose a scale such that the mean square value of the random process is equal to unity, then the Gaussian distribution takes the simple form

$$f(y) = \frac{1}{\sqrt{2\pi}} e^{-y^2/2} \tag{8.11}$$

Equation (8.11) is called the normalized Gaussian, or just the normal distribution. Next to the sine and cosine functions, it is probably the most important function in mathematical physics. When applied to turbulence-induced vibration, all the above discussions reduce to the following simple statement: Both the random pressure distribution and the distribution of the random response of the structure follow the normal distribution with the mean square values given by Equation (8.7). That is, the mean square pressure and vibration amplitudes are given by:

$$<p^2> = \int_{-\infty}^{\infty} p^2 f(p) dy$$

$$<y^2> = \int_{-\infty}^{\infty} y^2 f(y) dy$$

with $f(y)$ given by Equation (8.10).

8.3   Fourier Transform, Power Spectral Density and the Parseval Theorem

It was stated earlier that a random process such as turbulence or turbulence-induced structural response may not be easily interpreted in the time domain. For this reason, and for reasons of computational economy, one goes to the frequency domain to carry out the analysis. The "shuttle" that carries us from the time domain to the frequency domain is a mathematical process called Fourier transform (see also Chapter 13). If $y(t)$ is a function of the time variable t, the function $Y(\omega)$, defined as:

$$Y(\omega) = \int_{-\infty}^{\infty} y(t) e^{-j\omega t} dt , \quad j = \sqrt{-1} \tag{8.12}$$

is called the Fourier transform of $y(t)$ and is a function of ω. Conversely, $y(t)$ is the inverse transform of $Y(\omega)$ and is defined as

$$y(t) = \frac{1}{2\pi} \int_{-\infty}^{\infty} Y(\omega) e^{+j\omega t} d\omega$$

The following equation, the proof of which can be found in any book on statistics, is what makes the Fourier transform so useful[1]:

$$< y^2(t) >= \lim(T \to \infty) \int_{-T}^{T} y^2(t)dt = \lim(T \to \infty)\frac{1}{2\pi} \int_{-\infty}^{\infty} \frac{Y(\omega)Y*(\omega)}{2T}d\omega$$

This is known as the Parseval theorem. It enables us to find both the mean and the mean square values of a random variable in the frequency domain. If we define

$$S_y(\omega) = \frac{Y(\omega)Y*(\omega)}{4\pi T}$$ (8.13)

then the mean square of y is given by:

$$< y^2 >= \int_{-\infty}^{\infty} S_y(\omega)d\omega$$ (8.14)

$S_y(\omega)$ is the two-sided power spectral density (PSD) of y expressed as a function of radians/s. As defined, $S_y(\omega)$ is convenient for the mathematical development of the equations in turbulence-induced vibration analysis. In engineering applications, however, it is more convenient to express the power spectral density in terms of the frequency f in Hz, and define it only for positive values of f. From its definition, one can easily show that the PSD is an even function of the frequency. Therefore, we can define,

$$G_y(\omega)=2 S_y(\omega) \quad \text{for } \omega \geq 0$$ (8.15)
$$=0 \qquad \text{for } \omega < 0$$

But we want to express G as a function of the frequency f, in Hz, instead of ω. This can be achieved by noting that,

$$< y^2 >= \int_{0}^{\infty} G_y(\omega)d\omega = \int_{0}^{\infty} G_y(\omega)d(2\pi f) = \int_{0}^{\infty} G_y(f)df$$ (8.16)

It follows that:

---

[1] The placement of the factor $1/2\pi$ is arbitrary. Some authors put it in the forward transform while some put it in the inverse transform. Still others make it symmetrical by including $1/\sqrt{2\pi}$ in both the forward and the inverse transforms. As long as one is consistent, all these definitions lead to the same end results for any physically observable variable.

$$G_y(f) = 2\pi G_y(\omega) \quad \text{and,} \tag{8.17}$$

$$\begin{aligned} G_y(f) &= 4\pi S_y(\omega) \quad &for \quad f \geq 0 \\ &= 0 \quad &for \quad f < 0 \end{aligned} \tag{8.18}$$

Physically, the power spectral density is the energy distribution as a function of the frequency of the random variable y. If y is a voltage passing through a band of band-pass filters and collected in different bins, then G_y is the square of the voltage in each bin.

8.4 The Acceptance Integral in Parallel Flow over the Surface of Structures

In this chapter, we are concerned with the response of structures subject to turbulent flow over their surfaces (parallel flow). This is in contrast to what is covered in the following chapter, which deals with the responses of tubes and beams subject to turbulent cross-flow. The most widely used method to solve this type of problem is the acceptance integral method first formulated by Powell (1958). Chen and Wambsganss (1970) followed this method to estimate the parallel flow-induced vibration of nuclear fuel rods and Chyu and Au-Yang (1972) applied this method to estimate the response of panels excited by boundary layer turbulence. Au-Yang (1975) applied this method to estimate the response of reactor internal components excited by the coolant flow and again to cross-flow-induced vibration of a multi-span tube (Au-Yang, 1986; 2000). The following summarizes the essential steps leading to acceptance integral formulation of the response equation. The readers should refer to the cited references for more details.

To simplify the development, in this chapter we assume that the surface density of the structure is uniform, so that the mode-shape functions can be normalized to unity in accordance with Equation (3.30) in Chapter 3:

$$\int_A \psi_\alpha(\vec{x})\psi_\beta(\vec{x})d\vec{x} = \delta_{\alpha\beta} \tag{3.30}$$

We start with the modal response Equation (3.20) and (3.25) of Chapter 3, for a point mass-spring system (Figure 2.1). The equation of motion is, from Chapter 3,

$$m\ddot{y} + 2m\omega_0\zeta\dot{y} + ky = f$$

Fourier-transforming the above equation and using the upper case to represent the Fourier component of the corresponding variable, we get:

$$Y(\omega) = H(\omega)F(\omega), \quad \text{where} \tag{8.19}$$

$$H(\omega) = \frac{1}{m\,[(\omega_0^2 - \omega^2) + i2\zeta\,\omega_0\omega\,]} \tag{8.20}$$

and ω_0 is the natural frequency of the point mass-spring system. Multiplying Equation (8.19) by its own complex conjugate and using the definition of the PSD, Equation (8.13), we get,

$$S_y(\omega) = |H(\omega)|^2 S_f(\omega) \tag{8.21}$$

Engineers familiar with the fundamentals of modal analysis should recognized H as the familiar transfer function. Equation (8.21) indicates that the PSD of the response is equal to the PSD of the forcing function multiplied by the modulus squared of the transfer function. When the forcing function is random, Equation (8.19) cannot be solved by the usual deterministic method of structural dynamics. We give up the more ambitious attempt of solving the equation for the time history of response. Instead, we attempt to calculate the mean square value of the response. In the present case of a point mass-spring system subject to a random point-force, Equation (8.21) gives the solution to the mean square response. However, in the case of a structure of finite spatial extent excited by a spatially distributed random pressure, the situation is far more complicated because, in addition to the power spectral density of the forcing function, we have to consider its spatial distribution and how it matches the mode shape of the structure. This is achieved by the acceptance integral method formulated by Powell (1958). First, we give here again Equations (3.20) and (3.25) from Chapter 3:

$$y(\vec{x},t) = \sum_\alpha a_\alpha(t)\psi_\alpha(\vec{x}) \tag{8.22) =(3.20}$$

$$m_\alpha \ddot{a}_\alpha(t) + 2\omega_\alpha m_\alpha \zeta_\alpha \dot{a}_\alpha(t) + k_\alpha a_\alpha(t) = P_\alpha(t) \tag{8.23) = (3.22}$$

where

$$m_\alpha = \int_A \psi_\alpha(\vec{x}) m(\vec{x}) \psi_\alpha(\vec{x}) d\vec{x}$$
$$P_\alpha = \int_A \psi_\alpha(\vec{x}) p(\vec{x},t) d\vec{x} \tag{8.24}$$

are the generalized mass and generalized force defined in Chapter 3. However, since we are considering turbulent flow over the surface of a structure, $m(x)$ is a surface density while the force is now a pressure p. Fourier transforming Equation (8.23) and re-arranging,

$$A_\alpha(\omega) = H_\alpha(\omega) P_\alpha(\omega) \tag{8.25}$$

where $A_\alpha(\omega)$ is the Fourier transform of $a_\alpha(t)$ and must not be confused with the dynamic amplification factor A in Chapter 3, Equation (3.28). H is the *modal* transfer function (see Chapter 3),

$$H_\alpha(\omega) = \frac{1}{m_\alpha[(\omega_\alpha{}^2 - \omega^2) + i2\zeta_\alpha\omega_\alpha\omega \;]} \tag{8.26}$$

and

$$P_\alpha(\omega) = \int\limits_{-\infty}^{+\infty} P_\alpha(\vec{x},t)e^{-j\omega t}dt \tag{8.27}$$

is the Fourier component of the generalized pressure defined in Equation (8.24). Multiplying both sides of the above equation by their complex conjugates and using the definition of the PSD function, Equation (8.13), we get:

$$S_y(\omega) = |H_\alpha(\omega)|^2 S_p(\omega)$$

Substituting Equations (8.24), (8.27) into Equation (8.13), we get

$$S_y(\vec{x},\omega) =$$

$$\sum_\alpha \sum_\beta H_\alpha(\omega)H_\beta{}^*(\omega)\frac{1}{4\pi T}\int\limits_A d\vec{x}'\int\limits_A d\vec{x}''\psi_\alpha(\vec{x}')\int\limits_{-T}^{+T}\int\limits_{-T}^{+T}p(\vec{x}')p(\vec{x}'')e^{-i\omega(t''-t')}dt'\,dt''\psi_\beta(\vec{x}'')$$

in the limit as $T \to \infty$

$$\tag{8.28}$$

If the random pressure due to turbulence is completely uncorrelated, like raindrops on the roof, the structure will not vibrate. The only way a random pressure can excite a structure is if there is some correlation in the forcing function at different points on the structure. This correlation is measured by the cross-correlation function, defined as

$$R_p(\vec{x}',\vec{x}'',\tau) = \lim(T \to \infty)\frac{1}{2T}\int\limits_{-T}^{T}p(\vec{x}',t')p(\vec{x}'',t'+\tau)dt' \tag{8.29}$$

The correlation function is a time domain function. Its Fourier transform is by definition the cross-spectral density of the forcing function:

$$S_p(\vec{x}',\vec{x}'',\omega) = \frac{1}{2\pi}\int\limits_{-\infty}^{+\infty} R_p(\vec{x}',\vec{x}'',\tau)e^{-i\omega\tau}d\tau \tag{8.30}$$

When $x''=x'$, $R_p(\vec{x}', \vec{x}', \tau)$ is the autocorrelation function. It is shown in any book on signal analysis (Bendat and Piersol, 1971, for example) that the Fourier transform of the autocorrelation is just the PSD,

$$S_p(\vec{x}, \omega) = Lim(T \to \infty) \frac{1}{2\pi} \int_{-T}^{+T} R_p(\vec{x}, \vec{x}, \tau) e^{-i\omega\tau} d\tau = \frac{P(\omega)P^*(\omega)}{4\pi T}$$

The reason we have to go through all the above transformations is that presently there is no experimental technique to measure the cross-correlation function of the turbulent pressure as a time domain function, whereas there is at least some method to measure the cross-spectral density of the pressure as a frequency domain function. With these definitions, Equation (8.28) can be put into the form:

$$S_y(\vec{x}, \omega) = AS_p(\omega)\sum_\alpha \psi_\alpha(\vec{x})H_\alpha(\omega)H_\alpha{}^*(\omega)\psi_\alpha(\vec{x})\vec{J}_{\alpha\alpha}(\omega)$$
$$+ 2AS_p(\omega)\sum_{\alpha\neq\beta} \psi_\alpha(\vec{x})H_\alpha(\omega)H_\beta{}^*(\omega)\psi_\beta(\vec{x})\vec{J}_{\alpha\beta}(\omega) \tag{8.31}$$

α, β counted once, where

$$\vec{J}_{\alpha\beta}(\omega) = \frac{1}{A}\iint_{A\,A} \psi_\alpha(\vec{x}')[S_p(\vec{x}', \vec{x}'', \omega)/S_p(\vec{x}', \omega)]\psi_\beta(\vec{x}'')dx'dx'' \tag{8.32}$$

$\vec{J}_{\alpha\beta}$ is the acceptance integral. When $\beta = \alpha$, $\vec{J}_{\alpha\alpha}$ is known as the joint acceptance. Note that some authors write $\vec{J}_{\alpha\alpha}$ as $J_{\alpha\alpha}{}^2$. When $\alpha \neq \beta$, $J_{\alpha\beta}$ is known as the cross-acceptance. The terms $S_p(\vec{x}', \vec{x}'', \omega), S_p(\omega)$ are known as the cross-spectral density between two spatial points $\vec{x}'$ and $\vec{x}''$, and the power spectral density (PSD) at $\vec{x}'$ respectively. Equation (8.32) is general in that A can represent the surface of a two-dimensional structure, or the length of a one-dimensional structure. Similarly, $\vec{x}$ can be a vector coordinate in the case of a two-dimensional structure, or a scalar coordinate in the case of a 1D (beam or tube) structure. The integration is over the entire structure exposed to the flow excitation. Thus J is a double integral for one-dimensional structures, and a quadruple integral for a two-dimensional structure. It is customary to introduce the coherence function,

$$\Gamma(\vec{x}', \vec{x}'', \omega) = S_p(\vec{x}', \vec{x}'', \omega)/S_p(\vec{x}', \omega) \tag{8.33}$$

and rewrite Equation (8.32) in the following form:

$$\vec{J}_{\alpha\beta}(\omega) = \frac{1}{A}\iint_{A\,A} \psi_\alpha(\vec{x}')\Gamma(\vec{x}',\vec{x}'',\omega)\psi_\beta(\vec{x}'')d\vec{x}'d\vec{x}'' \tag{8.34}$$

Note that with the mode-shape functions normalized to unity in accordance with Equation (3.30), the acceptance integral defined in Equation (8.34) is dimensionless. The coherence function in general depends on the frequency as well as two spatial points. The mean square response is equal to the integral of the response PSD Equation (8.14) over the entire frequency range (from $-\infty$ to $+\infty$):

$$< y^2(\vec{x}) > = \int_{-\infty}^{+\infty} S_y(\vec{x},\omega)d\omega = \lim(f_{max}\to\infty)\int_0^{f_{max}} G_y(\vec{x},f)df \tag{8.35}$$

As it stands, the acceptance integral is dependent on any pair of spatial points, making it difficult to evaluate, especially in the days when it was first introduced by Powell (1958). Various simplifications, most of them dramatic and not quite justified, are made in an attempt to evaluate the acceptance integral and to estimate the rms response of the structure. This will be discussed in the following sections.

8.5   Factorization of the Coherence Function and the Acceptance Integral

In the general case in which the structure is two-dimensional, like a plate or a shell subject to boundary layer excitation, the coherence function Γ can be quite complicated. It is usually assumed that Γ is a product of its streamwise component and a cross-stream component, each depending only on its respective co-ordinate:

$$\Gamma(\vec{x}',\vec{x}'',\omega) = \Gamma_1(x_1',x_1'',\omega)\Gamma_2(x_2'',x_2'',\omega) \tag{8.36}$$

where x_1,x_2 are the streamwise and cross-stream coordinates and $\vec{x}',\vec{x}''$ are two different points on the surface of the structure. Even with this simplification, the procedure of determining the coherence functions and the subsequent evaluation of the acceptance integral in the general two-dimensional case can be extremely complicated and very often will involve experimental measurements and numerical techniques. The readers are referred to Chyu and Au-Yang (1972) for an example of such an application to estimate the response of a panel excited by turbulent boundary flow; and to Au-Yang (1975, 1977) for an application to estimate the response of the core support barrel of a nuclear reactor to the coolant flow.

<u>The One-Dimensional Coherence Function</u>

The coherence function must satisfy the following constraints in order to be consistent with the physics of its definition,

$$\Gamma(x',x'',f) = \Gamma*(x'',x',f)$$
$$\Gamma(x',x'',f) \to 1 \quad as \; x' \to x'' \tag{8.37}$$
$$\Gamma(x',x'',f) \to 0 \quad as \; |x'-x''| \to \infty$$

Based on these, some authors *assume* that the coherence function can be expressed in the following functional form

$$\Gamma(x',x'',f) = e^{-|x'-x''|/\lambda} e^{-i2\pi f(x'-x'')/U_{phase}}$$

where U_{phase} is the phase velocity of the wave front (Figure 8.2). In the general case, when the flow is inclined at an angle θ to the x_1 direction, the phase velocities in the x_1, x_2 directions are:

$$U_{phase1} = U_c/cos\theta, \quad U_{phase2} = U_c/sin\theta \tag{8.38}$$

In the subsequent discussion, we shall assume that the flow is in the x_1 direction (Figure 8.3), so that in the streamwise direction, the phase velocity is the same as the convective velocity. The coherence function then takes the form

$$\Gamma_1(x_1',x_1'',f) = e^{-|x_1'-x_1''|/\lambda_1} e^{-i2\pi f(x_1'-x_1'')/U_c} \tag{8.39}$$

which satisfies all the requirements in Equation (8.37). In Equation (8.39), the first factor governs the coherence range of the forcing function. The larger the λ, the *larger* the coherence range of the forcing function. When $\lambda \to \infty$, the forcing function is 100% coherent over the entire length of the structure. When $\lambda \to 0$, the forcing function is completely incoherent over the length of the structure. For this reason, λ is called the *correlation length* of the forcing function. It should be cautioned that some authors (e.g., Blevins, 1990) define the correlation length as ℓ_c with

$$\Gamma(x_1',x_1'',f) = e^{-2|x_1'-x_1''|/\ell_c} e^{-i2\pi f(x_1'-x_1'')/U_c}$$

Thus $\ell_c = 2\lambda$

The second factor in Equation (8.39) denotes the phase difference of the force at points x' and x''. As such, U_c is commonly called the *convective velocity*. It is a measure of how fast the turbulent eddies are carried down the stream. In the streamwise direction, since $(x_1'-x_1'')/Uc$ is the difference in time the front of the pressure wave arrives at the point x_1' and x_1'', and $2\pi f(x_1'-x_2'')/U_c$ is the phase difference in the forcing function between the two points x_1' and x_1''. It should be emphasized that the two factors in the coherence function are quite independent of each other. A completely coherent force may be

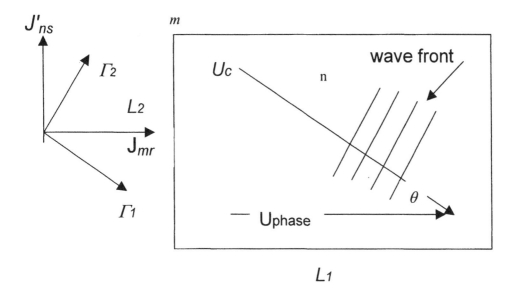

Figure 8.2 Convective and phase velocities

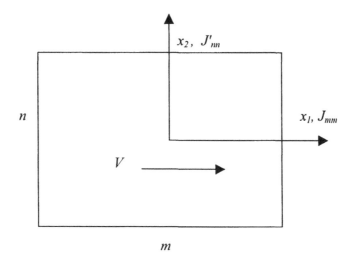

Figure 8.3 Velocity assumed in the x_1 direction

opposite in phase at two points on the structure. Conversely, a force that is in-phase at two points on the structure can be completely incoherent.

In the cross-stream (x_2) direction, since the wave front arrives at every point along the wave front at the same time, the phase angle is zero. The forcing function is always in-phase. The coherence function is equal to:

$$\Gamma_2(x_2',x_2'',f) = e^{-|x_2'-x_2''|/\lambda_2}$$
(8.40)

Equations (8.39) and (8.40) are the simplest representation of the coherence function. There are other more complicated empirical representations. One of those is used in computing the acceptance integrals given in the charts in Appendix 8A.

The Longitudinal and Lateral Acceptances

From Equation (8.34), the acceptance can also be factored into the product of a streamwise acceptance, J_{mr} (called longitudinal acceptance), and a cross-stream acceptance, J'_{ns} (lateral acceptance):

$$\bar{J}_{\alpha\beta} = J_{mr}J'_{ns}$$
(8.41)

For a rectangular plate, $\alpha=m,\ r\ldots;\ \beta=n,\ s\ldots$ are the modal indices in the streamwise and cross-stream directions (Figure 8.3). From Equations (8.34) and (8.39),

$$J_{mr} = \frac{1}{L_1}\iint_{L1}\psi_m(x_1')e^{-|x_1'-x_1''|/\lambda - i2\pi f(x_1'-x_1'')/U_c}\psi_r(x_1'')dx_1'dx_1''$$

The longitudinal acceptance is in general complex, with the real and imaginary parts given by

$$\text{Re } J_{mr} = \frac{1}{L_1}\int_0^{L_1}dx''\psi_m(x'')\int_0^{x''}\psi_r(x')e^{-(x'-x'')/\lambda}\cos\frac{2\pi f(x'-x'')}{U_c}dx' +$$
$$\frac{1}{L_1}\int_0^{L_1}dx''\psi_m(x'')\int_{x''}^{L_1}\psi_r(x')e^{-(x''-x')/\lambda}\cos\frac{2\pi f(x''-x')}{U_c}dx'$$
(8.42)

$$\text{Im } J_{mr} = \frac{1}{L_1}\int_0^{L_1}dx''\psi_m(x'')\int_0^{x''}\psi_r(x')e^{-(x'-x'')/\lambda}\sin\frac{2\pi f(x'-x'')}{U_c}dx' +$$
$$\frac{1}{L_1}\int_0^{L_1}dx''\psi_m(x'')\int_{x''}^{L_1}\psi_r(x')e^{-(x''-x')/\lambda}\sin\frac{2\pi f(x''-x')}{U_c}dx'$$
(8.43)

Likewise from Equations (8.34) and (8.40),

$$J'_{ns} = \frac{1}{L_2}\iint_{L2}\phi_n(x_2')e^{-|x_2'-x_2''|/\lambda}\phi_s(x_2'')dx_2'dx_2''$$
(8.44)

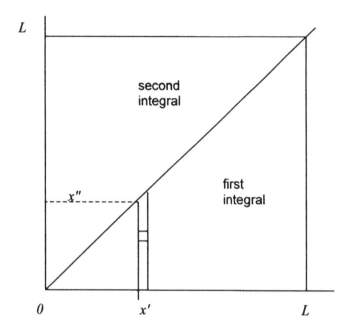

Figure 8.4 Domain of integration in the acceptance double integral

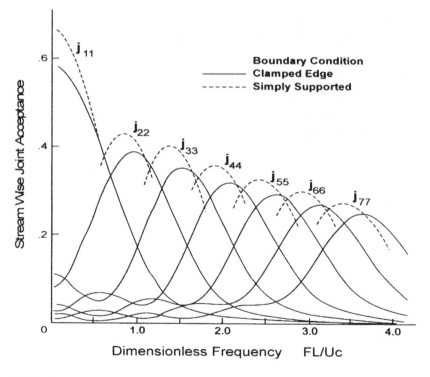

Figure 8.5 Joint acceptances for boundary layer turbulence over a flat rectangular panel
(Chyu and Au-Yang, 1972)

where ψ, ϕ are the mode-shape functions in the streamwise and cross-stream directions; and $\alpha=(m, n)$ and $\beta=(r, s)$ are the mode numbers, with m, r in the streamwise and n, s in the cross-stream directions. The acceptance integral in parallel flow usually has to be evaluated by numerical integration. Knowing the symmetry properties of the acceptance integral can eliminate over 50% of the computations.

The domain of integration is shown in Figure 8.4. In the majority of applications, the boundary conditions at $x=0$ and $x=L$ are the same. The first term is then equal to + or - the second term, depending on the symmetry of the mode shape. It is left for the readers to show that in this special case:

- The streamwise cross-acceptances are in general complex.
- The joint acceptances J_{mm}, J'_{nn} are always real.
- The cross-stream acceptances J'_{ns} are always real.
- For a rectangular panel with four edges simply-supported or clamped,

$Re\ J_{mr} = 0$ if $m+r$ is odd
$Im\ J_{mr} = 0$ if $m+r$ is even
$J'_{ns} = 0$ if $n+s$ is odd.

The numerical integration can then be carried out over only half of the domain of integration.

Figure 8.5 shows an example of the longitudinal joint acceptances for parallel flow over a flat rectangular panel. Charts for the acceptance integrals for turbulent boundary layer flow over rectangular panels with all four edges clamped are given in Appendix 8A. These acceptance integrals were computed by brute force numerical double integration, using Bull's representation (1967) of the coherence function given in Appendix 8A. As Figure 8.5 shows, the differences in the acceptance integrals between simply-supported and clamped boundary conditions are generally small. This is because the acceptance integral is a measure of the compatibility of the structural mode shapes to the forcing function distribution, and there is not much difference between the mode shapes of a rectangular panel with simply-supported or clamped boundary conditions. Therefore, the charts in the Appendix can also be used to estimate the response of simply-supported rectangular panels subject to turbulent boundary layer flow.

8.6   The Mean Square Response

From Equation (8.35), the mean square response is obtained by integrating the response PSD over the entire frequency range:

$$< y^2(\vec{x}) >= \int_{-\infty}^{+\infty} S_y(\vec{x}, \omega)d\omega = \int_{0}^{f_{max}} G_y(\vec{x}, f)df \qquad (8.45)$$

From Equations (8.31) and (8.26)

$$G_y(\vec{x}, f) = AG_p(f)\sum_\alpha \psi_\alpha(\vec{x})H_\alpha(f)H_\alpha *(f)\psi_\alpha(\vec{x})J_{\alpha\alpha}(f)$$
$$+ 2AG_p(f)\sum_{\alpha\neq\beta} \psi_\alpha(\vec{x})H_\alpha(f)H_\beta *(f)\psi_\beta(\vec{x})J_{\alpha\beta}(f)$$

(8.46)

$$H_\alpha(f) = \frac{1}{(2\pi)^2 m_\alpha[(f_\alpha{}^2 - f^2) + i2\zeta_\alpha f_\alpha f\ ]}$$

(8.47)

Provided that the forcing function is known, we can at this point proceed to evaluate the joint and cross-acceptances by numerical double integration, at different frequency intervals, then proceed to calculate the mean square response from Equations (8.45), (8.46) and (8.47) by numerical integration over f, to a cutoff frequency f_{max}. This approach will account for both the cross-term and off-resonance contributions to the response. With today's computers, this approach, though time-consuming, is absolutely feasible. Before the advent of today's high-speed and high-memory computers, such an approach was out of the question. Therefore, it is worthwhile to review the traditional way of simplifying Equations (8.45) to (8.47) to arrive at an approximate value for the mean square response. This would also give us some insight into one of the most commonly used equations in turbulence-induced vibration analysis.

The first assumption made to simplify Equation (8.35) with S_y given by Equation (8.31) is to ignore the cross-term $(\alpha \neq \beta)$ contribution to S_y. This is justified if damping is small and the normal modes are "well separated" in the frequency domain. The second assumption is that the turbulence is homogeneous, so that S_p is the same throughout the surface of the structure and Γ depends only on the separation distance between two points, $\Delta x = |\vec{x}' - \vec{x}''|$ (see Equations (8.39) and (8.40)). The third assumption is that $J_{\alpha\alpha}$ and S_p are slowly varying functions of ω in the vicinity of ω_α (Figure 8.6). The mean square response is given by integrating the response PSD over ω from $-\infty$ to $+\infty$. If the modal damping is small and the normal modes are "well separated," most of the contribution to the above integral will come from resonance peaks centered around the natural frequencies. If both S_p and $J_{\alpha\alpha}$ are slowly varying functions of ω around each of the natural frequencies, then

$$\bar{y}^2(\vec{x}) = \int_{-\infty}^{+\infty} S_y(\vec{x}, \omega)d\omega$$

$$= \sum_\alpha < y_\alpha{}^2(\vec{x}) > = \sum_\alpha AS_p(\omega_\alpha)\psi_\alpha{}^2(\vec{x})\bar{J}_{\alpha\alpha}(\omega_\alpha) \int_{-\infty}^{+\infty} |H_\alpha(\omega)|^2\ d\omega$$

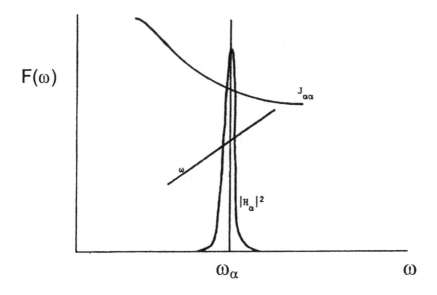

Figure 8.6 Variation of H and J in the neighborhood of the natural frequency

Since most of the contributions are from the vicinity of ω_α, we can multiply the integrand by ω and divide it by ω_α without affecting the results significantly:

$$<y_\alpha^2> = AS_p(\omega_\alpha)\psi_\alpha^2(\vec{x})\vec{J}_{\alpha\alpha}(\omega_\alpha)\frac{1}{\omega_\alpha}\int_{-\infty}^{+\infty}|H_\alpha(\omega)|^2\,\omega d\omega$$

Now change the variable of integration from ω to ω^2, and evaluate the resulting integral by contour integration and calculus of residues. Note that as ω goes from $-\infty$ to $+\infty$, ω^2 goes from $+\infty$ to 0 and back to $+\infty$. Therefore, the contour of integration in the complex ω^2 plane goes from $+\infty$ to 0, then from 0 back to $+\infty$. Figure 8.7 shows the contour of the integration. Substituting Equation (8.26) for H, we get:

$$<y_\alpha^2> = AS_p(\omega_\alpha)\psi_\alpha^2(x)\vec{J}_{\alpha\alpha}(\omega_\alpha)\frac{1}{2m_\alpha^2\omega_\alpha}\oint\frac{d(\omega^2)}{[(\omega_\alpha^2-\omega^2)^2+4\zeta_\alpha^2\omega_\alpha^4]} \qquad (8.48)$$

The integrand can be written as,

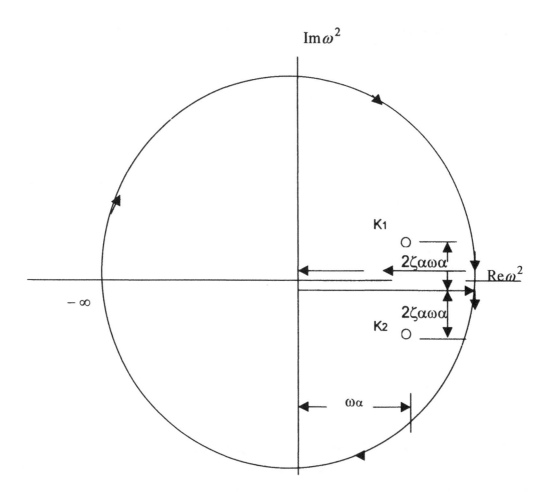

Figure 8.7 Contour integration in the complex ω^2 plane

$$\frac{-\dfrac{i}{4\zeta_\alpha\omega_\alpha^2}}{\omega^2 - \omega_\alpha^2(1 + i2\zeta_\alpha)} \quad + \quad \frac{\dfrac{i}{4\zeta_\alpha\omega_\alpha^2}}{\omega^2 - \omega_\alpha^2(1 - i2\zeta_\alpha)}$$

The integrand has two poles at $\omega^2 = \omega_\alpha^2(1 \pm i2\zeta_\alpha)$, with residues equal to

$K_j = \mp \dfrac{i}{4\zeta_\alpha\omega_\alpha^2}$. From calculus of residues,

$$\oint \quad = 2\pi i \sum_j K_j \tag{4.49}$$

The contour of integration encloses both poles. Substituting into equations (8.48) and (8.49) the values of the residues at the poles, we get

$$< y^2(\vec{x}) > = \sum_\alpha \frac{\pi A S_p(\omega_\alpha) \psi_\alpha^2(\vec{x}) \vec{J}_{\alpha\alpha}(\omega_\alpha)}{2 m_\alpha^2 \omega_\alpha^3 \zeta_\alpha}$$

At this point it is convenient to express the mean square response in terms of the single-sided fluctuating pressure power spectral density $G_p(f)$, which is more commonly used in engineering. From Equation (8.18), we get

$$< y^2(\vec{x}) > = \sum_\alpha \frac{A G_p(f_\alpha) \psi_\alpha^2(\vec{x}) \vec{J}_{\alpha\alpha}(f_\alpha)}{64 \pi^3 m_\alpha^2 f_\alpha^3 \zeta_\alpha} = \sum_\alpha \pi f_\alpha G_y(\vec{x}, f_\alpha) \qquad (8.50)$$

where

$$G_y(\vec{x}, f_\alpha) = \frac{A G_p(f_\alpha) \psi_\alpha^2(\vec{x}) \vec{J}_{\alpha\alpha}(f_\alpha)}{64 \pi^4 m_\alpha f_\alpha^4 \zeta_\alpha} \qquad (8.51)$$

is the *response* PSD at f_α.

Equations (8.50) and (8.51) are, up to this point, general. They can be applied to two-dimensional as well as one-dimensional structures. In the latter case, A is replaced by L, the length of the structure exposed to the flow excitation. In Equation (8.50), $\vec{x}$ is a vector coordinate on the surface of the structure and $J_{\alpha\alpha}$ must be obtained by integrating over the surface of the structure. In spite of the elegant way Equation (8.50) is derived, one must realized that it holds true only if:

- The cross-terms in Equation (3.17) are negligible,
- The turbulence is homogeneous so that S_p is independent of x and Γ depends only on Δx.
- The acceptance integral is a slowly varying function of ω in the vicinity of ω_α.

In the following section, the justifications of the above assumptions will be critically examined.

Validity of the Response Equation

Equation (8.50) is derived under the assumption that the cross-terms contributions to the total mean square responses are negligible. This is true if all of the following conditions are met:

- The cross-acceptances are small compared with the joint acceptances, or the transfer functions H from one mode to other different modes are small compared with that from one mode to the same mode.
- The acceptance integral is a slowly varying function of frequency near the natural frequencies (see Figure 8.6).
- The forcing function is homogeneous and isotropic. That is, not only is the power spectral density $G_p(\vec{x}, f)$ independent of $\vec{x}$, but the coherence function is dependent only on the separation distance $|\vec{x}' - \vec{x}''|$, not on the position vector $\vec{x}$ itself.

The charts in Appendix 8A, reproduced from Chyu and Au-Yang (1972), are plots of the joint and cross-acceptances for a rectangular panel subjected to turbulent boundary layer excitation. These values were computed by numerical integration of the acceptance integral. Bull's empirical correlation (1967), given in Appendix 8A, was used. Because of the symmetry of the coherence function, $J_{1m} = 0$ when m is odd. From these charts, it is obvious that the cross-acceptances in general are not negligible compared with joint acceptances.

The first two condition hold true when the natural frequencies of the structure are "well separated," as in a single span beam. It may not be true, for example, in a multiple-supported beam such as a heat exchanger tube. Cross-flow turbulence-induced vibration of a multi-span beam or tube will be discussed in the following chapter. In view of the wide availability of inexpensive high-speed computers over the last ten years, the necessity of ignoring cross-terms and off-resonance contribution to the total responses should be critically reviewed. With minimum programming effort, it is much more straightforward to directly compute the mean square response, after the acceptance integrals are computed, by numerical integration of the response PSD over the frequency, using Equations (8.45) to (8.47). In this way, not only are the cross-terms included, but the off-resonance contributions to the mean square responses are also included.

8.7   The Physical Meaning of the Acceptance Integral

Examination of Equation (8.31) shows that the factors involving the transfer function,

$$\psi_\alpha(x)H_\alpha(\omega)H_\beta{}^*(\omega)\psi_\beta(x)$$

and the acceptance integral, Equation (8.24) are very similar expressions. Indeed, they play similar roles in the response of the structure. For a point-mass-spring system subject to a random force, the acceptance integral reduces to unity. The acceptance integral plays no part in the response of the structure if it is a point. From this, it is obvious that the acceptance integral has something to do with the matching of the spatial distribution of the forcing function and the mode shapes of the structure. Indeed, while the transfer

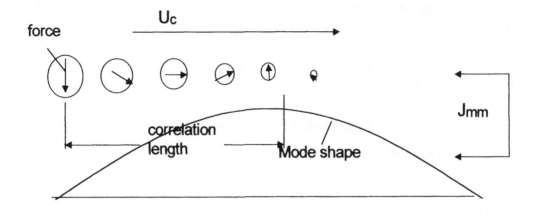

Figure 8.8 The joint acceptances as a measure of compatibility between the forcing function and the structural mode shape.

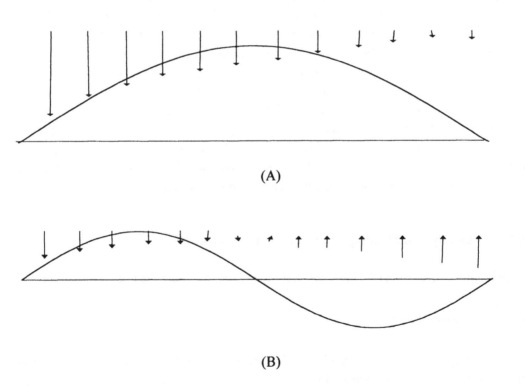

Figure 8.9 Parallel flow over a beam: (A) Phase relationship of forcing function favors the symmetric first mode; (B) favors the anti-symmetric second mode.

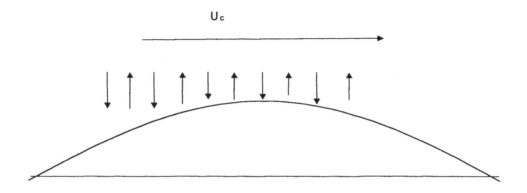

Figure 8.10 High frequency forcing function cannot excite any mode

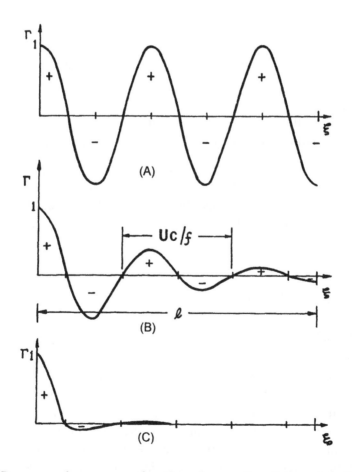

Figure 8.11 Pressure coherence as a function of separation distance $|x'-x''|$ in the special case when $L = U_c/4f$: (A) $\lambda \to \infty$; (B) $\infty > \lambda > U_c/4f$; (C) $\lambda \approx U_c/4f$

function Equation (8.26) is a measure of the matching in time domain between the periodicity of the forcing function and the natural period of vibration of the structure, the acceptance integral is a measure of the matching in space between the forcing function and the structural mode shape (Figure 8.8). Equation (8.26) shows that when the frequency of the excitation force is exactly the same as one of the natural frequencies of the structure, the transfer function reaches a maximum. This is the familiar phenomenon of resonant vibration. The joint acceptance, too, may contain maxima, as shown in Figure 8.5. Analogous to resonance in the time domain, when this happens, it is called coincidence.

It is easy to see why the joint acceptances exhibit maxima. Figure 8.9 shows a beam excited by turbulent parallel flow. As the eddies flow along the beam, the phase of the net force changes. How fast this changes depends on both the frequency of the forcing function and the convective velocity at which the eddies are being carried down the beam. If the phases of the eddies do not reverse as they are swept down the beam, clearly the symmetric mode will be preferentially excited, as shown in Figure 8.9A. On the other hand, if the phases of the eddies reverse as they are swept down the beam, the anti-symmetric mode will be preferentially excited while the first mode will be suppressed, as shown in Figure 8.9B. Figure 8.10 shows that the high-frequency component of the fluctuating pressure cannot excite the lower modes except at very high convective velocities. Figure 8.11 shows that in parallel flow-induced vibration, maximum response of the fundamental mode occurs when the correlation length is about ¼ of the "pressure wavelength" U_c/f.

Physicists would call the acceptance integral transition probability amplitude. $J_{\alpha\beta}$ is a measure of the probability that under the excitation of the forcing function $S_p(x', x'', \omega)$, a structure originally vibrating in the αth mode will change to the βth mode. The joint acceptance is a measure of the probability that a structure originally vibrating in the αth mode will remain in the same mode under the excitation of the force. From this, one can expect that there is an upper bound to the acceptance integral, as the transition probability obviously cannot exceed unity.

8.8   Upper Bound of the Joint Acceptance

From Equations (8.31) and (8.35), we see that the first step in estimating the structural response to turbulence is to evaluate the acceptance integral. Unfortunately, as given in Equations (8.42) to (8.44), the acceptance integral is difficult to calculate even if both of the necessary input parameters—the correlation length and the convective velocity—are known. Very often, mathematical expressions for these are not available. A very common way to get an approximate estimate of the response is to estimate an upper bound of the joint acceptance integral and ignore the cross-term contributions to the response. Hopefully, even this conservative estimate will give an acceptable structural response. It is often cited that,

$$\tilde{J}_{\alpha\alpha} \leq 1.0, \quad all \ \alpha$$

However, proof of the above statement is often lacking. Au-Yang (1986, 2000) showed that the above statement is correct only if the mode shapes are normalized to unity as in Equation (3.30) in Chapter 3:

$$\int\limits_{A} \psi_\alpha(\vec{x})\psi_\beta(\vec{x})d\vec{x} = \delta_{\alpha\beta} \qquad\qquad\qquad (8.52) = (3.30)$$

Equation (8.52) is the familiar orthonormality condition of the mode shapes. However, the mode shapes of a structure are orthogonal to each other *only* if the mass density, including the hydrodynamics mass (see Chapter 3), of the structure is uniform. The case when the total mass density is not uniform will be addressed in the following chapter. In addition, when we use the method of normal mode analysis, we intrinsically assume that any structural vibration form can be synthesized by a linear combination of its modal responses. This is the completeness condition and is expressed in the following equation:

$$\sum_\alpha \psi_\alpha(\vec{x}')\psi_\alpha(\vec{x}'') = \delta(\vec{x}'-\vec{x}'') \qquad\qquad\qquad (8.53)$$

Equation (8.53) holds ture *only* if Equation (8.52) is satisfied. Equations (8.52) and (8.53) are crucial to show that the upper bound of the joint acceptance integral is 1.0: Summing over all the joint acceptances, we get from Equation (8.34),

$$\sum_\alpha \tilde{J}_{\alpha\alpha}(\omega) = \sum_\alpha \frac{1}{A}\int\limits_{0}^{A}\int\limits_{0}^{A} \psi_\alpha(\vec{x}')\Gamma(\vec{x}',\vec{x}'',\omega)\psi_\alpha(\vec{x}'')d\vec{x}'\,d\vec{x}''$$

Interchanging the summation and integration, we have

$$\sum_\alpha \tilde{J}_{\alpha\alpha}(\omega) = \frac{1}{A}\int\limits_{A}\int\limits_{A}\sum_\alpha \psi_\alpha(\vec{x}')\psi_\alpha(\vec{x}'')\Gamma(\vec{x}',\vec{x}'',\omega)d\vec{x}'\,d\vec{x}''$$

$$= \frac{1}{A}\int\int \delta(\vec{x}'-\vec{x}'')\Gamma(\vec{x}',\vec{x}'',\omega)d\vec{x}'\,d\vec{x}'' = \frac{1}{A}\int\limits_{A}\Gamma(\vec{x}',\vec{x}',\omega)d\vec{x}'$$

But, by definition of the coherence function, Equation (8.37), $\Gamma(x, x, \omega) = 1.0$. Therefore,

$$\sum_\alpha \tilde{J}_{\alpha\alpha}(\omega) = A/A = 1.0 \qquad\qquad\qquad (8.54)$$

Since the sum of the joint acceptances *at a given frequency* is equal to 1.0, it follows that

$$\bar{J}_{\alpha\alpha}(\omega) \leq 1.0 \quad \textit{for all } \alpha \textit{ and at any } \omega \qquad\qquad (8.55)$$

It is seen that Equation (8.55) depends crucially on Equation (8.52) and (8.53), that is, on the way the mode shape is normalized. In practice, an analyst often uses a commercial finite-element computer program to solve for the natural frequencies and mode shapes. The mode shapes in these commercial computer programs are usually normalized with respect to the mass, by putting the generalized mass equal to unity (see Chapter 3, Equation (3.8)). Thus if the turbulence-induced vibration analysis is carried out with the help of a finite-element computer program in which the generalized mass is normalized to unity, it will be wrong to assume that the upper bound of the acceptance integral is unity.

8.9   The Turbulence PSD, Correlation Length and Convective Velocity

Here we return to the other half of the hybrid method for solving the turbulence-induced vibration problem—the determination of the turbulent forcing function. As can be see from the response Equations (8.45) to (8.47), (8.50) and the equations for the acceptance integrals, Equations (8.42) to (8.44), we need three parameters to characterize the turbulent forcing function: The convective velocity U_c, which determines the phase relationship of the forcing function at two different points on the surface of the structure; the correlation length λ, which determines the degree of coherence of the forcing function at two different points on the surface of the structure; and finally the power spectral density function, G_p, which determines the energy distribution as a function of the frequency of the forcing function. As discussed in the introductory section, these parameters are obtained by model testing and scaling. Designers who are facing the task of estimating the flow-induced vibration concern of "first-of-a-kind" structures should design scale model tests to estimate these parameters, as was done by Au-Yang and Jordan (1980), Au-Yang et al (1995), Chen and Wambsganss (1970), to quote just a few. Very often, however, the structure may be a tube, a pipe or the annular gap between concentric cylindrical shells for which test data already exist in the literature. In this section, existing data from the literature will be reviewed for application to turbulence-induced vibration estimates of power and process plant components.

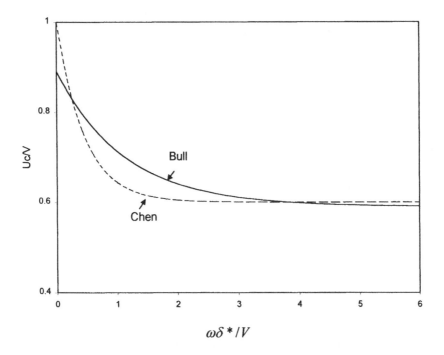

Figure 8.12 Comparison of convective velocities as predicted by
Chen and Wambsganss (1970) and by Bull (1967)

The Convective Velocity

Based on data obtained from turbulent flows, Chen and Wambsganss (1970) derived the
following empirical equation for the convective velocity as a function of frequency:

$$U_c / V = 0.6 + 0.4e^{-2.2(\omega\delta^*/V)} \tag{8.56}$$

Bull (1967) suggested a slightly different equation:

$$U_c / V = 0.59 + 0.30e^{-0.89\omega\delta^*/V} \tag{8.57}$$

where δ^* is the displacement boundary layer thickness for boundary layer flow, or the
"hydraulic radius" in confined internal flow. Figure 8.12 shows a comparison between
the convective velocities computed with Equations (8.56) and (8.57). Both equations
show that except at very low frequencies, the convective velocity is fairly independent of
the frequency, being equal to approximately 0.6 times the free stream velocity.

   The above equation apparently holds true only for boundary layer type turbulence. In
confined flow channels in which very high turbulence is generated by impingement or by
flow in 90-degree channels (e.g., pipe elbows), Au-Yang and Jordan (1980), Au-Yang et

al (1995) found, in two separate experiments, that the convective velocity is about the same as the mean free stream velocity. That is,

$$U_c \approx V \quad \text{for very turbulent confined flow} \tag{8.58}$$

The Displacement Boundary Layer Thickness

In Equations (8.56) and (8.57), $\delta*$ is the displacement boundary layer thickness—a fluid mechanical parameter that is discussed in books on fluid mechanics. For turbulent flow over a flat plate, a simple expression for estimating the displacement boundary layer thickness is (Streeter, 1966),

$$\delta* = \frac{0.37x}{\text{Re}^{1/5}}; \quad \text{Re} = \frac{Vx}{v} \tag{8.59}$$

where Re is the Reynolds number (see Chapter 6) based on the distance x from the leading edge of the flat plane. Equation (8.59) predicts that the turbulence boundary layer thickness will grow rapidly downstream from the leading edge of the flat plate. In most industrial applications, however, the flow is internal in long pipes or conduits. Obviously the boundary layer cannot grow indefinitely. In small pipes and narrow flow channels, the boundary layer ultimately will fill up the entire cross-section of the flow channel. In that case, $\delta*$ is the half-width, or hydraulic radius, of the flow channel,

$$\delta* = D_H / 2 = R_H, \quad \text{confined channel flow} \tag{8.60}$$

For flow between narrow parallel plates $2h$ apart,

$$\delta* = h \tag{8.61}$$

In large pipes or flow channels, the boundary layer thickness eventually will asymptotically reach a final value. This value depends on the Reynolds number and has been measured experimentally (see Duncan et al, 1960),

$$\delta* = \frac{D_H}{2(n+1)} \tag{8.62}$$

The value of n depends on the Renolds number, as shown in Figure 8.14.

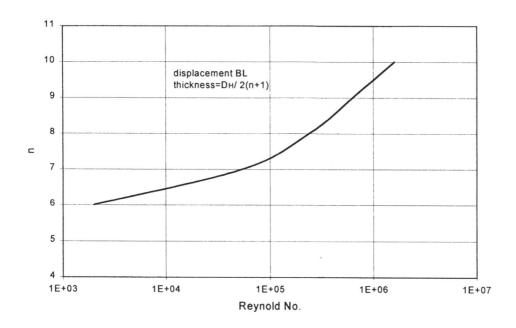

Figure 8.13 Dependence of n on the Reynolds' number
(Based on data from Duncan et al, 1960).

Correlation Length

Based on two different tests with annular axial flow, Au-Yang and Jordan (1980) and Au-Yang et al (1995) found that except at very low frequencies, at which the forcing function is dominated by acoustics, the correlation length λ is also fairly independent of frequency and is approximately equal to 0.4 times the annular gap width of the flow channel (Figure 8.14):

$$\lambda / g = 0.4 \tag{8.63}$$

where g is the annular gap width, which is equal to the hydraulic radius R_H. It should be mentioned once again that the correlation length λ is one half that of the correlation length ℓ_c, defined by authors such as Blevins (1990). Since ℓ_c is a more appropriate measure of the coherence range of the forcing function, the above experimental result agrees with our intuition that the coherence range of the turbulent forcing function is approximately equal to the characteristic length of the flow channel. Thus we can extrapolate Equation (8.63) to other flow channel geometry. For external axial flow through a tube bundle (see also Equation (5.11)),

$$\lambda = 0.2P(1 + P/2D) \tag{8.64}$$

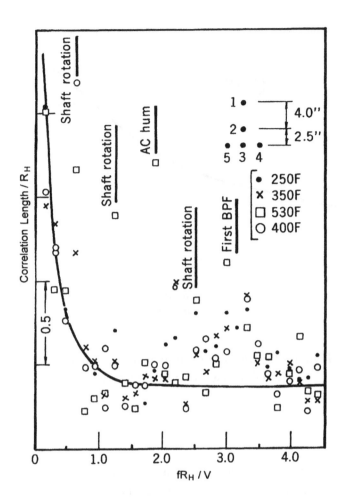

Figure 8.14 Correlation length as a function of dimensionless frequency in annular flow
(Au-Yang and Jordan, 1980)

where P is the pitch and D is the outside diameter of the tubes. For other flow channel cross-sections, we can take,

$$\lambda \approx 0.4 R_H \qquad\qquad (8.65)$$

where

$$R_H = D_H / 2 \qquad\qquad (8.66)$$

is the hydraulic radius and D_H is the hydraulic diameter.

Based on the discussions in Section 8.5, the correlation length for boundary layer flow is approximately equal to the displacement boundary layer thickness, which makes physical sense.

Turbulence Random Pressure Power Spectral Density

The last and the most important fluid mechanical parameter that characterized the turbulent forcing function is the power spectral density (PSD). Figure 8.15, reproduced from Chen (1985), shows the normalized PSD as a function of the circular frequency ω, plotted against the non-dimensional circular frequency $\omega\delta^*/V$. This set of data is for boundary layer turbulent flow over flat plates or in straight flow channels. The following should be noted when using this set of data: In Figure 8.15,

$$G_p(f) = 2\pi G_p(\omega) = 2\pi\Phi_{pp}(\omega) = 2\pi\rho^2 V^3\delta * \left\{\frac{\Phi_{pp}(\omega)}{\rho^2 V^3\delta *}\right\} \tag{8.67}$$

where δ^*, the displacement boundary layer thickness, is given in Equation (8.62) and is discussed at length in the sub-section on displacement boundary layer thickness. The quantity in { } is what is plotted in the ordinate in Figure 8.15. As pointed out by Chen (1985), the data in Figure 8.15 is unreliable in the low-frequency region marked "effective range." Figure 8.16, also reproduced from Chen (1985), shows the normalized boundary layer turbulence spectrum at low frequencies. Chen suggested the following empirical equation for the low-frequency PSD

$$\frac{G_p(f)}{\rho^2 V^3 D_H} = 0.272E - 5/S^{0.25}, \quad S < 5$$
$$= 22.75E - 5/S^3, \quad S > 5 \tag{8.68}$$

where

$$S = 2\pi f D_H / V \tag{8.69}$$

Equations (8.67) and (8.68) are applicable to straight flow channels, such as flow in nuclear fuel bundles and in long straight pipes. Industrial piping systems very often contain elbows and 90-degree turns and may have valves installed in them. Cavitation may occur downstream of these elbows and valves. The turbulence PSDs in such piping systems are generally much higher than what are given in Equations (8.67) and (8.68), as was shown by Au-Yang et al (1995), and by Au-Yang and Jordan (1980). In the 1995 test, light cavitation was observed; while in the 1980 test, there was no noticeable cavitation. Based on these two sets of data, the following empirical equations are suggested:

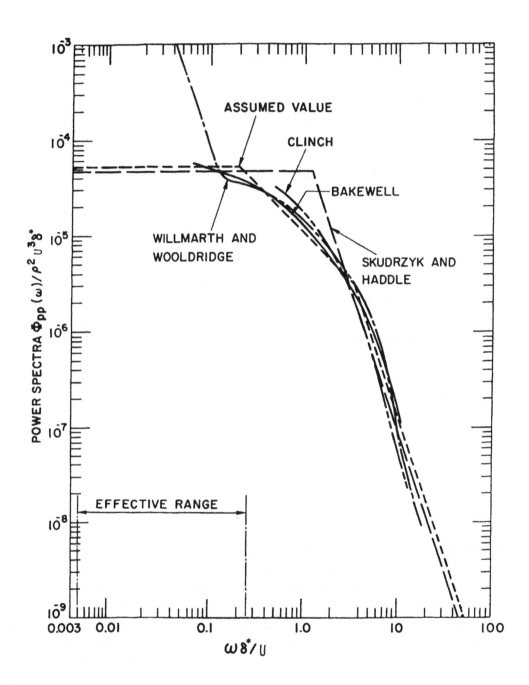

Figure 8.15 Boundary layer type of turbulence power spectral density
(Chen, 1985)

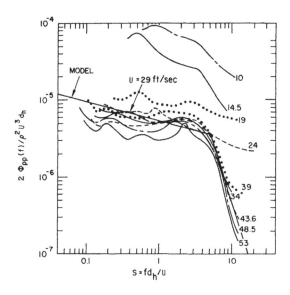

Figure 8.16 Normalized PSD for boundary layer type turbulence, low frequencies
(Chen, 1985)

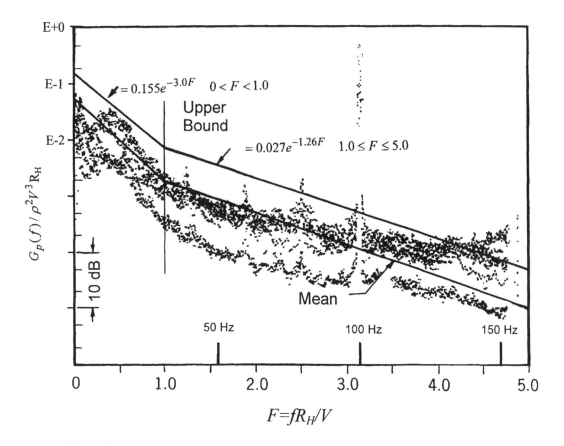

Figure 8.17 Comparison of empirical normalized PSD equation with field
measured data for confined annular flow (Au-Yang and Jordan, 1980)

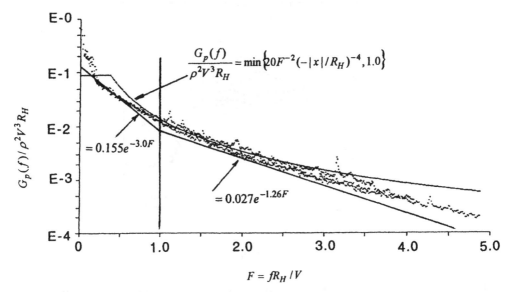

Figure 8.18 Comparison of empirical normalized PSD equation with data from confined, cavitating annular flow test (Au-Yang et al, 1995)

For turbulent flow without cavitation,

$$\frac{G_p(f)}{\rho^2 V^3 R_H} = 0.155e^{-3.0F}, \quad 0 < F < 1.0$$

$$= 0.027e^{-1.26F} \quad 1.0 \le F \le 5.0$$

(8.70)

where

$$F = fR_H / V$$

(8.71)

is the dimensionless frequency and $R_H = g$ is the annular gap width.

For turbulent flow with light cavitation,

$$\frac{G_p(f)}{\rho^2 V^3 R_H} = \min\{20F^{-2}(-|x|/R_H)^{-4}, 1.0\}$$

(8.72)

or values from Equation (8.70), whichever is larger, where $|x|$ is the absolute value of the distance from the cavitating source such as an elbow or a valve.

Figure 8.17 compares the normalized random pressure PSD, computed with Equation (8.70), and data reported in Au-Yang and Jordan (1980). This empirical equation, which was derived based on data from a scale model test, agrees reasonably well with the data obtained from measurement on the full-size prototype. Figure 8.18 compared the normalized random pressure PSD computed with Equations (8.70) and (8.72) with data reported in Au-Yang et al (1995).

Comparison between Figures 8.16, 8.17 and 8.18, 8.19 shows that in general, Au-Yang's data showed much higher turbulence intensity for the same velocity. This is not surprising as in Au-Yang's tests, the flow was allowed to impinge upon the flow channels perpendicularly before it flowed down the annular flow channel. In short, the flows were much more turbulent than those over flat plates or in straight pipes. Since industrial piping systems are seldom straight, Equation (8.70) is probably more representative of the turbulence PSD in practice.

8.10   Examples

In this section two examples are given to illustrate applications of the techniques discussed in this chapter to solve turbulence-induced vibration problems commonly encountered in the process and power industries. The first one is a quantitative calculation using acceptance integrals given in the charts in Appendix 8A. The second example is an application of the general concept for a "root-cause analysis" of a problem that occurs in the field, without getting into detailed quantitative calculations.

Example 8.1

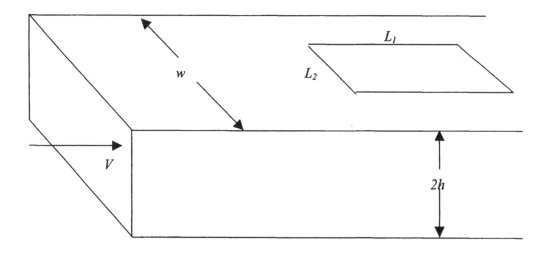

Figure 8.19 Turbulent flow over a flat plate

Air flows in a duct of rectangular cross-section. A rectangular cover plate is bolted on top of the duct. The dimensions of the duct, the cover plate and the density and viscosity of the air are given in Table 8.1. Estimate the turbulence-induced vibration amplitudes of the first four modes of the plate. Assume the cover plate is simply-supported at all four edges.

Table 8.1: Input for Problem 8.1

	US Unit	SI Unit
Air density, ρ	0.0763 lb/ft^3	1.222 Kg/m^3
	(1.143E-7 (lb-s^2/in) per in^3)	
Air kinematic viscosity, ν	0.000157 ft^2/s	1.459E-5 m^2/s
Flow velocity, V	200 ft/s (2400 in/s)	60.96 m/s
Channel height, $2h$	18 in	0.457 m
Channel width, w	24 in	0.610 m
Cover plate length L_1	18 in	0.457 m
Cover plate width, L_2	12 in	0.305 m
Cover plate thickness, t	1/8 in	0.00318 m
Plate material density, ρ_m	0.2831 lb/in^3	7863 Kg/m^3
	(7.327E-4 (lb-s^2/in) per in^3)	
Poison ratio, σ	0.29	0.29
Young's modulus, E	2.96E+7 psi	2.041E+11 Pa
Damping ratio, ζ	0.005	0.005

Solution

As always, we immediately convert all the given input in the US unit set into a consistent unit set with (lb-s^2/in), inch and second as the fundamental units. These are given in parentheses in Table 8.1. We shall use Equation (8.50) to estimate the response. As discussed before, a major part of the response calculation is in the acceptance integrals. So we must first decide how to calculate the acceptances. From the nature of the problem we expect the modal frequencies to be well separated so that cross-terms can be ignored. We can use Equations (8.42) to (8.44) together with the convection velocities and displacement boundary layer thickness discussed in Section 8.9 to compute the joint acceptances by numerical integration. However, since the problem is boundary layer turbulent flow over a flat plate, we can use the acceptance integrals in the Appendix. The following lists the step-by-step calculations of all the variables in the response Equation (8.50). We shall calculate the maximum response for each mode:

$$< y_\alpha^2(\vec{x}) >= \frac{AG_p(f_\alpha)\psi_\alpha^2(\vec{x})\vec{J}_{\alpha\alpha}(f_\alpha)}{64\pi^3 m_\alpha^2 f_\alpha^3 \zeta_\alpha} \tag{8.50}$$

Mode-Shape Function and Generalized Masses

Great care must be exercised to normalize the mode shape properly. Since the plate is of uniform thickness and we plan to use the charts in the Appendix, we can, and must, normalize the mode shape according to Equation (3.30):

$$\int_A \psi_\alpha(\vec{x})\psi_\beta(\vec{x})d\vec{x} = \delta_{\alpha\beta} \qquad (3.30)$$

For a simply-supported rectangular panel, the normalized mode shape functions in the longitudinal and lateral directions are:

$$\psi_m = \sqrt{\frac{2}{L_1}}\sin\frac{m\pi x}{L_1}, \quad \phi_n = \sqrt{\frac{2}{L_2}}\sin\frac{n\pi x}{L_2}$$

With these mode-shape functions, the maximum values of the mode shape for each mode is

$$\psi_{max} = \sqrt{2/L_1}, \quad \phi_{max} = \sqrt{2/L_2}$$

Note that the maximum rms vibration amplitudes for the four modes occur at different points on the panel. If a finite-element computer program is to be used to calculate the natural frequencies, then most likely the mode-shape normalization convention is different from Equation (3.30) above. In that case we must re-normalize the mode shape according to Equation (3.30) so that we can use the acceptance values in the Appendix. With this mode-shape normalization,

$$(A\psi_\alpha^2)_{max} = A(\psi_m\phi_n)^2 = A\frac{2}{L_1 L_2} = 2.0$$, and Equation (8.50) can be simplified to:

$$<y_\alpha^2(\vec{x})> = \frac{2G_p(f_\alpha)\vec{J}_{\alpha\alpha}(f_\alpha)}{64\pi^3 m_\alpha^2 f_\alpha^3 \zeta_\alpha} \qquad \alpha=(m,n)=(1,1),\ (2,1),\ (1,2),\ (2,2)$$

and the generalized mass for all the modes is just the surface mass density:

$$m_{mn} = \rho_m t = 9.159\text{E-5 (lb-s}^2\text{/in)} = 25.00 \text{ Kg/m}^2, \text{ all modes}$$

The Natural Frequencies

One can use a finite-element computer program to calculate the natural frequencies. For simply-supported, rectangular plates with uniform surface density, handbook formulas to calculate the frequencies are available. From Blevins (1979), Table 11-4, the natural frequencies are given by,

$$f_{mn} = \frac{\lambda_{mn}^2}{2\pi L_1^2} \left[\frac{Et^3}{12\rho_m t(1 - \sigma^2)} \right]^{1/2}$$

$$\lambda_{mn}^2 = \pi^2 \left[m^2 + n^2(\frac{L_1}{L_2})^2 \right]$$

Using these equations, the natural frequencies for the first four modes are found to be:

m	n	f_{mn} (Hz)
1	1	119
2	1	230
1	2	367
2	2	478

The natural frequencies, at least for the first four modes, are reasonably well separated. This is the justification for using Equation (8.50) without the cross-terms. We need only the joint acceptances, which are always real.

Displacement Boundary Layer Thickness and Δ*

We shall find the turbulence pressure PSD from Figure 8.15. For that we must know the displacement boundary layer thickness, which depends on the Reynolds number. Since this is an internal flow problem, we take the characteristic length as the half channel height, h (see Section 8.9), from which we get

$$\text{Re} = hV / v = 9.6E + 5$$

At this Reynolds number, from Figure 8.13, n=9.5. From Equation (8.62), the displacement boundary layer thickness is

$$\delta^* = \frac{h}{(n+1)} = 0.857 \text{ in } = 0.022 \text{ m}$$

$$\Delta_1^* = \delta^* / L_1 = 0.048, \qquad \Delta_2^* = \delta^* / L_2 = 0.072$$

The last two parameters are necessary to read off the acceptance values from charts in Appendix 8A.

The Convective Velocity

Since we are going to use the acceptance integrals given in the Appendix, which are computed with Bull's (1967) representation of the coherence function, we also use Bull's empirical formula for the convective velocity:

$$U_c = V(0.59 + 0.30e^{-0.89(2\pi f)\delta*/V}) \tag{8.57}$$

The convective velocities at the modal frequencies are given, in both US and SI units, in the second and third columns of Table 8.2a.

The Dimensionless Frequencies

With U_c known, the two dimensionless parameters, $4fL_1/U_c$, $4fL_2/U_c$, can be computed as a function of the modal frequencies. These are given in columns 4 and 5 in Table 8.2a. With the displacement boundary layer $\delta*$ computed, we now proceed to compute the dimensionless frequency $\omega\delta*/V$ which is necessary to calculate the pressure PSD. These are given in column 8 of Table 8.2a

Table 8.2a: U_c, Frequency Parameters and Joint Acceptances

(mode),(Hz)	U_c (in/s)	U_c (m/s)	$4fL_1/U_c$	$4fL_2/U_c$	J_{mm}	J'_{nn}	$\omega\delta*/V$
(1,1) , 119	1963	50.4	4.33	2.89	0.132	0.36	0.266
(2,1) , 230	1871	47.5	8.83	5.89	0.03	0.25	0.516
(1,2) , 367	1762	44.7	15.0	10.0	0.01	0.135	0.825
(2,2) , 478	1693	43.0	20.3	13.5	0.01	0.125	1.072

Table 8.2b: PSD and rms Responses

(mode),(Hz)	Normalized PSD	G_p (psi²/Hz)	G_p (Pa²/Hz)	y_{rms} (in)	y_{rms} (m)
(1,1) , 119	3.00E-5	2.918E-8	1.391	1.34E-4	3.54E-6
(2,1) , 230	2.50E-5	2.432E-8	1.159	1.90E-5	4.81E-7
(1,2) , 367	1.50E-5	1.459E-8	0.695	3.09E-6	8.11E-8
(2,2) , 478	1.00E-5	9.727E-9	0.464	1.64E-6	4.15E-8

The Joint Acceptances

We can proceed to calculate the longitudinal and lateral joint acceptances by numerical double integration, using either Bull's representation of the coherence function given in the Appendix, or the simpler expressions given in Section 8.5. Or we can use the charts in Appendix 8A. Here some engineering judgment must be made. The acceptances given in the Appendix are for a rectangular panel with all four edges clamped. Examination of Figure 8.5, however, shows that the differences in the joint acceptances between a rectangular panel with all four edges clamped or simply-supported are small, especially away from the "coincidence" peaks. Therefore, we can use the charts in the Appendix even though the panel in this example is simply-supported. With $\Delta_1^* = 0.048, \Delta_2^* = 0.072$ and the normalized frequencies computed above, we can just read off the longitudinal and lateral joint acceptances from the charts. These are given in columns 6 and 7 of Table 8.2a. Because the modal frequencies are well separated, we can ignore the cross-terms. No cross-acceptances are needed in the calculation.

The Random Pressure PSD

Knowing $\omega\delta^*/V$, the normalized fluctuating pressure PSD can now be read off from Figure 8.15. These are given in the second column of Table 8.2b. Finally, the random pressure PSD is obtained from the normalized PSD from Equation (8.67),

$$G_p(f) = 2\pi\rho^2 V^3 \delta^* \{normalized\ PSD)$$

These are given, in both US and SI units, in columns 3 and 4 of Table 8.2b.

The rms Responses

With all the variables in Equation (8.50) computed, we can now compute the mean square response for each mode, from which the modal rms modal responses can be computed. These are given, in both US and SI units, in the last two columns of Table 8.2b. Note that these maximum vibration amplitudes occur at different points on the panel. The following should be noted:

- The joint acceptances generally are small.
- The longitudinal joint acceptances are very small at high frequencies.
- The mean square vibration amplitude decreases rapidly at higher modes, which justifies using the normal mode analysis method.

To find the total rms response, we must add the mean square responses *at the same point* on the panel from all the modes, then take the square root of the sum.

Example 8.2 and Case Study

In a pressurized water reactor, the core is suspended by a flange at the top (Figure 2.20). This flange provides the core with the necessary clamping force and stiffness. If this clamping force is maintained so that the system stiffness and the fundamental modal frequency remain sufficiently high as designed, turbulence-induced vibration due to the coolant flow in the downcomer annulus surrounding the reactor core would not be large enough to cause excessive stresses in the connecting bolts. However, if due to thermal expansion, material degradation or other reasons, the clamping force is reduced, resulting in a loss of system stiffness and a corresponding lowering of the fundamental modal frequency, turbulence-induced vibration may increase by a factor much larger than that caused by the decrease in system stiffness alone.

Such an example occurred in the early 1970s. During a pre-operational test, the vibration amplitude of the core of a pressurized water reactor was measured to be excessively high. Apparently the clamping force was not high enough to provide the core with enough stiffness, resulting in a decrease of the fundamental beam mode frequency. Closely related problems again occurred in the early 1980s. Bolts attaching thermal shields coaxially to the cores of pressurized water nuclear reactors were found to be broken in all three different pressurized water reactor designs in the United States (Sweeney and Fry, 1986). Extensive vibration analysis in the latter case showed that if the material of the connecting bolts had maintained its properties, neither turbulence nor coolant pump acoustically-induced vibration could have fatigued the bolts. It was concluded that the material in the bolts had degraded over time, resulting in loss of clamping forces and, as a result, decreases in the fundamental beam mode frequencies of the thermal shields. Once initiated, this process became self-propagating until the bolts eventually failed by fatigue. The decreases in the fundamental beam mode frequency were detected by ex-core vibration detectors based on measuring the energy spectra of the neutrons escaping from the core of the nuclear reactor. These bolts were replaced with new designs made of different kinds of materials. The problems do not recur after another 15 years of operations.

The above two problems are very similar in nature. In the following, we carry out a qualitative study based on the reactor core model in Example 4.3 (Chapter 4). This example shows that very often, we can gain insight into a problem without carrying out the detailed quantitative numerical calculations. Since this is a semi-quantitative analysis, only simple calculations in US unit are shown.

Consider the pressurized water reactor in Example 4.3. We found before that the natural frequency of the pendulous mode of the core is 21.9 Hz. The length of the core is 300 in. (7.62 m), its diameter is 157 in. (3.988 m) and the annular gap width is 10 in. (0.254 m), as shown in Example 4.3. The reactor core is subjected to turbulence-induced excitation due to the coolant flow in the downcomer annulus (Figure 8.20). Assuming the mean stream velocity of the coolant flow is 20 ft/s, we proceed to qualitatively assess the increase in turbulence-induced vibration amplitude if, because of thermal expansion, the clamping force is reduced and the fundamental pendulous modal frequency decreases

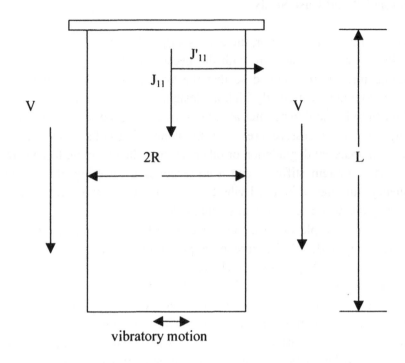

Figure 8.20: Turbulence-induced vibration of nuclear reactor core

to 2.0 Hz.   Here we take the hydraulic radius of the flow channel as equal to the annular gap width, $R_H = g = 10$ in. (0.254 m). At 21.9 Hz,

$F = f R_H / V = 0.91$ and $\omega R_H / V = 5.7$

For annular flow without a well-defined boundary layer, from Equation (8.58),

$U_c = V = 240$ in/s

and the normalized PSD is, from Equation (8.70),

$$\frac{G_p(f)}{\rho^2 V^3 R_H} = 0.155 e^{-3*0.91} = 0.01$$

At 2.0 Hz, $F = 2*10/240 = 0.083$,

The normalized PSD is, from Equation (8.70) again,

$$\frac{G_p(f)}{\rho^2 V^3 R_H} = 0.155 e^{-3*0.083} = 0.12$$

There is a fourteen-fold increase in the excitation energy because the system is now subject to the turbulence spectrum at a lower frequency. Next, let us look at the acceptance integrals. The correlation length is, from Equation (8.65),

$\lambda = 0.4 R_H = 4$ in

Since we are interested only in the fundamental beam mode,

$$\vec{J}_{11,11} = J_{11} J'_{11} \qquad \text{(See Figure 8.20)}$$

and there are no cross-terms involved. We shall omit the indices from now on, with the understanding that we are dealing with only the 1,1 mode. It is shown in the following chapter that the lateral joint acceptance is

$$J' \approx 2\lambda_1 / \pi R = 0.032$$

This does not change with the frequency. From the discussions in Section 8.4, the joint acceptance is always real. Thus, the integral involves only the real part of the streamwise coherence function. From Equation (8.39), the real part of the streamwise coherence function is:

$$\text{Re}\,\Gamma_1 = e^{-|x'-x''|/\lambda_1} \cos\frac{2\pi f(x'-x'')}{U_c}$$

At 21.9 Hz, $U_c = 240$ in/s, the "pressure wavelength" is

$$\ell = U_c / f = 11 \text{ in}$$

There are 300/11 = 27.38 complete pressure waves over the length of the reactor core. That is, the coherence function reverse signs 27.38 times over the length of the reactor core. Because of this frequent sign reversal, the net contribution to the joint acceptance is very small, in the order

$$\approx const * 0.38 * 11 / 300 = 0.014 * const$$

At 2.0 Hz, $U_c = 240$ in/s, the "pressure wave length" is

$$\ell = U_c / f = 120 \text{ in}$$

There are only 2.5 pressure waves over the length of the core. This leaves a much larger residual contribution to the acceptance integral, in the order of,

$$\approx const * 0.5 * 120 / 300 = 0.2 * const$$

The decrease in the fundamental pendulous frequency has resulted in a fourteen-fold increase in the joint acceptance. Finally, from Equation (8.4), the mean square response is proportional to f^{-3}. This gives another factor of $(21.9/2)^3 = 1,313$ to the mean square response. To summarize, as a result of loosing the clamping force, the mean square response goes up by a factor of approximately $14*14*1,313 = 257,000$ times, or a rms response of 500 times. Of this, the smaller system stiffness accounts for a factor of 36. It was this huge increase in vibration amplitude that rapidly failed the connecting bolts by fatigue.

The above example shows that very often, some insight into turbulence-induced vibration can be obtained without carrying out lengthy numerical analyses.

Appendix 8A: Charts of Acceptance Integrals

Bull's Representation of the Coherence Function

Equations (8.39) and (8.40) are the simplest but by no means the only empirical equations proposed for the coherence functions. Based on turbulent boundary layer flow over flat surfaces, Bull (1967) suggested the following empirical equations for the streamwise and cross-stream coherence functions.

$$\Gamma_1(x_1', x_1'', f) = e^{-2\pi f \alpha_1 |x_1' - x_1''|/U_c} e^{-i2\pi f(x_1' - x_1'')/U_c} \quad \text{if} \quad 2\pi f \Delta_1^* \geq 0.37$$

$$= e^{-\alpha_2 |x_1' - x_1''|/\delta^*} e^{-i2\pi f(x_1' - x_1'')/U_c} \quad \text{if} \quad 2\pi f \Delta_1^* < 0.37 \tag{8.72}$$

$$\Gamma_2(x_2', x_2'', f) = e^{-2\pi f \alpha_3 |x_2' - x_2''|/U_c}$$

$$\text{if} \quad \left| \frac{x'_2 - x''_2}{L_2} \right| \geq -\Delta_2^* \left[9.1 \log(\frac{2\pi f \delta^*}{U_c}) + 5.45 \right] \tag{8.73}$$

$$= c + de^{-\alpha_4 |x'_2 - x''_2|/\delta^*}$$

$$\text{if} \quad \left| \frac{x'_2 - x''_2}{L_2} \right| < -\Delta_2^* \left[9.1 \log(\frac{2\pi f \delta^*}{U_c}) + 5.45 \right]$$

where $\alpha_1 = 0.1,$ $\alpha_2 = 0.037,$ $\alpha_3 = 0.715,$ $\alpha_4 = 0.547$

$c = 0.28,$ $d = 0.72,$ $\Delta_1^* = \delta^* / L_1,$ $\Delta_2^* = \delta^* / L_2$

with the convective velocity given by,

$$U_c = V(0.59 + 0.30e^{-0.89(2\pi f)\delta^*/V})$$

and δ^* is the displacement boundary layer thickness discussed before.

In this Appendix, charts of acceptance integrals for a rectangular flat plate with uniform surface density and with all four edges clamped are given. Since the acceptance is dimensionless and the integrals were evaluated based on non-dimensional parameters, these charts are applicable to rectangular panels subject to boundary layer type of turbulent excitation. It is assumed that the flow is in the longitudinal direction of the panel. The acceptances were reproduced from Chyu and Au-Yang (1972), which gives the acceptances up to the (7,7) mode. For brevity, acceptances up to only the (3,3) mode are reproduced in this Appendix. These values were obtained by numerical double integration of the acceptance integrals based on Bull's correlation, Equations (8.72) and (8.73), and with the mode shape normalized to unity according to Equation (3.30). Due to the symmetry of the mode shape functions,

J'_{nn} is always real
Re $J_{12} = 0$,   Re $J_{23} = 0$
Im $J_{13} = 0$

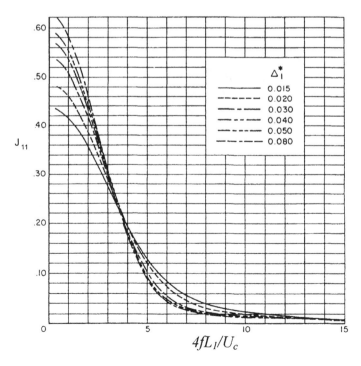

Figure 8.21 Longitudinal joint acceptance J_{11}

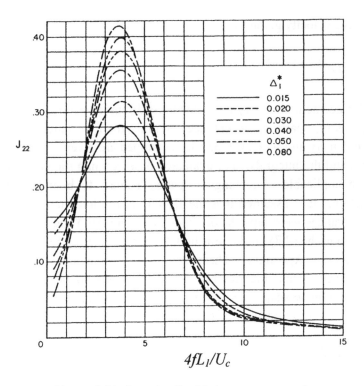

Figure 8.22 Longitudinal joint acceptance J_{22}

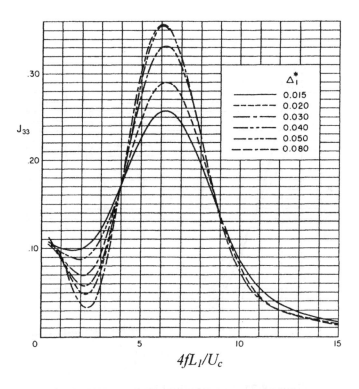

Figure 8.23 Longitudinal joint acceptance J_{33}

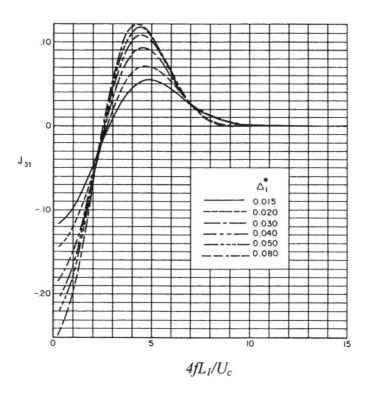

Figure 8.24 Real part of longitudinal cross-acceptance J_{31}

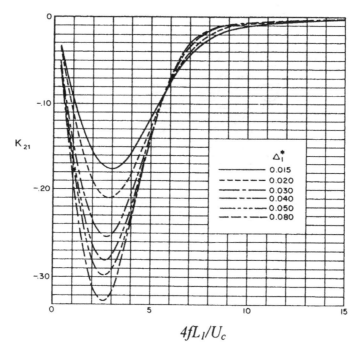

Figure 8.25 Imaginary part of longitudinal cross-acceptance J_{21}

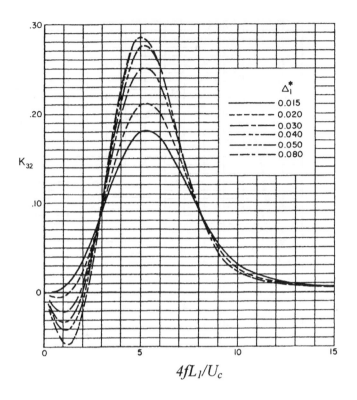

Figure 8.26 Imaginary part of longitudinal cross-acceptance J_{32}

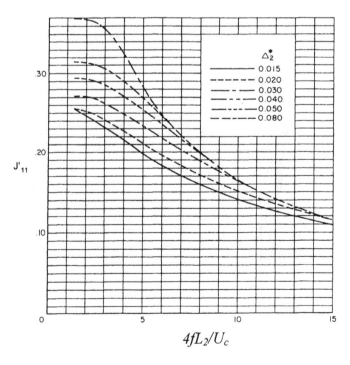

Figure 8.27 Lateral joint acceptance J'_{11}

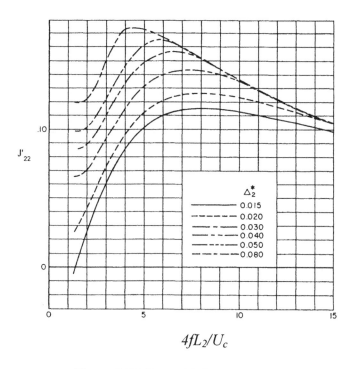

Figure 8.28 Lateral joint acceptance J'_{22}

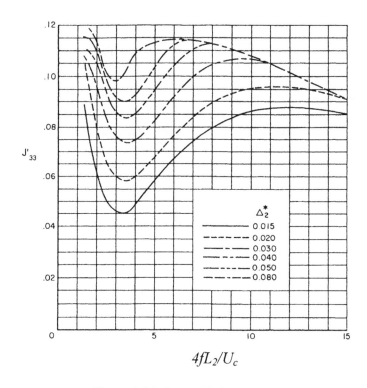

Figure 8.29 Lateral joint acceptance J'_{33}

References

Au-Yang, M. K., 1975, "Response of Reactor Internals to Fluctuating Pressure Forces," Journal Nuclear Engineering and Design, Vol. 35, pp. 361-375.

Au-Yang, M. K. and Connelly, W. H., 1977, "A Computerized Method for Flow-Induced Random Vibration Analysis of Reactor Internals," Journal Nuclear Engineering and Design, Vol. 42, pp. 257-263.

Au-Yang, M. K. and Jordan, K. B., 1980, "Dynamic Pressure Inside a PWR—A Study Based on Laboratory and Field Test Data," Journal Nuclear Engineering and Design, Vol. 58, pp. 113-125.

Au-Yang, M. K., 1986, "Turbulent Buffeting of a Multi-Span Tube Bundle," ASME Transaction, Journal Vibration, Stress, Reliability in Design, Vol.108, pp. 150-154.

Au-Yang, M. K., Brenneman, B. and Raj, D., 1995, "Flow-Induced Vibration Test of an Advanced Water Reactor Model, Part I--Turbulence Induced Forcing Function," Journal Nuclear Engineering and Design, Vol. 157, pp. 93-109.

Au-Yang, M. K., 2000, "The Joint and Cross-Acceptances in Cross-Flow Induced Vibration, Part I—Theory and Part II—Charts and Applications," ASME Transaction, Journal of Pressure Vessel Technology, Vol. 122, pp. 349-361.

Blevins, R. D., 1979, Formulas for Natural Frequencies and Mode Shape, Van Nostrand Reinhold, New York.

Blevins, R. D., 1990, Flow-Induced Vibration, Second Edition, Van Nostrand Reinhold, New York.

Bendat, J. S. and Piersol, A. G., 1971, Random Data: Analysis and Measurement Procedures, Wiley-Interscience, New York.

Bull, M. K., 1967, "Wall-Pressure Associated with Subsonic Turbulent Boundary Layer Flow," Journal of Fluid Mechanics, Vol. 28, part 4, pp. 719-754.

Chen, S. S. and Wambsganss, M. W., 1970, "Response of a Flexible Rod to Near Field Flow Noise," in Proceeding of Conference on Flow-Induced Vibration in Reactor System Components, Argonne National Laboratory Report ANL-7685, pp. 5-31.

Chen, S. S. and Wambsganss, M. W., 1972, "Parallel Flow-Induced Vibration of Nuclear Fuel Rods," Journal Nuclear Engineering and Design, Vol.18, pp. 253-278.

Chen, S. S., 1985, Flow-Induced Vibration of Circular Cylindrical Structures, Report No. ANL-85-51, Argonne National Laboratory.

Chyu, W. J. and Au-Yang, M. K., 1972, Random Response of Rectangular Panels to the Pressure Field Beneath a Turbulent Boundary Layer in Subsonic Flow, NASA TN D-6970.

Duncan, W. J., Thom, A. S. and Young, A. D., 1960, An Elementary Treatise on the Mechanics of Fluids, Arnold, London.

Hurty, W. C. and Rubinstein, M. F., 1964, Dynamics of Structures, Prentice Hall, Englewood Cliffs.

Powell, A., 1958, "On the Fatigue Failure of Structures Due to Vibrations Excited by Random Pressure Fields," Journal Acoustical Society of America, Vol. 30 No. 12, pp. 130-1135.

Sweeney, F. J. and Fry, D. N., 1986, "Thermal Shield Support Degradation in Pressurized Water Reactors, in Flow-Induced Vibration-1986, ASME Special Publication PVP-Vol.104, Edited by S. S. Chen, pp. 59-66.

Streeter, V. L., 1966, Fluid Mechanics, Fourth Edition, McGraw Hill, New York.

CHAPTER 9

TURBULENCE-INDUCED VIBRATION IN CROSS-FLOW

Summary

The response of one-dimensional structures such as beams and tubes to cross-flow turbulence has important applications to heat exchanger design, operation and maintenance. In cross-flows, because the front of the pressure wave reaches all points on the structure at the same time, the forces along the length of the structure are always in phase, contrary to the case of parallel flow-induced vibration. This greatly simplifies the acceptance integral method of turbulence-induced vibration analysis discussed in the previous chapter. By making the additional assumption that the random pressure is fully coherent across the width of the structure, the equation for the mean square response of the structure derived in the last chapter is simplified to:

$$< y^2 >= \sum_n \frac{L G_F(f_n)\phi_n^{\,2}(x)}{64\pi^3 m_n^{\,2} f_n^{\,3}\zeta_n} J_{nn} \quad + \quad cross\text{-}terms \tag{9.4}$$

Here $G_F = D^2 G_p$ is the random force PSD, expressed in (force/length)^2/Hz and J_{nn} are the joint acceptance integrals in the axial (cross-stream) direction. Because of the absence of the phase angle in the coherence function, J_{nn} are much easier to compute in cross-flow over 1D structures compared with those for parallel flows discussed in the previous chapter. Using the finite-element technique, the joint and cross-acceptances for cross-flow over beams and tubes with arbitrary boundary conditions are computed as a function of λ/L and are given as design charts in Appendix 9B. In particular, it is shown that as long as the correlation length is small compared with the half flexural wavelength of the structure, the following relationship:

$$J_{nn} = 2\lambda / L \qquad as \ n\pi\lambda / L \rightarrow 0 \tag{9.14}$$

is true irrespective of the boundary conditions of the beam.

Knowing the correlation length λ and the random pressure power spectral density G_F, Equations (9.4) and (9.14) can be used to calculate the responses of a single-span beam or tube with uniform linear mass density. For a beam with many spans and with non-uniform linear mass density, the acceptance integral method is generalized to

$$< y^2 >= \sum_n \frac{\phi_n^{\,2}(x)}{64\pi^3 m_n^{\,2} f_n^{\,3}\zeta_n} \sum_i \ell_i G_F^{(i)}(f_n) j_{nn}^{(i)} + cross \ terms \tag{9.34}$$

where

$$G_F^{(i)} = \int_0^{\ell i} G_F(x)\phi_n^2(x)dx \tag{9.35}$$

is the "generalized random force PSD" for the i-th span. In Equation (9.34), the summation is over all the modes n; within each mode, the summation is over all the spans i that are subjected to cross-flow. If we can further assume that within each span of the multi-span beam, the linear density is approximately constant, then the span acceptance $j_{nn}^{(i)}$ takes the same form as the single-span acceptance integrals given in the charts in Appendix 9B.

As mentioned in Chapter 8, the cross-terms contributions to the mean square response, Equation (9.34), can be ignored only if the normal modes are well separated in the frequency domain. Since this is usually not the case in a beam or tube with multiple supports, cross-modal contributions to the total mean square response should be included in the analysis. The span joint and cross-acceptance integrals can be used to assess the cross-modal contributions to the total response, as is discussed in Section 9.7.

The present state-of-the-art in turbulence-induced vibration analysis requires that the parameters describing the forcing function: the correlation length, λ, and the random force PSD, G_F, be measured experimentally. Based on data from several experiments, it is estimated that in a tube bundle, the correlation length is around 0.2 times the hydraulic (or effective) diameter,

$$\lambda \approx 0.2P(1 + P/2D) \tag{9.18}$$

for a tube bundle with pitch P and tube diameter D. Upper bound values of three to four times the tube diameter have been mentioned by several authors; the latter is more representative of the correlation length for cross-flow over a single tube.

The simplest formula for the random force PSD is given by Pettigrew and Gorman (1981):

$$G_F = \{C_R D(\frac{1}{2}\rho V_p^2)\}^2 = D^2 G_p \tag{9.20}$$

where the random lift coefficient, C_R, is given as function of frequency in Figure 9.4, or approximately by the equation:

$$\begin{aligned} C_R &= 0.025 \quad s^{1/2} & 0 < f < 40 \; Hz \\ &= 0.108*10^{-0.0159f} & f \geq 40 \; Hz \end{aligned} \tag{9.21}$$

As defined in Equation (9.20), the random lift coefficient has the dimension of time while the random pressure power spectral density G_F has the dimension of (force/length)^2/Hz. The following alternate empirical equation, which better fits the more recent test data, is suggested for estimating the normalized random pressure PSD:

$$\begin{aligned}
\overline{G}_p &= 0.01 \quad for \quad F < 0.1 \\
&= 0.2 \quad for \quad 0.1 \le F \le 0.4 \\
&= 5.3E - 4 / F^{7/2} \quad for \quad f > 0.4
\end{aligned} \qquad (9.25)$$

In a heat exchanger, the fluid very often is a two-phase mixture. This greatly complicates the forcing function and the energy dissipation mechanism of the vibrating structure. Two-phase flow-induced vibration is very much a developing science. Several tests pointed to the evidence that two-phase flows with void fractions between 60% and 85% add 2% to 3% critical damping to the structure.

Random pressure PSDs for two-phase flow are even scarcer. Of the small amount of data available, evidence shows that two-phase forcing functions are smaller compared with the corresponding single-phase spectra with the same dynamic pressure head. This, together with the higher damping ratios in two-phase flows, indicates that turbulence-induced vibration is probably less severe in two-phase regions.

Acronyms

1D   One dimensional
PSD   Power Spectral Density

Nomenclature

C_{ij} Element of the coherence matrix
C_R Random lift coefficient
d_B Characteristic void length
D Outside diameter of tube
f Frequency, in Hz
F Dimensionless frequency, defined by Equation (9.23) in single-phase flow and Equation (9.42) in two-phase flow
G_F Single-sided random force PSD, in (force/unit length)^2/Hz, $= D^2 G_p$
$G_F^{(i)}$ Generalized random force PSD for the i-th span
G_p Single-sided random pressure PSD, in (force/unit area)^2/Hz
$\overline{G}_p$ Normalized random pressure PSD (dimensionless)
$j_{nn}^{(i)}$ Joint acceptance for the i-th span
J'_{mm} Joint acceptance in the direction orthogonal to tube or beam axis

J_{ns} Acceptance in the direction of the tube or beam axis

L ·   Length of tube or beam

L_e Length of tube subject to cross-flow

ℓ_c $=2\lambda$, alternative definition of correlation length

ℓ_i Length of one span

m_n Generalized mass

P Pitch of a tube bundle

U_c Convective velocity

U_{phase} Phase velocity

v_f Specific volume of liquid phase

v_g Specific volume of vapor phase

v_{fg} $v_g - v_f$

V Cross-flow velocity

V_p Pitch velocity

x Void fraction, or position coordinate along the axis of the tube or beam

Δx Finite-element length

$\langle y^2 \rangle$ Mean square vibration amplitude

α Void fraction

δ Dirac delta function

θ Angle between velocity vector and axis of tube or beam

ρ Density of fluid

ρ_f Mass density of liquid phase

ρ_g Mass density of vapor phase

ρ_{fg} $=1/v_{fg}$

ϕ Mode-shape function along the axis of the tube or beam

$\bar{\phi}$ Mode-shape function normalized to one span

λ Correlation length, cross-flow direction

λ_2 Correlation length, cross-stream direction

Φ Adjusted two-phase PSD, defined by Equation (9.41)

Γ_1 Coherence function in the streamwise direction

Γ_2 Coherence function in the cross-stream direction

ω Frequency, in radian/s

ζ Damping ratio

Indices

m, r Modal index, perpendicular to axis of tube or beam

n, s Modal indices, along axis of tube or beam

9.1   Introduction

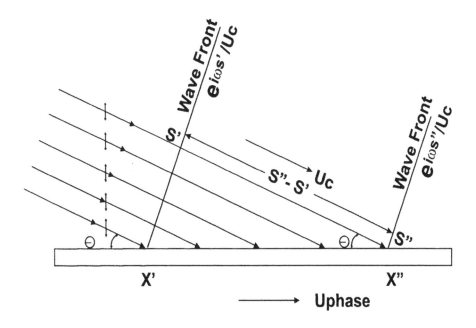

Figure 9.1 Cross-flow over a 1D structure

In Chapter 8, turbulence-induced vibration due to parallel flow over the surfaces of structures is considered. In this chapter, we shall consider turbulence-induced vibration due to cross-flow over a one-dimensional (1D) structure such as a tube or a beam, as shown in Figure 9.1 (also compare with Figure 8.2). In the most general case, the flow impinges upon the axis of the structure obliquely at an angle θ. The wave front propagates along the length of the structure at a velocity equal to

$$U_{phase} = U_c / \cos\theta$$

Points along the axis of the structure thus respond to the same wave front of the forcing function at different times. In other words, there are phase differences between the forcing function at different points along the structure. Now consider the special case when the flow is perpendicular to the axis of the structure. In this case, $\theta = 90$ deg. and U_c is infinite. The wave fronts arrive at all the points along the structure at the same time; the structure therefore responds to a forcing function that is always in-phase. We recall that this is the situation for points along the cross-stream direction in the previous chapter, except that there, the cross-stream direction corresponds to the lateral acceptance

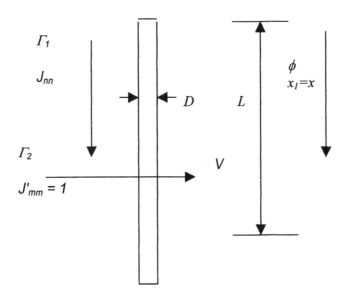

Figure 9.2 Notations for cross-flow over 1D structures

integral (in the x_2 direction). In this chapter, since the structure under consideration is one-dimensional, we take x_1 as the direction along the structure (and perpendicular to the direction of flow). When the flow is inclined to the structural axis, we just resolve the velocity into two directions: a cross-flow component that is considered in this chapter; and an axial flow component that will be considered in the next chapter. The acceptance integral method of cross-flow turbulence-induced vibration analysis discussed in this chapter closely follows the non-mandatory ASME Boiler Code, Sec. III Appendix N-1300 (ASME, 1998).

Using the same logic as in Chapter 8, we establish the corresponding notations for cross-flow-induced vibration over 1D structures as shown in Figure 9.2:

- The axis of structure is in the x_1 direction, which is the cross-stream direction.
- The mode-shape function in the x_1 direction will be denoted as ϕ_n.
- The streamwise direction is x_2, which is the lateral direction of the structure.
- The longitudinal acceptance is along the structural axis and is in the cross-stream direction, and will be denoted by J_{ns}.

Finally, we make a simplifying assumption that the dimension of the structure in the x_2 direction is small compared with the correlation length of the forcing function, so that the latter is 100% coherent and in-phase across the width of the structure. From Equations (8.39) and (8.40),

$$\Gamma_1(x_1',x_1'',f) = e^{-|x_1'-x_1''|/\lambda}$$ (9.1)

$$\Gamma_2(x_2',x_2'',f) = e^{-|x_2'-x_2''|/\lambda_2} = 1$$ (9.2)

From Equations (8.42) to (8.44),

$$J_{ns} = \frac{1}{L}\iint_L \phi_n(x_1')e^{-|x_1'-x_1''|/\lambda}\phi_s(x_1'')dx_1' dx_1''$$ (9.3)

and

$$J'_{mr} = 1$$

Based on these assumptions, for the special case of cross-flow over a 1D structure (we shall suppress the subscript 1 from now on), equation (8.50) simplifies to:

$$<y^2> = \sum_n \frac{LG_F(f_n)\phi_n^2(x)}{64\pi^3 m_n^2 f_n^3 \zeta_n} J_{nn} \quad + \quad cross\text{-}terms$$ (9.4)

Here $G_F = D^2 G_p$ is the random force PSD, expressed in (force/length)^2/Hz, and J_{nn} is the longitudinal joint acceptance. The readers are cautioned that G_F is often called the random pressure PSD in the literature dealing with heat exchanger tube dynamics. This term is reserved for G_p in this book. From the discussions in Section 8.8, it also follows here that if we normalize the mode-shape function to unity, as in Equation (3.30), which is possible only if the beam is of uniform linear mass density, then

$$\sum_n J_{nn} = 1.0$$ (9.5)

9.2   Exact Solution for the Joint Acceptance Integrals

With the greatly simplified expression for the coherence function, we can proceed to derive an expression for the acceptance integral using Equation (9.3), for the simplest possible 1D structure—that of an uniform beam spring-supported at both ends. In this case, one can normalize the mode shape according to Equation (3.30),

$$\int_0^L \phi_n(x)\phi_r(x)dx = \delta_{ns}$$ (9.6)

The mode-shape functions that satisfies Equation (9.6) are

$$\phi_1 = 1/\sqrt{L}$$

$$\phi_2 = \sqrt{\frac{12}{L}}(x - L/2) \qquad (9.7)$$

Substituting into Equation (9.3), we get

$$J_{11} = \frac{1}{L}\int_0^L \frac{1}{\sqrt{L}} e^{-|x'-x''|/\lambda} \frac{1}{\sqrt{L}} dx' \, dx''$$

By separating the region of integration into two—one for $x' > x''$ and the other for $x' < x''$, as shown in Figure 8.5, we get by straightforward integration:

$$J_{11} = \frac{2\lambda}{L}\left[1 - \frac{\lambda}{L}(1 - e^{-L/\lambda})\right] \qquad (9.8)$$

When the correlation length λ is small (e.g., 1/10 times the length of the beam), the above equation reduces to the well-known equation,

$$J = 2\lambda/L, \quad \lambda \ll L \qquad (9.9)$$

At this point, we have derived Equation (9.9) only for a rigid beam spring-supported at both ends. For applications to industrial systems, a much more common case is that of a beam simply-supported at both ends. If the linear mass density is uniform, as in the case of the above spring-supported beam, again we can normalized the mode shape according to Equation (9.6), and the mode-shape functions are:

$$\phi_n = \sqrt{\frac{2}{L}}\sin\frac{n\pi x}{L} \qquad (9.10)$$

The derivation of the joint acceptance is far more tedious. In Appendix 9A, it is shown that, for a simply-supported beam,

$$J_{nn} = \frac{2(\lambda/L)^2}{1+(n\pi)^2(\lambda/L)^2}\left[\frac{2(n\pi)^2(\lambda/L)^2\{(-1)^{n+1}e^{-L/\lambda}+1\}}{1+(n\pi)^2(\lambda/L)^2} + \frac{L}{\lambda}\right] \qquad (9.11)$$

It can be readily derived from Equation (9.11), that

$$J_{nn} = 8/(n\pi)^2 \qquad \text{for odd values of } n \text{ and } \lambda \gg L \qquad (9.12)$$

$$J_{nn} = 0 \qquad \text{for even values of } n \text{ and } \lambda \gg L \qquad (9.13)$$

and that

$$J_{nn} = 2\lambda / L \qquad \text{as } n\pi\lambda / L \to 0 \tag{9.14}$$

Summing Equation (9.12) over all n, we have,

$$\sum_n J_{nn} = \frac{8}{\pi^2}(1 + \frac{1}{3^2} + \frac{1}{5^2} + \frac{1}{7^2} + ...) = \frac{8}{\pi^2}(\frac{\pi^2}{8}) = 1.0$$

which is a special case of Equation (8.54). Notice that Equation (9.14) is valid only if the correlation length is small compared with each flexural wavelength of the structure. Thus, a simply-supported tube may have a length that is ten times larger than the correlation length of the forcing function over the length of the tube; Equation (9.14) cannot be used to evaluate, e.g., $J_{10,10}$. This is different from the usual small correlation length conclusion and has deep effects on the calculation of the response of a beam with many spans.

At this point, it is proper to point out once again that some authors (e.g., Blevins, 1990, equation 7-47) define the correlation length ℓ_c as,

$$\Gamma(x', x'', f) = e^{-2|x'-x''|/\ell_c}$$

That is,

$$\ell_c = 2\lambda \tag{9.15}$$

For an isolated tube, it was found that ℓ_c varies between 6 to 8 diameters of the tube. That is, λ varies between 3 to 4 diameters of the tube. As will be discussed later, in a tube bundle, λ is much closer to 1 tube, instead of 3 tube diameters. However, if the vortex shedding frequency is close to one of the natural frequencies of the tube, the correlation length can dramatically increase. Thus, the method developed in this chapter can be used to assess the lock-in vortex-induced vibration amplitude deep inside a tube bundle, as will be demonstrated in Examples 9.1 and 9.3.

With the above measured values of the correlation length, Equation (9.14) is widely used together with Equation (9.4), *with the cross-terms ignored*, to estimate the response of tubes subjected to cross-flow. However, the very restrictive assumptions under which these equations are derived must be remembered. In addition to the three assumptions listed in Section 8.6 under which Equation (8.51) and consequently Equation (9.4) is derived, Equation (9.14) is further restricted to cross-flows over tubes or beams when $n\pi\lambda$ is small compared with the length L of the structure.

9.3   Finite-Element Method of Evaluating the Acceptance Integrals

The acceptance integral, Equation (9.3), can be readily evaluated with the help of techniques developed in finite-element structural analysis. In the finite-element method, the structural mode-shape function is approximated by a number of discretized "elements." The higher the number of elements used to represent the structure, the more accurate the representation. Structural properties and external forces are integrated over each element first, then the structural responses are synthesized from the individual elements. Au-Yang and Connelly (1977) followed this method to estimate the response of reactor internal components to inhomogeneous turbulent flows. In their approach, the surface of the structure was discretized into elements and the forcing function was assumed to be constant over each of these elements. The acceptance integrals were then evaluated for each pair of these elements. Brenneman (1987) derived closed-form expressions for the integrated effect of the coherence function over each element, assuming that the mode shape was constant over each element. In the special case of cross-flow, these expressions, called coherence integrals by Brenneman, reduce to:

For $x_j > x_i$,

$$C_{ij} = 4\lambda^2 \sinh^2(\Delta x / 2\lambda) e^{-(xj - xi)/\lambda}$$

For $i=j$,

$$C_{ii} = 2\lambda^2 \{\Delta x / \lambda - 2\sinh(\Delta x / 2\lambda) e^{-\Delta x/2\lambda}\} \tag{9.16}$$

Brenneman proceeded to solve directly for the structural responses using these "coherence integrals," thus bypassing the intermediate steps of computing the acceptance integrals. Brenneman's approach is the most direct. However, it requires specialized computer programs to handle the large coherence matrices. The objective of this chapter is to present a method to calculate the acceptance integrals using standard commercial structural-analysis computer programs. Once calculated and tabulated, these acceptance integrals can be used to calculate the response of a wide variety of beam structures to random forces.

The acceptance integral for the entire beam can be obtained by summation over the contribution from all the elements. From Equations (9.3) and (9.16),

$$J_{ns} = \frac{1}{L} \Sigma_i \Sigma_j \phi_n(x_i) C_{ij} \phi_s(x_j) \tag{9.17}$$

where the mode shapes ϕ can be obtained from closed form solutions or numerically from a finite-element computer program.

To verify the accuracy of the finite-element solution to the acceptance integral using Equations (9.16) and (9.17), the joint acceptances for cross-flow-induced vibration in the

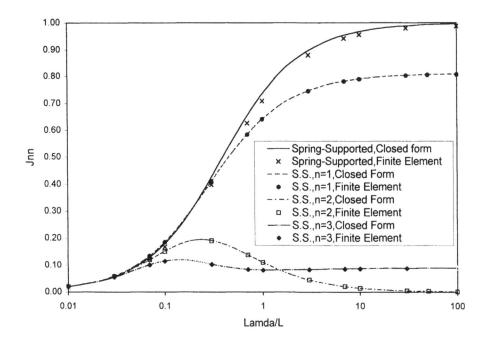

Figure 9.3 Comparison of Joint Acceptances Computed with Closed-Form
and Finite-Element Solutions (Au-Yang, 2000)

two special cases for which closed-form solutions are derived earlier are calculated by the
two different methods.

Case 1: Spring-Supported Rigid Beam with Uniform Mass Density

First, the joint acceptance J_{11} is computed with the closed-form Equation (9.8), then it is
numerically computed with a spreadsheet, using the finite-element Equations (9.16) and
(9.17), with 30 elements and for ratios of λ / L ranging from 0.01 to 100. The results are
plotted in Figure 9.3.

Case 2: Simply-Supported Beam with Uniform Mass Density

The joint acceptances J_{11}, J_{22}, J_{33} are first computed with the closed-form solution
Equation (9.11), and then with the finite-element Equations (9.10), (9.16) and (9.17),
again with 30 elements and using a spreadsheet. The results are also plotted in Figure
9.3.   It can be seen from Figure 9.3 that agreement between the finite-element and the
exact solutions are excellent in these two special cases.

Using the finite-element method, Equations (9.16), (9.17) together with mode-shape functions from a standard commercial finite-element structural analysis computer program, Au-Yang (2000) computed the joint and cross-acceptances for single-span beams of uniform linear mass density but with different boundary conditions. These are reproduced in Appendix 9B. These charts can be used together with Equation (9.4) to compute the responses of single-span beams with uniform linear mass density to turbulent cross-flow. Application of these single-span acceptance integrals to calculate the responses of multiple-supported beams will be discussed in a later section.

9.4   Correlation Length and Power Spectral Density for Single-Phase Cross-Flows

The response Equation (9.4) involves two experimentally determined parameters for defining the forcing function: the random pressure power spectral density, G_p, and the correlation length; the latter is necessary to evaluate the acceptance integral. This is very similar to the situation in parallel flow-induced vibration, except that in cross-flow, the convective velocity is not required. In Section 8.9, the random pressure PSD and correlation length for parallel flow are discussed. In this section, these functions for cross-flow will be given.

Correlation Length

Direct measurement of the correlation length λ would require the use of arrays of dynamics pressure transducers to measure the space-time coherence of the forcing function, as in the case of turbulent parallel flow discussed in Section 8.9. Unfortunately such data is rare. Very often the correlation length is "backed out" from the measured vibration amplitudes of, or the forces acting on the tubes instead. However, since tube vibration also depends on the random pressure PSD and the damping ratio, both of which also carry large experimental uncertainties, the determination of the correlation length based on the measured tube responses understandably carries a large margin of uncertainty. Based on very limited data, Blevins et al (1981) suggested a correlation length of approximately 3.0 tube diameters. On the other hand, Axisa et al (1988) suggested a correlation length of 4.0 tube diameters but stated that it was probably much less than that in a tube bundle. These suggested values are consistent with the correlation lengths for cross-flow over an isolated tube. For a tube inside a bundle, physical intuition suggests that the hydraulic diameter, or the effective diameter, should govern the correlation length of the tube bundle. It is suggested in Section 8.9 that based on direct measurement of the coherence of the random pressure in annular flows, the correlation length is about 0.2 times the hydraulic or effective tube diameter, or (See Equation (8.64)),

$$\lambda = 0.2P(1 + P/2D) \qquad (9.18)$$

Since most heat exchangers have P/D ratios between 1.3 and 1.5, application of the above equation shows that in a tightly packed tube bundle characteristic of what is in practical heat exchangers, the correlation length is approximately

$$\lambda \approx 0.5D \tag{9.19}$$

This also agrees with our physical intuition that the correlation length in cross-flow over tube bundles is comparable to the gap clearance between the tubes. Computed rms vibration amplitudes using this value of correlation length together with the single-phase PSD suggested by Pettigrew and Gorman (1981), discussed below, seem to give overall reasonable agreement with what are observed in the field.

The Random Force PSD for Single-Phase Cross-Flow

Based on a large number of data from laboratory and field tests, Pettigrew and Gorman (1981) suggested the following empirical equation for the random force PSD for turbulent single-phase cross-flow over tube bundles:

$$G_F = \{C_R D(\frac{1}{2}\rho V_p^2)\}^2 = D^2 G_p \tag{9.20}$$

where V_p, the pitch velocity, is the velocity of flow through the gaps between the tubes and C_R is what Pettigrew and Gorman called the random lift coefficient. However, since C_R has the unit of time, it is not a true aerodynamic coefficient. Note that G_F has the unit of (force per unit length)^2/Hz.

The random lift coefficient is not only a function of the frequency, but also depends on the tube locations. Figure 9.4, reproduced from the above-cited reference, shows typical values of the random lift coefficient over the range of frequencies commonly encountered in practical heat exchanger tubes. The following equation fits the upstream curve reasonably well,

$$\begin{aligned} C_R &= 0.025 \quad s^{1/2} & 0 < f < 40\ Hz \\ &= 0.108*10^{-0.0159f} & f \geq 40\ Hz \end{aligned} \tag{9.21}$$

For the downstream curve, the value is about half that given by Equation (9.21). Because of the limited frequency range based on which the curve was derived, the data should not be extrapolated beyond 200 Hz.

The more common way of presenting tube bundle random pressure data is to plot the normalized random pressure PSD against the dimensionless frequency, similar to Figures 8.17, 8.18 and 8.19 in Section 8.6, except it is more customary to define the normalized random pressure PSD as:

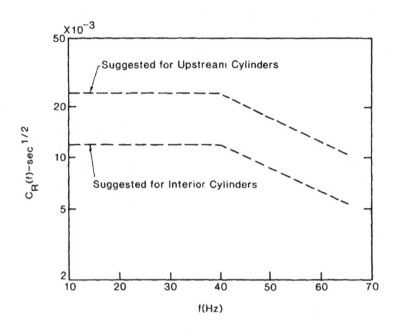

Figure 9.4 Random lift coefficient for cross-flow over tube bundles
(Pettigrew and Gorman, 1981)

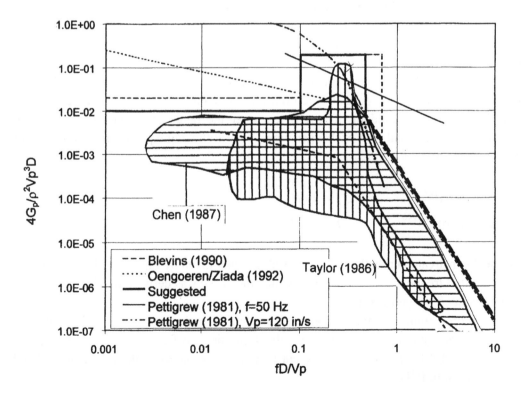

Figure 9.5 Suggested normalized random pressure PSD versus dimensionless frequency

$$\overline{G}_p = \frac{G_p}{(\frac{1}{2}\rho V_p^2)^2 (\frac{D}{V_p})} = \frac{4G_p}{\rho^2 V_p^3 D} \tag{9.22}$$

and the dimensionless frequency,

$$F = \frac{fD}{V_p} \tag{9.23}$$

From Equations (9.21) and (9.22), we get

$$C_R^2 = \frac{D}{V_p}\overline{G}_p \tag{9.24}$$

Based on data from Chen and Jendrizejczyk (1987) and Taylor et al (1986), Blevins (1990) suggested the following bound for the normalized PSD:

$$\overline{G}_p = 0.02 \quad for \quad 0 < F < 0.1$$
$$= 0.2 \quad for \quad 0.1 \le F \le 0.7$$

Beyond F=0.7, $\overline{G}_p$ decreases exponentially to $\overline{G}_p$=1.0E-7 at F=10.

Based on test data with tube bundles with P/D ratios of 1.95, 1.5 and 3.0, as well data from Chen and Jendrizejczyk (1987) and Taylor et al (1986), Oengoeren and Ziada (1992) suggested the following bound for the normalized random pressure PSD,

$$\overline{G}_p = 8.0E - 3/F^{1/2} \quad for \quad F < 0.4$$
$$= 5.3E - 4/F^{7/2} \quad for \quad F > 0.4$$

Blevins' (1990) suggested equation is probably slightly over conservative since there seems to be no test data lying above $\overline{G}_p$=0.01 for F<0.1. For F>0.4, Onegoeren and Ziada (1992) suggested formula bounds the test quite well. However, the "hump" between F=0.1 and F=0.4 represents possible vortex-shedding excitation and thus should be regarded as real. Based on all the above considerations, it is suggest that the following empirical equation be used as an upper bound for the normalized random pressure PSD in tube bundles,

$$\overline{G}_p = 0.01 \quad for \quad F < 0.1$$
$$= 0.2 \quad for \quad 0.1 \le F \le 0.4 \tag{9.25}$$
$$= 5.3E - 4/F^{7/2} \quad for \quad f > 0.4$$

Figure 9.5 plots the empirical equations suggested by the different authors against a backdrop of test data by Chen and Jendrzejczyk (1987) and Taylor et al (1986). For comparison purpose, the normalized random pressure PSD computed with Equation (9.20) together with the random lift coefficient from Figure 9.4, for the two special cases when (i) the tube frequency is fixed at 50 Hz, and (ii) when the pitch velocity is fixed at V_p=120 in/s (3 m/s) is also shown on the same plot. It is seen that Pettigrew and Gorman's (1981) empirical equation is generally conservative and almost bounds the vortex-shedding spectrum.

Vortex-Induced Response

As observed by Oengoeren and Ziada (1992), in the range between F=0.1 and F=0.4, there is evidence of "vorticity" activities. This may explain the "hump" in the normalized PSD in this range. Thus, Equation (9.4) together with Equation (9.25) can be used to estimate the vortex-induced response of a tube inside a bundle, when the dimensionless frequency (Strouhal number) is between F=0.1 and F=0.4. However, it must be realized that when vortex lock-in occurs, not only does the random pressure PSD increases, but the forcing function becomes much more coherent over the entire length of the tube. To estimate the vortex-induced vibration amplitude of the tube during lock-in, the acceptance integral J in Equation (9.4) should be set to 1.0 to be conservative or less conservatively, λ should be set to the length of the tube.

Example 9.1

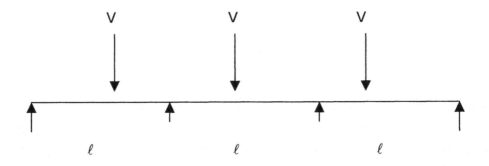

Figure 9.6 Response of a three-equal-span tube to single-phase cross-flow

Figure 9.6 shows a tube pin-supported at four equally spaced points. The tube is subject to uniform cross-flow over all of its spans. The structural dimensions, material properties, cross-flow velocities and fluid densities as well as the density of the water inside the tube are given in Table 9.1 in both US and SI units. Calculate the maximum rms vibration amplitude of the tube due to cross-flow assuming (i) $\lambda = 3 * D_o$, (ii) $\lambda = L$, the total length of the tube.

Table 9.1: Input Parameters for Example 9.1

	US Unit	SI Unit
Total length, L	150 in	3.81 m
Length of each span	50 in	1.27 m
OD, D_o	0.625 in	0.01588 m
ID, D_i	0.555 in	0.01410 m
Young's modulus, E	29.22E+6 psi	2.02 E+11 Pa
Tube material density, ρ_m	0.306 lb/in^3	8470 Kg/m^3
	(7.919E-4 (lb-s^2/in)/in^3)	
Damping ratio, ζ	0.01	0.01
Cross-flow velocity, V	40 ft/s	12.19 m/s
	(480 in/s)	
Shell-side fluid density, ρ_s	1.82 lb/ft^3	29.15 Kg/m^3
	(2.734E-6 (lb-s^2/in)/in^3)	
Tube-side water density, ρ_i	43.6 lb/ft^3	698.4 Kg/m^3
	(6.550E-5 (lb-s^2/in)/in^3)	

Solution

Following our standard practice, the first step is to convert all the units in the US unit system into a consistent unit set. This means the velocity V in in/s and densities in (lb-s^2/in)/in^3. Those who are still not familiar with this process must refer to Chapter 1 on units and dimensions. The converted units are given in parenthesis in Table 9.1.

The support plates do not separate the three spans of this tube; Equation (9.4) can be used. Since the tube has uniform mass density, we can normalize the mode shape to unity (Equation (9.6)):

$$\int_0^L \phi_n(x)\phi_s(x)dx = \delta_{ns}$$

The *first* mode-shape function for this multi-span tube is,

$$\phi_1(x) = \sqrt{\frac{2}{L}} \sin\frac{3\pi x}{L}$$

With this mode-shape normalization, the generalized mass is just the total uniform mass per unit length. This includes the material mass of the tube, the mass of water contained in the tube and the mass of fluid displaced by the tube (see Chapter 4 on hydrodynamic mass),

$$m_1 = \rho_m A_m + \rho_s A_o + \rho_i A_i .$$

The computed parameters—such as the tube cross-sectional, inside and outside areas, the generalized mass above, as well as other computed intermediate values—are given in Table 9.2 in both US and SI units.

Table 9.2: Computed Parameters for Example 9.1

	US Unit	SI Unit
Outside cross-sectional area A_o	0.307 in^2	1.99E-4 m^2
Inside cross-sectional area A_i	0.242 in^2	1.56E-4 m^2
Tube cross-sectional area A_m	0.0649 in^2	4.19E-5 m^2
Moment of inertia $= I = \pi(D_o^4 - D_i^4)/64$	0.00283 in^4	1.18E-9 m^4
Generalized mass $m_1 = \rho_m A_m + \rho_s A_o + \rho_i A_i$	6.81E-5 (lb-s^2/in)/in	0.469 Kg/m
$(\rho_s V^2 /2)$	0.317 psi	219 Pa
C_R (at 21.9 Hz, Figure 9.4)	0.025	0.025
$G_F = (D_o C_R (\rho_s V^2 /2))^2$	2.42 (lb/in)^2/Hz	0.739(N/m)^2/Hz
Max $\{\phi_1(x) = \sqrt{\frac{2}{L}} \sin\frac{3\pi x}{L}\}$	0.115	0.725
$<y^2>/J_{33}$	5.01E-2 in^2	3.21E-5 m^2

The first mode of this three-span beam is the same as the third mode of a simply-supported beam of total length L. Its modal frequency is given by (see Chapter 3, Table 3.1),

$$f_1 = \frac{9\pi^2}{2\pi} \sqrt{\frac{EI}{m_1 L^4}}$$

Using the given values of E and L in Table 9.1 and the computed values of I in Table 9.2, we get:

$f_1 = 21.9$ Hz

The random force PSD is, from Equation (9.20) and Figure 9.4,

$$G_F = (DC_R(\rho_s V^2 / 2))^2$$

Its values in both the US and SI units are given in Table 9.2.

The maximum responses occur when the mode shape assumes its maximum value of

$$\phi_{max} = \sqrt{\frac{2}{L}} * 1.$$

This is different in the two different unit systems because the mode-shape function has dimensions. The maximum values of the mode shape occurs at the mid-point of each span and are given in both the US and SI units in Table 9.2. Finally, using Equation (9.4), the first modal rms response is found to be

$$< y^2 >= 0.0501 J_{33} \text{ in}^2 \quad \text{or} \quad 3.21E\text{-}5 J_{33} \text{ m}^2$$

(i)   With $\lambda/L = 3*0.625/150 = 0.0125$, from Figure 9.17 in Appendix 9B or from Figure 9.3,

$$J_{33} \approx 2\lambda / L = 2*3*0.625/150 = 0.025$$

The maximum first modal contribution to the response is

$y_{rms} = 0.035$ in or 8.95E-4 m   This occurs at the mid-point of each span.

(ii)   With $\lambda/L = 1.0$, from the chart, Figure 9.17 in Appendix 9B,

$$J_{33} \approx 0.09$$

$y_{rms} = 0.067$ in or 1.7E-3 m

which is about twice as large as in case (i) when the correlation length is small.

9.5   The Acceptance Integral for a Structure with Non-uniform Mass Density

Very often in practice, the mass density and the stiffness of a structure are not uniform. In a nuclear steam generator, for example, the secondary (or shell) side fluid density varies along the length of the tube, very often from sub-cooled water to superheated steam with widely different densities, as shown in Figure 9.12. This gives rise to a non-uniform hydrodynamic mass (see Chapter 4), which may be much higher than the mass of fluid displaced by the tube. In many cases, a leaking tube may have been repaired by a sleeve, or be plugged and stabilized with a rod or a cable (Au-Yang, 1987). These sleeves or stabilizers seldom span the entire length of the tube. Thus, practicing engineers are often confronted with dynamics analyses of tubes with non-uniform total mass densities and stiffness. In this case the mode shapes are no longer orthogonal to each other, although they are still orthogonal to each other with respect to the mass density. The corresponding orthonormality equation is (compare with Equation (3.30)):

$$\int_0^L m(x)\phi_n(x)\phi_s(x)dx = m_{ns}\delta_{ns} \tag{9.26}$$

Morse and Feshbach (1953, Chapter 6) shows that the completeness relation for mode shapes normalized according to Equation (9.26) is,

$$\sum_n \phi_n(x')\phi_s(x'')/m_n = \delta(x'-x'')/m(x') \tag{9.27}$$

In most commercial finite-element structural analysis computer programs, the generalized mass m_n is normalized to 1.0. Based on Equations (9.26) and (9.27), Au-Yang (1986) suggested a generalized definition for the acceptance integral when the mass of the beam or tube is not uniform:

$$J_{ns}(f) = \frac{1}{L(m_n m_s)^{1/2}} \int_0^L \int_0^L m^{1/2}(x')\phi_n(x')\Gamma(x',x'',f)\phi_s(\vec{x}'')m^{1/2}(x'')dx'\,dx'' \tag{9.28}$$

As defined in Equation (9.28), the acceptance integral is independent of the mode-shape normalization, as it should be if we want to preserve the acceptance integral's original meaning as a transition probability amplitude (see Chapter 8), which is a physically observable variable. Using the orthonormality and completeness Equations (9.26) and (9.27) and following the same procedure as in Section 8.8, one can easily show again that

$$\sum_n J_{nn}(f) = 1.0$$

It follows from the above derivation that the relationship

$$\frac{J_{\alpha\alpha}}{\int \phi_n^2(x)dx} \le 1.0 \tag{9.29}$$

as is often cited in the literature, is true only if the mass of the beam or tube is uniform.

9.6 Practical Heat Exchanger Tubes with Multiple Supports—The Span Acceptance

Industrial heat exchanger tubes or nuclear fuel bundles usually have many spans with highly non-uniform section properties along the tube length. As discussed before, under these conditions, the normal modes of the tube are no longer orthogonal to each other. That is, Equations (9.6) and (8.53) can no longer be met. One must normalize the mode shapes according to Equation (9.26). Many commercial finite-element computer programs are available for modal analysis of multiple-supported tubes with non-uniform cross-sectional properties and arbitrary constraint conditions at the support plates. Almost without exception, these computer programs normalize the mode shapes according to Equation (9.26) and thus intrinsically assume that Equation (9.27) is true. Furthermore, all of these computer programs at least offer the option of normalizing the mass matrix to an identity matrix, or

$$\{\phi\}^T [m]\{\phi\} = [I] \tag{9.30}$$

in finite-element notations. To use the acceptance integral method together with modal analysis results from finite-element computer programs to estimate the cross-flow turbulence-induced vibration amplitudes of multi-span, industrial heat exchanger tubes, Au-Yang (1986) introduced the concept of "span joint acceptance" defined as:

$$j_{nn}^{(i)}(f) = \frac{1}{\ell_i m^{(i)}} \int_0^{\ell i}\int_0^{\ell i} m^{1/2}(x')\overline{\phi}_n(x')\Gamma(x',x'',f)\overline{\phi}_n(x'')m^{1/2}(x'')dx'\,dx'' \tag{9.31}$$

where

$$m_n^{(i)} = \int_0^{\ell i} m(x)\overline{\phi}_n^{\,2}(x)dx \tag{9.32}$$

and $\overline{\phi}_n(x)$ and $\phi_n(x)$ are the same mode-shape function, except $\overline{\phi}_n(x)$ is normalized to one span while $\phi_n(x)$ is normalized with respect to the entire tube. That is,

$$\overline{\phi}_\alpha(x) = k\phi_\alpha(x) \tag{9.33}$$

where k is a constant. Note that as defined in Equation (9.31), $j_{nn}^{(i)}$ is independent of mode-shape normalization. Au-Yang (1986) showed that if the linear mass density can be approximately regarded as constant within each span, then the response at any point on a multi-span tube separated by tube support plates, so that there is no correlation across different spans, can be written as (compared with Equations (9.4)),

$$<y^2> = \sum_n \frac{\phi_n^2(x)}{64\pi^3 m_n^2 f_n^3 \zeta_n} \sum_i \ell_i G_F^{(i)}(f_n) j_{nn}^{(i)} + cross-terms \qquad (9.34)$$

where[1]

$$G_F^{(i)} = \int_0^{\ell i} G_F(x)\phi_n^2(x)dx \qquad (9.35)$$

is the "generalized random force PSD" for the i-th span. In Equation (9.34), the summation is over all the modes n and within each mode, the summation is over all the spans i that are subjected to cross-flow. The mode-shape function ϕ and the modal generalized mass m_n are obtained from the same finite-element model. In most computer programs, $m_n=1.0$ for all modes.

Since in Equation (9.34), the joint acceptance as defined in Equation (9.31) has exactly the same form as that of a one-span tube, the results derived in Sections 9.2 and 9.3 can be used. If we can assume the tube is simply-supported at each of the intermediate supports, we have from Equation (9.11),

$$j^{(i)}_{nn} = \frac{2(\lambda/\ell_i)^2}{1+(n\pi)^2(\lambda/\ell_i)^2}\left[\frac{2(n\pi)^2(\lambda/\ell_i)^2\{(-1)^{n+1}e^{-\ell i/\lambda}+1\}}{1+(n\pi)^2(\lambda/\ell)^2}+\frac{\ell_i}{\lambda}\right] \qquad (9.36)$$

and

$$j_{nn}^{(i)} = 2\lambda/\ell_i \quad \text{for } n\pi\lambda/\ell_i \to 0 \qquad (9.37)$$

Alternatively, the charts given in Appendix 9B can be used to evaluate the acceptance for each span. Since in practice one seldom encounters more than three half waves within one span, the acceptance integrals given in Appendix 9B are enough to cover all the cases likely to be encountered in practice.

---

[1]As originally formulated (Au-Yang, 1986), the generalized random force PSD was defined as $G_F^{(i)} = \int_0^{\ell i} G_F(x)\phi_n^2(x)dx / \int_0^{\ell i}\phi_n^2(x)dx$ and the mean square response, $<y^2>$, contains an extra factor $\int_0^{\ell i}\phi_n^2(x)dx$ to cancel the integral in the denominator of $G_F^{(i)}$. This preserves the physical meaning of the generalized force PSD but adds unnecessary computation.

It should be noted that the very acceptance integral concept was developed based on the assumption of homogeneous turbulence. That is, until Equation (9.34), G_F has been assumed to be constant. Equation (9.35) is introduced at the last step as an engineering approximation to address the practical problem of non-uniform loading on industrial heat exchanger tubes. In the special case when all the spans are of equal length, ℓ_0, and the random force PSD G_0 is the same over the entire length of the tube, Equation (9.35) together with Equation (9.37) reduces to:

$$<y^2> = \sum_n \frac{\ell_0 G_0 \ (f_n)\phi_n^2(x)}{64\pi^3 m_n^2 f_n^3 \zeta_n}(2\lambda/\ell_0)\sum_i \int_0^{\ell_i} \phi_n^2(x)dx$$

The last factor is of course just the integral of the mode shape over all the spans. Thus,

$$<y^2> = \sum_n \frac{\ell_0 \ (f_n)\phi_n^2(x)}{64\pi^3 m_n^2 f_n^3 \zeta_n}(2\lambda/\ell_0)G_0 \int_0^L \phi_{nn}^2(x)dx \qquad (9.38)$$

for $\lambda << \ell_0$

If, in addition, the mass density is uniform, then

$$m_n = m\int_0^L \phi_{nn}^2(x)dx$$

Equation (9.34) reduces to,

$$\bar{y}^2 = \sum_n \frac{\ell_0(f_\alpha)\phi_\alpha^2(x)}{64\pi^3 m^2 f_n^3 \zeta_n}G_0 \ [(2\lambda/\ell_0)/\int_0^L \phi_{nn}^2(x)dx] \qquad (9.39)$$

Partial Span Loading

It should be pointed out that in Equations (9.34) and (9.37), ℓ_i refers to the length of the entire span even if only part of the span is subject to cross-flow. For $\lambda << \ell_i$, $j_{ns}^{(i)} \approx 2\lambda/\ell_i$ even if only $\ell_e < \ell_i$ of the span is loaded. The span length ℓ_i in the joint acceptance cancels out the span length ℓ_i in the sum $\sum_i \ell_i G_F^{(i)} j_{ns}^{(i)}$. The effect of partial loading is completely accounted for in the generalized random force PSD $G_F^{(i)}$.

Example 9.2

This example is given without specifically using any unit system so that the physics of the problem is not masked by the complications caused by the units.

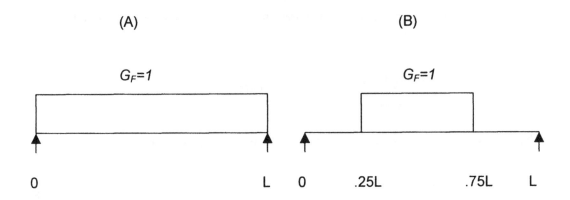

Figure 9.7 Totally loaded and partially loaded simply-supported beams

Figure 9.7 shows two rigid beams that are both spring-supported at the ends. In Figure 9.7(A) the tube is uniformly loaded with a constant random force PSD equal to (1.0 per unit length)^2/Hz, while in Figure 9.7(B) the load is over only the middle half of the span. In both cases $\lambda<<L$. Assume the total linear mass density is uniform and is equal to m per unit length; find the first modal response in each case in terms of L, m, f_1 and ζ_1.

Solution

(i)   Suppose we normalize the mode shape in the finite-element tradition with $m_1=1.0$, the mode shape function will be $\phi_1 = 1/\sqrt{mL}$. Since the span joint acceptance is independent of mode-shape normalization, we have

$$j_1^{(1)} = 2\lambda/L \quad \text{and} \quad G_a = \int_0^L 1.\phi^2 dx = \frac{1}{m}$$

From Equation (9.34), ignoring the cross-terms, we have,

$$< y_a^2 >= \frac{2\lambda}{64\pi^3 m^2 f_1^3 \zeta_1 L}$$

(ii)   Using the same mode-shape normalization we have, since the joint acceptance is independent of partial span loading, again,

$$j_1^{(1)} = 2\lambda / L, \quad \text{but,} \quad G_b = \int_{L/4}^{3L/4} 1.\phi^2 dx = \frac{1}{2m}$$

It follows that

$$< y_b^2 > = \frac{\lambda}{64\pi^3 m^2 f_1^3 \zeta_1 L} = < y_a^2 > / 2$$

as one would expect intuitively. If we normalize the mode shape according to

$$\int_0^L \phi^2 dx = 1$$

then $m_1=m$. The span joint acceptance in both cases are still equal to $j=2\lambda/L$, but then

$$G_a = \int_0^L 1.\phi^2 dx = 1 \qquad \text{and} \qquad G_b = \int_{L/4}^{3L/4} 1.\phi^2 dx = \frac{1}{2}$$

We get the same responses as computed before. This example shows that in a properly formulated problem, the results for any physically observable variable should not be dependent on the mode shape normalization.

Example 9.3

Let us return to Example 9.1, except in this case the tube is a heat exchanger tube supported by plates that completely separate the random pressure fields, so that the pressures are completely uncorrelated across the spans. Find the maximum rms response of the first mode again assuming again (i) $\lambda = 3 * D$, (ii) $\lambda = \ell$.

Part (ii) of this problem has some practical importance in this case of a multi-span tube. Deep inside a tube bundle, even if lock-in vortex-induced vibration (see Chapter 5) occurs, the response spectrum would not have a sharp classic resonance peak as in the case of an isolated tube. The question whether we can use the empirical equations in Chapter 5 to predict the lock-in vortex-induced vibration amplitudes is very much a controversial subject. However, it is well known that when the vortex-shedding frequency is close to one of the natural frequencies of the tube, the correlation length of the random pressure field increases tremendously. Thus, one way to estimate the lock-in vortex-induced vibration inside a tube bundle is to use the method developed in this chapter for cross-flow turbulence-induced vibration, and assume that the correlation

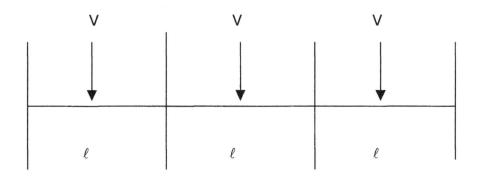

Figure 9.8 Response of a three-equal-span heat exchanger tube to cross-flow

length is about the length of the entire span (or, more conservatively, $j_{nn}^{(i)} = 1.0$), as in case (ii) of this example.

(i)   When the correlation length is small,

$j_{11}^{(i)} \approx 2\lambda / \ell_i = 2*3*0.625/50 = 0.075$ for each span.

$$< y_1^2 >= \frac{\phi_1^2(x)}{64\pi^3 m_1^2 f_1^3 \zeta_1} \sum_i \ell_i G_F^{(i)}(f_1) j_{11}^{(i)}$$

$$= \frac{\phi_1^2(x)}{64\pi^3 m_1^2 f_1^3 \zeta_1} \sum_i 2\lambda G_F \int_0^{\ell i} \phi_1^2 dx$$

Since $\sum_i \int_0^{\ell i} \phi_1^2 dx = \int_0^{\ell} \phi_1^2 dx = 1$

by our normalization for constant linear mass density, we get the same answer as in Example 9.1. This example shows that when the correlation lengths are small compared with the span lengths, the support plates have little effect on breaking up the coherence of the pressure field. The response of the tube is the same whether the supports are point supports or plate supports separating the flow regions. If, however:

(ii)   When the correlation length is equal to the span length for each of the spans, then from Figure 9.17 in Appendix 9B,

$j_{11}^{(i)} = 0.64$ $i=1,2,3$

$$< y_1^2 >= 0.0501 * (3 * 0.64) = 0.0962 \text{ in}^2$$

$$y_{rms} = 0.31 \text{ in}$$

compared with 0.067 in. when the spans are not separated by the support plates as in Example 9.1. The above example shows that it is not necessarily true, even in cross-flow, that smaller correlation length will result in smaller responses. Very often in practice, disks are installed along beams to "break up" the correlation length and thus reduce their responses to cross-flow. The above example shows that these disks should not be installed at the support points of the beam.

9.7   Cross-Modal Contributions to Responses

So far we have been ignoring the cross-terms contributions to the responses in Equation (9.4) or Equation (9.34). As discussed in Section 8.6, this is justifiable if either the cross-acceptance is negligible, or if the modal frequencies are well separated. In the latter case, the cross-terms that contain the factor $H_n(f)H_s(f)$, $n \neq s$, are small compared with the joint terms. However, if $f_n \approx f_s$ and J_{ns} is not small, then there will be cross-modal contribution to the mean square response. The "cross-term contribution" in Equation (9.34) is,

$$< y^2 >= \sum_{n \neq s} \frac{2\phi_n(x)\phi_s(x)}{64\pi^3 m_n m_s f_{ns}^3 \zeta_{ns}} \sum_i \ell_i G_F^{(i)}(f_{ns}) j_{ns}^{(i)}$$

$$G_F^{(i)} = \int_0^{\ell i} G_F(x)\phi_n(x)\phi_s(x)dx \tag{9.40}$$

with n, s counted once for all n and s such that $f_n \approx f_s$. In Equation (9.40), m_n, m_s, ϕ_n and ϕ_s are global generalized masses and mode shapes. These values are normally printed out from the finite-element computer program used in the modal analysis. f_{ns} and ζ_{ns} are respectively, the average natural frequencies and damping ratios of the two modes involved. As an example, Figure 9.9 shows the first 10 modes of a heat exchanger tube with 17 spans. In this particular example, the span of interest is the second from the top (16th span), which is subject to high-velocity steam cross-flow. Examination of Figure 9.9 shows that the modes that contribute most to the response of this span are modes 3 and 4, with frequencies of 37.4 and 38.0 Hz. Thus, these two modes are not "well separated" in the frequency domain. From Equation (9.39), the cross-acceptance between modes 3 and 4 involves the 16th span cross-acceptance between these two modes. Examination of the mode shape plot shows that both modes 3

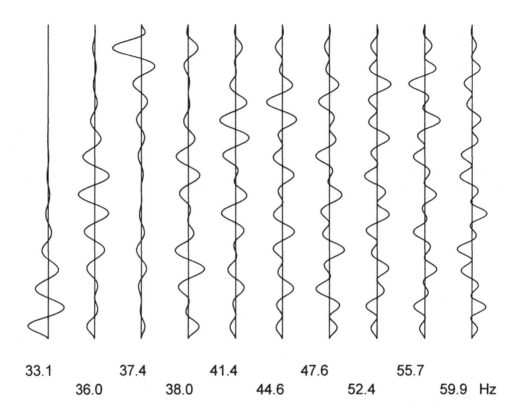

33.1   37.4   41.4   47.6   55.7

36.0   38.0   44.6   52.4   59.9 Hz

Figure 9.9 First 10 modes of a heat exchanger tube with 17 spans

and 4 have one half-wave in the 16th span. The span cross-acceptance j_{34}^{16} is therefore the same as the joint acceptance for a single-span tube with span length ℓ_{16}. In view of the insensitivity of the joint acceptance to the exact mode shape, especially at small correlation lengths, one can find the value of j_{34}^{16} from either Figure 9.17 or 9.18 in Appendix 9B, even though the mass density is only approximately uniform within this span. When the correlation length λ is small compared with ℓ_{16}, $j_{34}^{16} = 2\lambda / \ell_{16}$. Assuming that the only significant forcing function is over the 16th span, the cross-modal (between modes 3 and 4) contribution to the response is, from Equation (9.39), equal to:

$$<y_{34}^{2}> = \frac{\phi_3 (x)\phi_4(x)}{64\pi^3 m_3\ m_4 f_{34}^{3}\zeta_{34}}\{\ell_{16}G_F^{(16)}(f_{34})\}\{2\lambda / \ell_{16}\}$$

$$G_F^{16} = \int_0^{\ell16} G_F(x)\phi_3(x)\phi_4(x)dx$$

where f_{34} and ζ_{34} are the average natural frequency and damping ratio for modes 3 and 4.

Similarly, $< y_{43}^2 > = < y_{34}^2 >$

The cross-modal contributions from the other pairs of closed-space modes can be estimated similarly. Obviously, in the general case when all the spans are subject to cross-flow excitation and there are many closely-spaced modes, the above process can be very tedious. Specialized computer programs have been developed to carry out the analysis in more efficient ways. In this example at least, the cross-modal contribution to the responses from modes 3 and 4 are about the same as the direct contributions from these two modes. Thus, ignoring the cross-terms in Equation (9.34) can lead to unconservative results.

Example 9.4

To highlight the essential topic of discussion in this chapter: The use of commercial finite-element structural analysis computer programs to solve practical flow-induced vibration problems, the following example is given without specific mention of the units, except that it is given in *consistent* units. Readers who are familiar with the dimensions of heat exchanger tubes will recognize from the given data, that the units used are those of the US system. However, in going through the sample, one does not have to pay attention to which unit system is used.

Figure 9.10 shows a heat exchanger tube with three spans. The tube is clamped at one end and simply-supported at the other and at two intermediate tube support plates. These support plates completely divide the shell into three compartments so that the pressure field is not cross-correlated across different spans. The tube is subject to cross-flow over its spans, with the cross-flow velocities given in the last column of Table 9.3 and plotted in Figure 9.11. The density of the fluid in each span can be assumed to be approximately constant. Other relevant data are given below, all in a consistent unit set.

OD of tube, D_o=0.625 length units
ID of tube, D_i=0.555 length units
Young's modulus E=2.9228E+7 units
Poison ratio=0.3
Equivalent material mass density, including mass of water contained, ρ_m=7.9257E-4 (see Chapter 4 for hydrodynamic mass calculations)
Tube-side fluid density=6.532E-5 mass unit /unit vol.
Shell-side fluid densities:
    Span 1: 2.816; Span 2: 2.711; Span 3: 2.726 mass unit/unit vol.
Damping ratio 0.03
Correlation length λ = 3*OD of tube

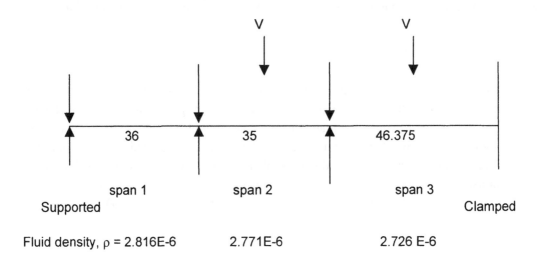

Figure 9.10 A three-span heat exchanger tube

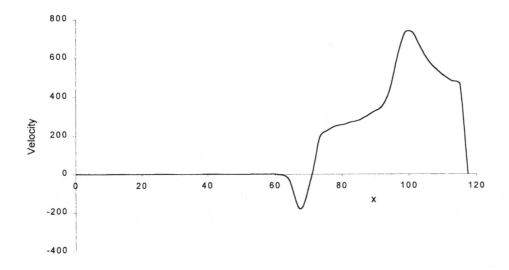

Figure 9.11 Cross-flow velocity distribution over length of tube

Compute the first modal vibration amplitude at the point of maximum deflection. What would be the vibration amplitude if the pressure becomes fully correlated over each span?

Solution

As in Examples 9.1 and 9.3, the first step is to calculate the natural frequencies and mode shapes. We shall use a commercial finite-element structural analysis computer program, NASTRAN, to solve this problem. A finite-element model is set up, with 10 elements each in the first two spans and 20 elements in the third span. The node numbers, their locations, boundary conditions (if any) are shown in the first three columns of Table 9.3. The mean square response for the first mode is, from Equation (9.40),

$$< y_1^2 > = \frac{\phi_1^2(x)}{64\pi^3 m_1^2 f_1^3 \zeta_1} \sum_{i=1}^{3} G_F \ell_i j_{11}^{(i)}$$

We shall compute the rms displacement for the first mode at the maximum point of deflection, i.e., at node 30009 (near mid point of 3rd span). From Table 9.3, at this point,

$$\phi_1(30009) = 19.17$$

Other known values are:

$m_1 = 1.0$ from NASTRAN normalization

$f_1 = 41.37$ Hz from NASTRAN
$\zeta_1 = 0.03$ (given)
$\ell_1 = 36, \quad \ell_2 = 35 \quad \ell_3 = 46.375$ (given)

Other variables are discussed below.

The span acceptance $j_{11}^{(i)}$

Since the span linear mass density is constant, we can use the single-span joint acceptance given in the charts in Appendix 9B. Furthermore, since the tube is quite slender even in each span, we can use the approximate result:

$$j_{11}^{(i)} \approx 6D_o / \ell_i \quad \text{for each span}$$

From this: $j_{11}^{(1)} = 0.104; \quad j_{11}^{(2)} = 0.107; \quad j_{11}^{(3)} = 0.08$

Table 9.3: Noding Scheme in the Finite-Element Model

Boundary condition	Node	x	Mode Shape ϕ_1	Velocities
Clamped	30020	117.380	0	0
	30019	115.061	0.4754	465
	30018	112.742	1.765	482
	30017	110.423	3.666	508
	30016	108.104	5.98	543
	30015	105.785	8.51	586
	30014	103.466	11.07	655
	30013	101.147	13.5	732
	30012	98.828	15.64	732
	30011	96.509	17.36	603
	30010	94.190	18.56	431
	30009	91.871	19.17	353
	30008	89.552	19.12	327
	30007	87.233	18.43	302
	30006	84.914	17.09	280
	30005	82.595	15.15	271
	30004	80.276	12.7	258
	30003	77.957	9.835	250
	30002	75.638	6.668	228
	30001	73.319	3.441	194
Supported	30000	71.000	0	0
	20009	67.500	-4.758	-181
	20008	64.000	-8.846	-24.4
	20007	60.500	-11.95	0
	20006	57.000	-13.84	0
	20005	53.500	-14.38	0
	20004	50.000	-13.54	0
	20003	46.500	-11.42	0
	20002	43.000	-8.241	0
	20001	39.500	-4.308	0
Supported	20000	36.000	0	0
	10009	32.400	4.391	0
	10008	28.800	8.324	0
	10007	25.200	11.43	0
	10006	21.600	13.41	0
	10005	18.000	14.08	0
	10004	14.400	13.37	0
	10003	10.800	11.36	0
	10002	7.200	8.252	0
	10001	3.600	4.337	0
Supported	10000	0.000	0	0

The Span-Random Force PSD $G_F^{(i)}(f = 41.37Hz)$

From Equation (9.35),

$$G_F^{(i)} = (OD * C_R / 2)^2 \sum_j (\rho_i V_j^2)^2 \phi_j^2 \Delta x_j$$

(Note: j is nodal index; i is span index)

From Figure 9.4, at f=41.37 Hz, $C_R = 0.025$. Using the nodal values in Table 9.4 for the mode shape and velocities, we calculate by hand or use a spreadsheet to find:

$$G_F^{(1)} = 0$$
$$G_F^{(2)} = 3.68E - 5$$
$$G_F^{(3)} = 0.22$$

all in units of (force/length)^2/Hz

The Response

From Equation (9.40) with $G_F^{(i)}$ given above,

$$< y^2 (node\ 30009) >= \frac{19.17^2}{64\pi^3 * 1^2 * 41.37^3 * 0.03} * \{35 * 3.68E - 5 * *0.107 +$$

$$46.375 * 0.22 * 0.08\} = 7.07E - 5$$

$y_{rms} = 8.4E - 3$ length units at node 30009.

When the pressure PSD is fully correlated over each span:

The span-joint acceptances, $j_{11}^{(i)} = 1$ for each span. The response is then,

$$\bar{y}^2_{max} = \frac{19.17^2}{64 * \pi^3 * 1^2 * 41.37^3 * 0.03} * \{35 * 3.68E - 5 * 1 + 46.375 * 0.22 * 1\} = 8.88E\text{-}4$$

$y_{rms} = 29.8E - 3$ length units

This response is an approximate estimate of what would happen if lock-in vortex-induced vibration occurs inside a tube bundle. Inside a closely packed tube bundle, even if lock-in vortex-induced vibration occurs, the tube response would not exhibit the classic sharp peak as in an isolated cylinder. The responses, too, would be less than those given by the classic empirical lock-in vortex-induced vibration of an isolated cylinder. However, if lock-in occurs, the pressure field will be fully correlated along the span in which the lock-in condition is satisfied. Thus, an estimate of the lock-in vortex-induced response can be obtained using turbulence-induced vibration analysis technique, but with $j_{nn}^{(i)} = 1.0$.

9.8   Vibration due to Turbulent Two-Phase Cross-Flow

In a heat exchanger, very often the shell-side fluid is a two-phase mixture of steam and water. Figure 9.12 shows a nuclear steam generator of the once-through design. The tube-side fluid is primary water from the nuclear reactor. Heated to more than 600 deg. F (315.6 deg. C), this water flows from the top to the bottom of the tubes. The shell-side fluid enters the bottom of the steam generator in the form of sub-cooled water and flows upward. As it rises, it is heated up by the tube-side water and changes from sub-cooled water to a two-phase steam-water mixture and finally into superheated steam before it exits the steam generator. This particular design is called a once-through steam generator (OTSG). In this design, the cross-flow regions are the bottom and top spans of the tube bundle, where the fluid is either single-phase water or superheated steam. This is not true in the case of most recirculation steam generators (RSG, see Figure 9.13). While the shell-side fluid generally still enters the tube bundle at the bottom in the form of sub-cooled water, it does not change to super-heated steam as it rises to the critical U-bend region. The U-bends in RSGs are subject to two-phase turbulent cross-flow.

Two-Phase Flow Regimes

As shown in Figures 9.12 and 9.13, as the shell-side fluid changes from single-phase water to single-phase, super-heated steam, it undergoes different two-phase flow regimes. Not only are the forcing functions in these regimes different from those derived from Equations (9.20) and (9.21) or Equation (9.25) and Figure 9.4 for single-phase flows, but also they are different from one another. In addition, the thermodynamics of a two-phase fluid is quite different from that of a single-phase fluid, resulting in significantly different energy transfer mechanisms between the vibrating structure and the flowing fluid. This will be discussed below.

Two-Phase Parameters

Since a two-phase mixture contains two components, it is understandably much more complicated to characterize a two-phase fluid than a single-phase one. The general thermodynamics of a two-phase fluid is well beyond the scope of this book. Interested

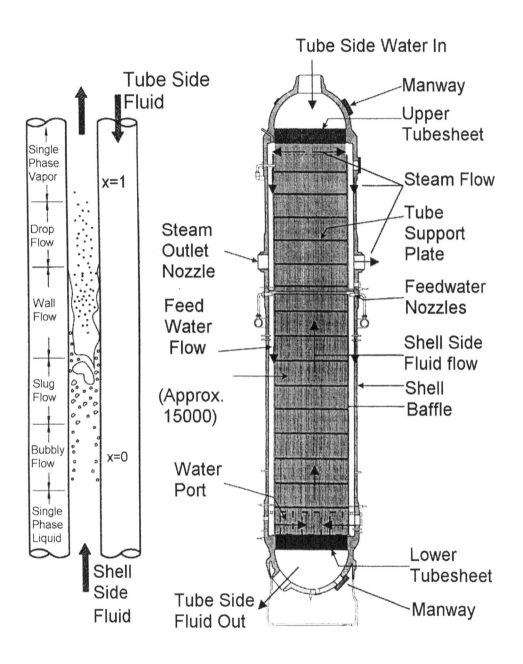

Figure 9.12 Shell-side flow regimes in a nuclear steam generator of the
once-through design

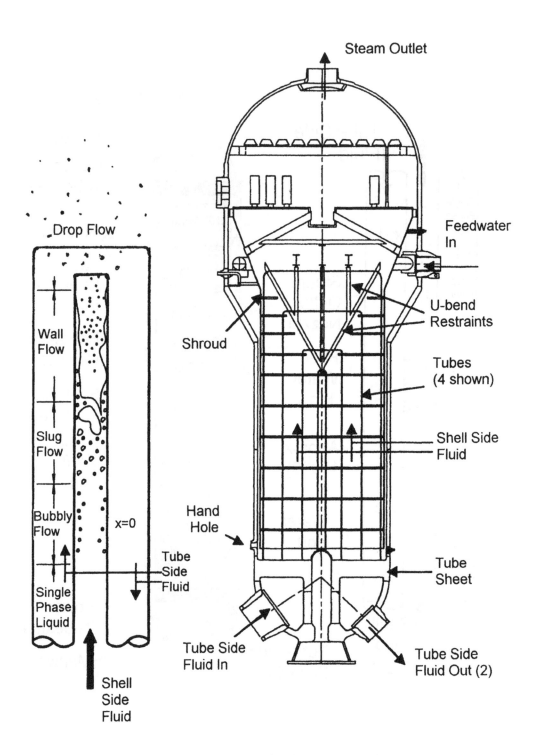

Figure 9.13 Shell-side flow regimes in a nuclear steam generator of the
recirculatory design

readers are referred to Tong (1966) or El-Wakil (1962). The following defines the terminology absolutely required for two-phase flow-induced vibration analysis. In what follows, we always assume the two phases form a homogeneous mixture. We shall follow as closely as possible terminology and notations in the ASME Steam Table (1979) and assume that the readers are familiar with the meaning of saturated pressure and temperature:

ρ_f is the mass density of liquid phase of the mixture

ρ_g is the mass density of the vapor phase of the mixture

$v_f = 1/\rho_f$ is the specific volume of the liquid

$v_g = 1/\rho_g$ is the specific volume of the vapor

$v_{fg} = v_g - v_f$

$\rho_{fg} = 1/v_{fg}$

x is the quality of two-phase mixture (percent mass of steam in the mixture)

At a given temperate and pressure, the above parameters can be obtained from the ASME Steam Table which we have used in several work examples. However, two-phase parameters pertaining to flow-induced vibration are expressed not in terms of the above parameters, but in terms of another parameter called the void fraction. We shall use the notation α to denote the void fraction. The void fraction is a measure of the gaseous volume of the two-phase mixture, and is related to the quality x by (El-Wakil, 1962):

$$\alpha = \frac{xv_g}{v_f + xv_{fg}} \tag{9.41}$$

Two-Phase Damping

Tests conducted both in Canada and in Europe conclusively showed that the damping ratios in a tube vibrating in a two-phase mixture are higher than those in a tube vibrating in a single-phase fluid. The exact origin of this additional energy dissipation is not known and the subject is beyond the scope of this book. Suffice to state that the suggestion that it originates from phase transition between the two states of the fluid does not seem to hold, as similar increases in damping ratios were observed even in air-water mixtures. Figure 9.14, reproduced from Pettigrew and Taylor (1993), shows the measured fluid damping ratio of a tube vibrating in an air-water "two-phase" mixture with different void fractions. Assuming that 1% of the damping is from the material damping of the tube and damping from the single-phase fluid, it can be seen from Figure 9.14 that two-phase mixtures can add as much as 3% to the damping of the tube, which is significant and may explain why vortex-shedding has never been observed in two-phase flows. Pettigrew and Taylor's observation is generally consistent with that of Axisa et al (1986), who used a steam-water mixture in their test. Base on these two sets of data, it is conservative to assume that between a void fraction of 60% and 85% which is applicable to the shell-side fluid conditions in the U-bend regions of most RSGs, two-phase fluid

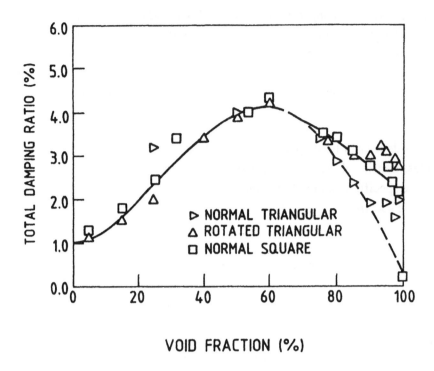

Figure 9.14 Two-phase damping (Pettigrew and Taylor, 1993)

adds to the damping ratio of a typical steam generator tube by at least 2% of its critical damping value (see Chapter 4) over its value due to material and tube-support plate interaction. Since the latter is typically about 0.5% for a tightly supported tube and about 3% to 5% for a loosely supported, multi-span tube, we can see that two-phase damping should be included in the vibration analysis.

Random Pressure PSD in Two-Phase Flows

With so many different flow regimes and so many parameters to describe just the physical state of a two-phase mixture, one can imagine the magnitude of deriving empirical equations for the random pressure PSD for two-phase flows. Two-phase flow-induced vibration is still very much an ongoing research topic. Here we quote only the more successful results. For "wall-type" flow, which is applicable to the U-bend regions of most nuclear steam generators of the recirculation designs, Pettigrew and Taylor (1993) reasonably collapse data from several tests by expressing the adjusted PSD Φ as a function of the reduced frequency, F, defined as,

$$\Phi = \frac{G_F(f)}{D^2}$$

(9.42)

$$F = fd_B / V_p \tag{9.43}$$

where

$$d_B = 0.00163 \sqrt{\frac{V_p}{1-\alpha}} \quad m \tag{9.44}$$

is what Taylor called the characteristic void length expressed in *meters*. Pettigrew and Taylor (1993) called Φ a normalized PSD—a term reserved to mean "dimensionless spectrum" in this book. Examination of Equation (9.42) indicates that Φ has the same dimension as random pressure PSD defined in Chapter 8. Based on this, it does not appear that Φ is a true power spectral density of the two-phase random pressure field. In spite of all these shortcomings, the proposed scheme seems to collapse data from several different tests reasonably well, as shown in figure 9.15. The following equation bounds all the data in Figure 9.15:

$$\begin{aligned} \Phi &= 10^{-4(\log_{10} F + 1)}, \quad F \geq 0.1 \\ &= 1000 \qquad\qquad F < 0.1 \end{aligned} \tag{9.45}$$

with Φ in $(N/m^2)^2/Hz$. It must be cautioned that Equation (9.45) is based on very limited data from tests on tubes with typical heat exchanger dimensions. Therefore, it is applicable only to wall-type two-phase flow-induced vibration analysis of typical heat exchanger tubes.[1] From Equation (9.42), we get,

---

[1] As presented by Pettigrew and Taylor (1993), Equation (9.42) was expressed as,

$$\Phi = \frac{G_F(f)(L_e / L_0)}{D^2}$$

and L_e is the length of the tube subject to the cross-flow, expressed in terms of a reference length L_0. The authors then chose $L_0 = 1.0$ m as the "reference" length. This presentation has inherent conceptual difficulties as it implies that two identical tubes subject to the same cross-flow would experience different forces *per unit length*, if the lengths subject to cross-flow are not the same. Apparently the authors chose this convention because they defined the joint acceptance as $J = 2\lambda / L_e$ instead of $J = 2\lambda / L$. The latter convention is followed in this book with the effect of partial span loading completely accounted for in the generalized pressure PSD, Equation (9.35). Following the approach used in this book, it is best to ignore the factor L_e/L_0 in the original Pettigrew and Taylor presentation. Since Figure 9.15 depends on how the authors normalized the source data, Equation (9.42) should be used with caution. Extrapolation much beyond the 1.0 meter length where most of the heat exchanger span lengths lie is not recommended.

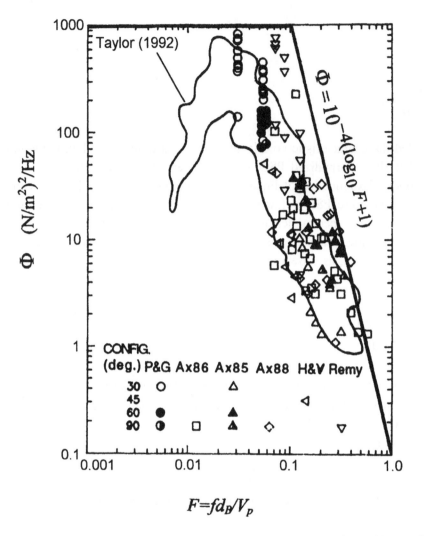

Figure 9.15 Two-phase random pressure PSD for wall-type flow
with L_0=1.0 m. (Pettigrew and Taylor, 1993)

$$G_F(f) = D^2 \Phi \quad (\text{N/m})^2/\text{Hz} \tag{9.46}$$

Consistent SI units must be used in Equation (9.46).

Example 9.5

The entire span of a heat exchanger tube is subject to two-phase cross-flow. The
temperature, pressure and velocity of the flow, as well as the dimensions of the tube are

given in Table 9.4. Estimate: (i) the void fraction; (ii) the added damping due to the two-phase flow; (iii) the random forcing function G_p, G_F in both US and SI units; and (iv) the corresponding G_p, G_F in both US and SI units if the single-phase equation for the random forcing function, Equation (9.20), together with Figure 9.4 are used instead.

Table 9.4: Input Parameters for Example 9.5

	US Unit	SI Unit
OD of tube, D_o	0.875 in	0.0222 m
Length of tube, L_e	40 in	1.016 m
Cross-flow pitch velocity, V_p	15 ft/s	4.572 m/s
Density of fluid, ρ	11 lb/ft^3	176.2 Kg/m^3
Temperature, T	517.27 deg. F	315.6 deg. C
Tube natural frequency	20 Hz	20 Hz

Solution:

(i) To find the void fraction from Equation (9.41), we must know the specific volumes of the gas and liquid phase at the stated saturation temperature of the two-phase mixture. From the ASME Steam Table (1979), at a temperature of 517.27 deg. F (315.6 deg. C) and a density of 11 lb/ft^3 (176.2 Kg/m^3), the specific volumes of the vapor, liquid and the quality of the two-phase mixture are given in the first 3 rows of Table 9.5. From these, $v_{fg} = v_g - v_f$ can be computed. This is given in the 4th row of Table 9.5. The void fraction α can then be computed from Equation (9.41) and is equal to 0.80.

Table 9.5: Computed Parameters for Example 9.5

	US Unit	SI Unit
Specific volume of vapor, v_g	0.5753 ft^3/lb	0.03592 m^3/Kg
Specific volume of liquid, v_f	0.02084 ft^3/lb	0.0013 m^3/Kg
Quality of mixture, x	0.126	0.126
$v_{fg} = v_g - v_f$	0.5545 ft^3/lb	0.0346 m^3/Kg
Void fraction, α	0.80	0.80
Two-phase G_p at 20 Hz	2.07E-5 psi^2/Hz	984 Pa^2/Hz
Two-phase G_F at 20 Hz	1.58E-5 (lb/in)^2/Hz	0.486 (N/m)^2/Hz
Single-phase G_p at 20 Hz	4.50E-5 psi^2/Hz	2139 Pa^2/Hz
Single-phase G_F at 20 Hz	3.45E-5 (lb/in)^2/Hz	1.056 (N/m)^2/Hz

(ii) From Figure 9.14, at a void fraction of 0.80, the total damping ratio of the tube is about 3.0% of critical. Of this, about 1% is from the tube vibrating in single-phase water. Thus, the two-phase mixture adds about 2.0% of critical damping to the tube.

(iii) From Equation (9.44) and the given pitch velocity, the characteristic void length can be calculated. Here, we must be careful that the empirical equation is given in SI units. Therefore, we must calculate d_B with the pitch velocity in m/s. From Equation (9.44),

$$d_B = 0.00163\sqrt{\frac{4.572}{1-0.80}} = 0.0078 \text{ m}$$

At 20 Hz, the dimensionless frequency F is, from Equation (9.43),

$F=20*0.0078/4.572=0.034<0.1$

Therefore,

$\Phi=1000$ Pa^2/Hz.

From Equation (9.46),

$$G_F(f = 20\ Hz) = \frac{0.02222^2}{(1.016/1.0)} * 1000 = 0.486 \ \ (N/m)^2/Hz, \text{ or}$$

$$G_p(f = 20\ Hz) = \frac{1}{(1.016/1.0)} * 1000 = 984 \ \ Pa^2/Hz$$

U.S. engineers should convert the units back into the US unit system at this point:

$$G_F(f = 20\ Hz) = 0.486*(0.225/39.37)^2 = 1.58 \text{ E-5 } (lb/in)^2/Hz$$

$$G_p(f = 20\ Hz) = 984/(6.895E+3)^2 = 2.07 \text{ E-5 } psi^2/Hz$$

(iv) If the single-phase equation is used, then from Figure 9.4, at 20 Hz, $C_R=0.025$. From Equation (9.20), one can readily compute the corresponding random pressure and random force PSD. These are given in the last two rows of Table 9.5. In this particular example, the two-phase excitation force is less than the corresponding single-phase excitation force while the damping is higher.

Appendix 9A: Derivation of Equation (9.11) (By B. Brenneman)

From Equation (9.3),

$$J_{nn}(\omega) = \frac{1}{L}\int_0^L\int_0^L \phi_n(x')\Gamma(x',x'',\omega)\phi_n(x'')dx'\,dx''$$

with

$$\Gamma(x',x'',f) = e^{-|x'-x''|/\lambda} \quad \text{and} \quad \phi_n = \frac{1}{\sqrt{L}}\sin\frac{n\pi x}{L}$$

From this, we get

$$J_{nn} = \frac{2}{L}\int_{x''=0}^L [\int_{x'=0}^{x''} \sin\frac{n\pi x'}{L}e^{-(x''-x')/\lambda}dx' + \int_{x'=x''}^L \sin\frac{n\pi x'}{L}e^{-(x'-x'')/\lambda}dx']\sin\frac{n\pi x''}{L}dx''$$

Now change the variable of integration to

$$u = \sin\frac{n\pi x'}{L}, \quad du = \frac{n\pi}{L}\cos\frac{n\pi x'}{L}dx'$$
$$v = \lambda e^{-(x''-x')}, \quad dv = e^{-(x''-x')}dx'$$

for the first integral, and

$$u = \sin\frac{n\pi x'}{L}, \quad du = \frac{n\pi}{L}\cos\frac{n\pi x'}{L}dx'$$
$$v = -\lambda e^{-(x'-x'')}, \quad dv = e^{-(x'-x'')}dx'$$

for the second integral. We have for the first integral,

$$I_1 = \lambda\sin\frac{n\pi x'}{L}e^{-(x''-x')/\lambda}\Big|_{x'=0}^{x''} - \frac{n\pi\lambda}{L}\int_{x'=0}^{x''}\cos\frac{n\pi x'}{L}e^{-(x''-x')/\lambda}dx'$$

Change the variable of integration again,

$$u = \cos\frac{n\pi x'}{L}, \quad du = -\frac{n\pi}{L}\sin\frac{n\pi x'}{L}dx'$$

$$v = \lambda e^{-(x''-x')/\lambda}, \quad dv = \lambda e^{-(x''-x')/\lambda}dx'$$

We get for the first integral,

$$I_1 = \lambda \sin\frac{n\pi x''}{L} - n\pi\lambda^2 \cos\frac{n\pi x''}{L} + n\pi\lambda^2 e^{-x''/\lambda} + (n\pi\lambda)^2 I_1$$

$$I_1 = [\lambda^{-1} \sin\frac{n\pi x''}{L} - \frac{n\pi}{L}\cos\frac{n\pi x''}{L} + \frac{n\pi}{L}e^{-x''/\lambda}]/[\frac{1}{\lambda^2} + (\frac{n\pi}{L})^2]$$

Similarly; for the second integral;

$$I_2 = [\lambda^{-1} \sin\frac{n\pi x''}{L} + \frac{n\pi}{L}\cos\frac{n\pi x''}{L} - (-1)^n \frac{n\pi}{L}e^{-(L-x'')/\lambda}]/[\frac{1}{\lambda^2} + (\frac{n\pi}{L})^2]$$

Therefore,

$$J_{nn}L^2[\frac{1}{\lambda^2} + (\frac{n\pi}{L})^2]/2 =$$

$$\int_{x''=0}^{L} [\frac{2}{\lambda}\sin^2\frac{n\pi x''}{L} + \frac{n\pi}{L}\sin\frac{n\pi x''}{L}e^{-x''/\lambda} - (-1)^n \frac{n\pi}{L}\sin\frac{n\pi x''}{L}e^{-(L-x'')/\lambda}]dx''$$

$$= x''/\lambda |_{x''=0}^{L} + \frac{n\pi}{L}(I_3 + I_4)$$

Changing the variable of integration again,

$$u = \sin\frac{n\pi x''}{L}, \quad du = \frac{n\pi}{L}\cos\frac{n\pi x''}{L}dx''$$

$$v = -\lambda e^{-x''/\lambda}, \quad dv = e^{-x''/\lambda}dx''$$

and

$$v = +\lambda e^{-(L-x'')/\lambda}, \quad dv = e^{-(L-x'')/\lambda}dx''$$

we get,

$$I_3 = -\lambda \sin\frac{n\pi x''}{L}e^{-x''/\lambda} |_{x''=0}^{L} + \frac{n\pi\lambda}{L}\int_{x''=0}^{L} \cos\frac{n\pi x''}{L}e^{-x''/\lambda}dx''$$

The last integral is evaluated by changing the variables of integration to:

$$u = \cos\frac{n\pi x''}{L}, \quad du = -\frac{n\pi}{L}\sin\frac{n\pi x''}{L}dx''$$

$$v = -\lambda e^{-x''/\lambda}, \quad dv = e^{-x''/\lambda}dx''$$

From which,

$$I_3 = -\frac{n\pi\lambda^2}{L}[(-1)^n e^{-L/\lambda} - 1] - (\frac{n\pi\lambda}{L})^2 I_3$$

$$I_3 = [-(-1)^n e^{-L/\lambda} + 1][(\frac{n\pi}{L})/\{\lambda^{-2} + (\frac{n\pi}{L})^2\}]$$

Similarly,

$$I_4 = [e^{-L/\lambda} - (-1)^n][(\frac{n\pi}{L})/\{\lambda^{-2} + (\frac{n\pi}{L})^2\}]$$

Therefore,

$$J_{nn} = \frac{2(\lambda/L)^2}{1 + (n\pi)^2(\lambda/L)^2}[\frac{2(n\pi)^2(\lambda/L)^2\{(-1)^{n+1}e^{-L/\lambda} + 1\}}{1 + (n\pi)^2(\lambda/L)^2} + \frac{L}{\lambda}]$$

$$J_{nn}(\lambda/L \to \infty) \to 8/(n\pi)^2 \quad \textit{for n odd}$$
$$\to 0 \quad \textit{for n even}$$
$$J_{nn}(n\pi\lambda/L \to \varepsilon) \to 2\lambda/L \quad \textit{for all n}$$

Appendix 9B: Charts of Acceptance Integrals

In this appendix, the joint and cross-acceptance integrals for uniform, single-span beams with different boundary conditions are given. In each case, the plot is given in two ranges: in linear scale from $\lambda/L = 0$ to $\lambda/L = 0.2$, and then in semi-log scale up to $\lambda/L = 10$. These acceptance integrals are computed either from the closed-form solutions (for spring-supported and simply-supported beams) or with the finite-element method. These charts are reproduced from Au-Yang (2000). Notice that the acceptance integrals are not very sensitive to the boundary conditions. Thus, for a beam that is simply-supported at one end and clamped at the other, the average value between a beam simply-supported at both ends and one clamped at both ends can be used.

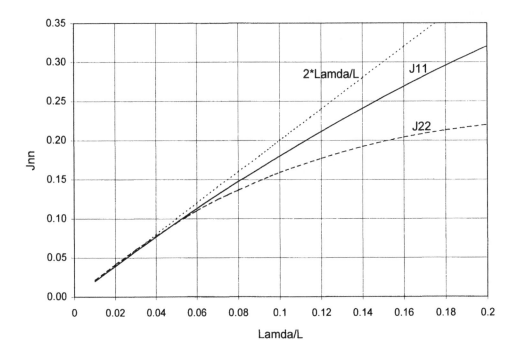

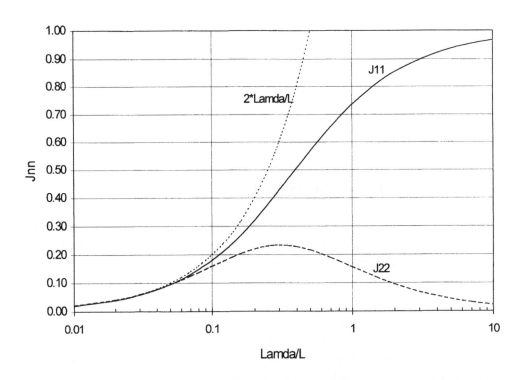

Figure 9.16 Joint acceptances for a rigid beam spring-supported at both ends

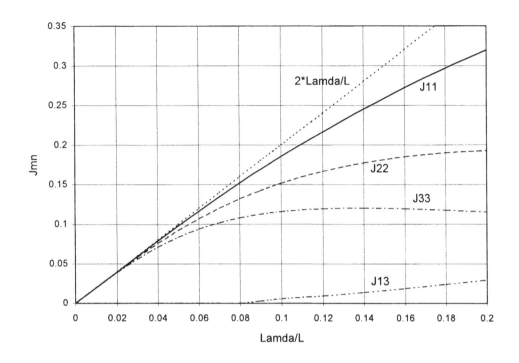

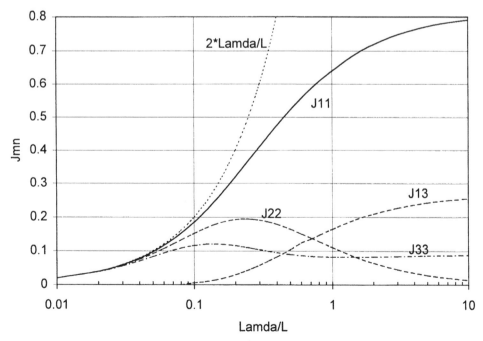

Figure 9.17 Joint and cross-acceptances for a beam
simply-supported at both ends

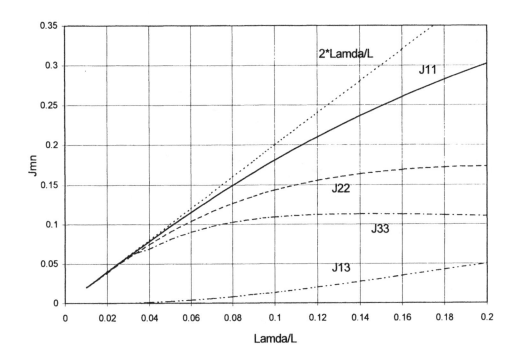

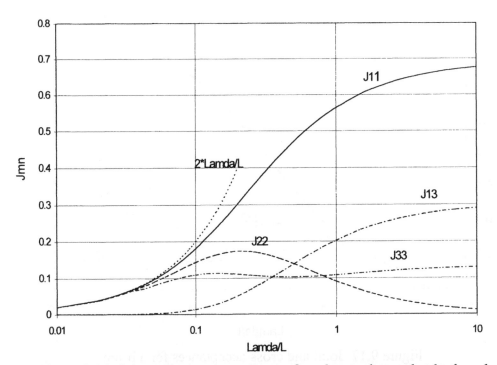

Figure 9.18 Joint and cross-acceptances for a beam clamped at both ends

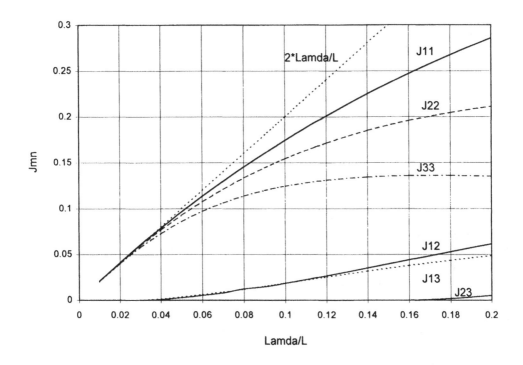

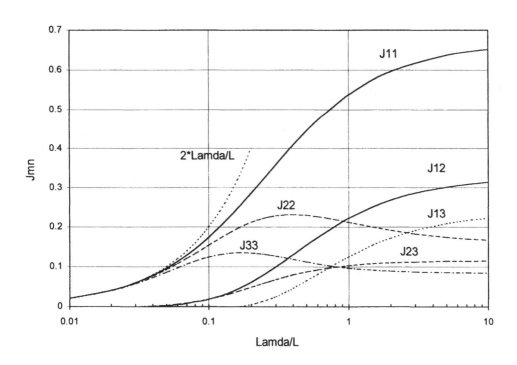

Figure 9.19 Joint and cross-acceptances for a cantilever beam

References:

ASME, 1979, Steam Table, Fourth Edition.

ASME, 1998, Boiler Code Sec III Appendix N-1300 Series.

Au-Yang, M. K. and Connelly, W. H., 1977, "A Computerized Method for Flow-Induced Random Vibration Analysis of Nuclear Reactors Internals," Journal Nuclear Engineering and Design, Vol. 42, pp. 257-263.

Au-Yang, M. K., 1986, "Turbulent Buffeting of a Multi-Span Tube Bundle," ASME Transaction, Journal of Vibration, Acoustics Stress and Reliability in Design, Vol. 108, pp. 150-154.

Au-Yang, M. K., 1987, "Development of Stabilizers for Steam Generator Tube Repair", Journal Nuclear Engineering and Design, Vol. 103, pp. 189-197.

Au-Yang, M. K., 2000, "The Joint and Cross-Acceptances in Cross-Flow Induced Vibration, Part I--Theory and Part II--Charts and Applications," ASME Transaction, Journal of Pressure Vessel Technology, Vol. 122, pp. 349-361.

Axisa, F., et al, 1986, Flow-Induced Vibration of Steam Generator Tubes, Electric Power Research Institute Report EPRI NP-4559.

Axisa, F., et al., 1988, "Random Excitation of Heat Exchanger Tubes by Cross-Flow," in Symposium on Flow-Induced Vibration and Noise, Vol. 2, ASME Special Publication, edited by M. P. Paidoussis, pp. 23-46.

Blevins, R. D., Gilbert, R. J. and Villard, B., 1981, "Experiments on Vibration of Heat Exchanger Tube Arrays in Cross-Flow," Transaction of the Sixth International Conference on Structural Mechanics in Reactor Technology, Paper B6/9.

Blevins, R. D., 1990, Flow-Induced Vibration, Second Edition, Van Nostrand Reinhold, New York.

Blevins, R. D., 1993, "Turbulence-Induced Vibration," in Technology for the 90s, , edited by M. K. Au-Yang, ASME Press, New York, pp. 681-709.

Brenneman, B., 1987, "Random Vibrations Due to Small-Scale Turbulence With the Coherence Integral Method", ASME Journal of Vibration, Acoustics, Stress, and Reliability in Design, Vol.109, pp. 158-161.

Chen, S. S. and Jendrezejczyk, J. A., "1987, "Fluid Excitation Force Acting on a Square Tube Array," ASME Transaction, Journal of Fluid Engineering, Vol. 109, pp. 415-423.

El-Wakil, M. M., 1962, Nuclear Power Engineering, Chapter 11, McGraw Hill, New York.

Morse, P.M. and Feshbach, H., 1953, Method of Theoretical Physics, Chapter 6. McGraw Hill, New York.

Oengoeren, A. and Ziada, S., 1992, "Unsteady Fluid Force Acting on a Square Tube Bundle in Air Cross-Flow," in Proceedings, 1992 International Symposium on Flow-Induced Vibration and Noise, Volume 1, edited by M. P. Paidoussis, ASME Press, pp. 55-74.

Pettigrew, M. J. and Gorman, D. J., 1981, "Vibration of Heat Exchanger Bundles in Liquid and Two-Phase Cross-Flow," in Flow-Induced Vibration Guideline, ASME Special Publication PVP-52, edited by P. Y. Chen, pp. 89-110.

Pettigrew, M. J. and Taylor, C. E., 1993, "Two-Phase flow-Induced Vibration," in Technology for the 90s, edited by M. K. Au-Yang, ASME Press, pp. 811-864.

Taylor et al, 1986, "Experimental Determination of Single and Two-Phase Cross-Flow-Induced Forces on tube Rows," in Flow-Induced Vibration--1986, ASME Special Publication PVP Vol. 104, edited by S. S. Chen, pp. 31-39.

Tong, L. S., 1966, Boiling Heat Transfer and Two-Phase Flow, Wiley, New York.

CHAPTER 10

AXIAL AND LEAKAGE-FLOW-INDUCED VIBRATIONS

Summary

Under normal conditions, axial flow-induced vibrations of power and process plant components are of much less concern than cross-flow-induced vibrations with comparable flow velocities and fluid densities. Because of this and because industry research effort over the past 20 years has been focused on cross-flow-induced vibration, axial flow-induced vibration is often overlooked in the industry. A detailed review of the operating history of commercial nuclear power, in which detailed documentation of every flow-induced vibration incident is required by law and is available to the public, revealed that in the last 40 years, axial flow-induced vibration has caused as much monetary loss to the industry as cross-flow-induced fluid-elastic instability and vortex-induced vibration combined.

In the absence of narrow flow channels, axial flow-induced vibration can be estimated by three different methods:

(1) The acceptance integral method developed in Chapter 8 for parallel-flow-induced vibration.
(2) Equation of Wambsganss and Chen (1971) which estimates the *minimum* response of a rod or a tube subject to axial flow,

$$y_{rms}(x) = \frac{0.0255\kappa\gamma D^{1.5}D_H^{1.5}V^2\psi(x)}{L^{0.5}f^{1.5}m_t\zeta} \tag{10.1}$$

(3) Paidoussis' (1981) equation, which estimate the upper bound responses,

$$\frac{y_{max}}{D} = (5E-4)K\alpha^{-4}\{\frac{u^{1.6}\varepsilon^{1.8}\,\text{Re}^{0.25}}{1+u^2}\}\{\frac{D_h}{D}\}^{0.4}\{\frac{\beta^{2/3}}{1+4\beta}\} \tag{10.2}$$

Under normal conditions, all three equations yield results that are low compared with cross-flow-induced vibration with comparable flow velocities and fluid densities. However, in the presence of very narrow flow channels surrounded by flexible structures, a phenomenon called leakage-flow-induced vibration may occur in which the structure extracts energy from the potential pressure energy and gets into spontaneous vibration, often with very detrimental results. Leakage-flow-induced vibration had caused the nuclear industry substantial financial losses over the last 40 years. In spite of this, there is no industry-accepted, simple equation to estimate the critical flow velocity at which leakage-flow-induced instability may occur. The following are guidelines to avoid the situations that favor leakage-flow-induced instability:

- Avoid, if at all possible, flow in very narrow gaps bounded by flexible structural boundaries.
- In flow channels bounded by flexible structural boundaries, avoid the situation when a higher-pressure drop occurs in the upstream end of the channel than that in the down-stream end.
- In the case of a cantilever rod surrounded by a narrow fluid gap, avoid the condition when the flow direction is from the free end to the clamped end of the cantilever.
- Avoid massive end pieces in components in narrow flow channels.

Under normal operating conditions, axial flow-induced instability is usually not a problem in industrial piping systems designed to withstand seismic and accidental pipe sever events.

Nomenclature

A, A_o	Outside cross-sectional area
A_i	Inside cross-sectional area
D, D_o	Outside diameter of the pipe
D_i	Inside diameter of the pipe
D_H	Hydraulic diameter
E	Young's modulus
f	Modal frequency, Hz
I	Moment of inertia
J_{11}	First modal joint acceptance
K	= 1.0 for "controlled," very quiet flow
	= 5.0 for turbulent flow in industrial environment
L	Length of rod or tube
m	Structural mass per unit length
m_t	Total (structural and hydrodynamic) mass per unit length
Re	Reynolds number
u	$= VL\sqrt{\rho A / EI}$
V	Flow velocity
y_{0-p}	0-peak response of a tube or a rod
y_{rms}	Root mean square response of a tube or a rod
α	$= \pi$ for simply-supported rods,
	$= 4.73$ for rods clamped both ends
β	$= \rho A / (\rho A + m)$
γ	Specific gravity of fluid

Γ_1	Coherence function in the streamwise direction
Γ_2	Coherence function in the cross-stream direction
ε	$= L/D$
ζ	Modal damping ratio
κ	Numerical constant in Equation (10.1)
v	Kinematic viscosity
ρ	Fluid density
ψ	Mode-shape function

10.1   Introduction

In the absence of narrow flow passages, axial flow-induced vibration is a much smaller problem than cross-flow-induced vibration. The exception is axial flow in steam nozzles and in cavitating venturis. In at least one incident, high-velocity unsteady steam flow exiting steam nozzles had caused severe vibration problems in the attached piping system, requiring expensive in-situ modifications. In other incidents, cavitating venturis designed to limit the mass flow in piping systems by its "choking" action, had caused large amplitude vibrations in the attached piping systems, causing handwheels to back off and nozzle joints to crack. Other than these isolated examples, axial flow-induced vibration is usually not a concern in the power and process industries, although it can be a serious problem in rocket and jet engine exhaust nozzles.

The situation dramatically changes when flexible structures inside narrow flow passages are involved. When the conditions are right, a phenomenon called leakage-flow-induced instability can occur, where the structure can be set into spontaneous vibration. The term "leakage-flow" was derived from the fact that this type of flow-induced vibration usually involves only a very small volumetric flow rate. The structure is not excited by such dynamic forces as turbulent eddies or vortex shedding. Rather, it extracts the potential energy form the static pressure in the surrounding fluid, and becomes unstable as the negative "fluid damping" becomes larger than the total system damping.

A review of flow-induced vibration problems in the 40 years of commercial nuclear power operation revealed that leakage-flow-induced vibration repeatedly caught the industry by surprise. Aside from fretting wear caused by turbulence-induced vibration, which by now is almost accepted as a normal operational problem, leakage-flow-induced vibration may be the second mostly costly flow-induced vibration mechanism in the nuclear industry, especially in the early days of commercial nuclear power. Yet, in spite of this, there have been very little research activities on this flow-induced vibration phenomenon. Today, the exact mechanism of leakage-flow-induced instabilities is still not well known, and there is no industry-accepted method to predict the condition under which leakage-flow-induced instability will occur. Even published codes and standards and design guides rarely discuss this issue. Engineers who are aware of the potential of leakage-flow-induced vibration usually resolve to qualitative methods to avoid the situation in which the potential of leakage-flow-induced instability may occur, instead of designing the components to operate below the instability threshold. In the following sections, simplified expressions for assessing the vibration amplitudes due to steady axial flows are given together with numerical work examples and qualitative discussion of leakage-flow-induced vibration. Excitation of piping systems due to unsteady fluid forces such as pulsation flow, water or steam hammer, pipe whip, as well as axial flow-induced vibration problems in very flexible structures are not discussed. The readers are referred to the work by Chen (1985) and Paidoussis (1998) for these specialized topics.

10.2 Turbulence-Induced Vibration in Axial Flow

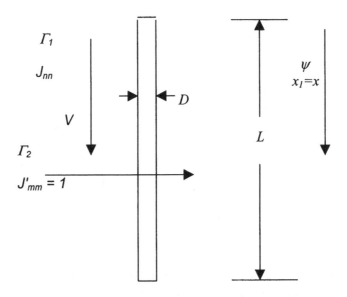

Figure 10.1 Notations for axial flow over 1D structures
(Compare with Figure 9.2)

The acceptance integral method for parallel-flow-induced vibration discussed extensively in Chapter 8 applies to turbulence-induced vibration in axial flows. Indeed, much of the data on the power spectral density, such as those by Clinch (1969), Au-Yang and Jordan (1980), Au-Yang et al., (1995), were for axial flows. In this chapter, we limit ourselves to a sub-set of axial flow-induced vibration problems in which the structure is one-dimensional, with the flow along the axis of the structure (Figure 10.1; also compare with Figure 9.2). The longitudinal acceptance is in the direction of the flow and along the axis of the structure, while the lateral acceptance is perpendicular to the flow and the axis of the structure. As in the case of cross-flow over 1D structures, we assume in the present case that the fluctuating pressure is fully coherent across the structure and, as a result, the lateral acceptance is equal to unity. Based on this assumption, only the longitudinal acceptances have to be computed. The calculation can be carried out following the examples given in Chapter 8 and in Example 10.2 (given later in this Chapter). However, alternative simplified expressions are available to estimate the vibration amplitudes in these special cases.

10.3　Some Simplified Equations for External Axial Flow

For vibration due to external axial flow, Wambsganss and Chen (1971) theoretically derived a simplified equation for estimating the *lower bound* response of a flexible cylindrical structure to turbulence[1]:

$$y_{rms}(x) = \frac{0.0255 \kappa \gamma D^{1.5} D_H^{1.5} V^2 \psi(x)}{L^{0.5} f^{1.5} m_t \zeta} \tag{10.1}$$

where

κ　= 2.56E-3　lbf-s$^{2.5}$-ft $^{-5.5}$ = 7.85 N-s$^{2.5}$-m$^{-5.5}$
γ　= Specific gravity of fluid (density of fluid/density of water)
D　= Outside diameter (OD) of the cylinder

―――――――――――――――

[1] As published by Chen (1985), γ was not in the Equation (10.1). The original equation was given as:

$$y_{rms}(x) = \frac{0.018 \kappa D^{1.5} D_H^{1.5} V^2 \psi(x)}{L^{0.5} f^{1.5} m_t \zeta}$$

and was applicable to water flow only. The unit of κ was quoted to be in lb-s$^{2.5}$ft $^{-5.5}$ and m_t was in mass per unit length. If we carry out a dimensional check as in Example 1.2, we would find in order that the right-hand side of Equation (10.1) has the length unit as the left-hand side, then the mode-shape function must be dimensionless and the lb in κ must be in lb force. In some other publications, Chen (1988) quoted the mode-shape function was equal to $\sqrt{2} \sin(\pi x / L)$ which is consistent with a dimensionless mode shape function, but was a very unconventional way of normalizing the mode shape. In some other publications, Chen gave κ=0.244 N-s$^{2.5}$m$^{-5.5}$, which is not consistent with 2.56E-3 lbf-s$^{2.5}$-ft $^{-5.5}$. Since Chen originally derived the equation in US units, κ=7.85 N-s$^{2.5}$m$^{-5.5}$ is correct. In Equation (10.1), $\sqrt{2}$ =1.414 is combined with 0.018 in Chen's original equation so that the mode-shape function can be normalized to maximum equal to unity (such as a sine function). In this revised form, Equation (10.1) gives results that agree better with those obtained from the acceptance integral method.

D_H = Hydraulic diameter = 4*cross-sectional area/wetted perimeter
    = OD for rods = diametral gap width for annulus
V = Flow velocity
ψ = Mode shape function, normalized to maximum value = 1.0
L = Length of rod
f = Modal frequency, Hz
m_t = Total mass (structural+hydrodynamic) per unit length
m = Structural mass per unit length
ζ = Modal damping ratio

Equation (10.1) theoretically applies to any mode of vibration. In actual applications, Equation (10.1) suffers from vaguely defined notations. Paidoussis' (1981) equation,

$$\frac{y_{max}}{D} = (5E-4)K\alpha^{-4}\{\frac{u^{1.6}\varepsilon^{1.8}\,Re^{0.25}}{1+u^2}\}\{\frac{D_h}{D}\}^{0.4}\{\frac{\beta^{2/3}}{1+4\beta}\} \qquad (10.2)$$

is more often used for hand calculations. Equation (10.2) is a simplified empirical equation for the response of flexible structures to external axial flow, and is applicable only to the fundamental mode. The variables in Equation (10.2) are defined as follows:

α $=\pi$ for simply-supported rods
    =4.73 for rods clamped at both ends
K = 1.0 for "controlled," very quiet flow = 5.0 for turbulent flow
u $=VL\sqrt{\dfrac{\rho A}{EI}}$
ρ = Fluid density
A = Cross-sectional area
E = Young's modulus
I = Moment of inertia
ε $= L/D$
Re = Reynolds number = VD/v
v = Kinematic viscosity
β $= \rho A/(\rho A + m + hydrodynamic\ mass)$

Other notations are defined under the Wambsganss and Chen equation.

Example 10.1

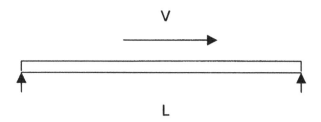

Figure 10.2 Axial flow over a simply-supported tube

Take the tube in Example 9.1, and consider only one span of the tube. Assume this tube is simply-supported at both ends and subject to the same external flow velocity as in Example 9.1, except that the flow is along the tube axis instead of perpendicular to it (Figure 10.2), and the pressure and temperature of the flow are as given in Table 10.1. All the other parameters are the same as in Example 9.1 and are repeated in Table 10.1. What is the response of this tube?

Table 10.1: Input Parameters for Example 10.1

	US Unit	SI Unit
Ambient pressure	1200 psi	8.27E+6 Pa
Ambient temperature	600 deg. F	315 deg. C
Length of tube, L	50 in	1.27 m
OD, D_o	0.625 in	0.01588 m
ID, D_i	0.555 in	0.01410 m
Young's modulus, E	29.22E+6 psi	2.02 E+11 Pa
Tube material density, ρ_m	0.306 lb/in^3	8470 Kg/m^3
	(7.919E-4 (lb-s^2/in)/in^3)	
Damping ratio, ζ	0.01	0.01
Axial flow velocity, V	40 ft/s	12.19 m/s
	(480 in/s)	
Shell-side fluid density, ρ_s	1.82 lb/ft^3	29.15 Kg/m^3
	(2.734E-6 (lb-s^2/in)/in^3)	
Tube-side water density, ρ_i	43.6 lb/ft^3	698.4 Kg/m^3
	(6.550E-5 (lb-s^2/in)/in^3)	

Table 10.2: Computed Parameters for Example 10.1

	US Unit	SI Unit
Outside cross-sectional area, A_o	0.307 in^2	1.99E-4 m^2
Inside cross-sectional area, A_i	0.242 in^2	1.56E-4 m^2
Tube cross-sectional area, A_m	0.0649 in^2	4.19E-5 m^2
Moment of inertia, $I = \pi(D_o^4 - D_i^4)/64$	0.00283 in^4	1.18E-9 m^4
Total mass density, $m_t = \rho_m A_m + \rho_s A_o + \rho_i A_i$	6.81E-5 (lb-s^2/in)/in	0.469 Kg/m
Sp. gr. of shell-side fluid, $\gamma = \rho_s / \rho_w$	0.0292	0.0292
Reynolds number, Re	3.47E+5	3.47E+5
$u=$	0.07627	0.07627
$\varepsilon = L/D =$	80	80
$\beta = \rho_s A_o / m_t$	0.0123	0.0123

Solution

Method 1: Using Paidoussis' Equation

Paidoussis' Equation (10.2) is the simplest one to use for this problem. Most of the necessary intermediate parameters have been computed in Example 9.1. These are repeated in Table 10.2 for easy reference. There is one additional parameter that is necessary in Equation (10.2) but is not calculated in Example 9.1—the Reynolds number. From the ASME Steam Table (1979), at 1,200 psi and 600 deg. F, the kinematic viscosity is ν=6.0E-6 ft^2/s. Since this is given in feet unit, we must use the velocity in ft/s and the diameter of the tube in ft when we calculate the Reynolds number, Equation (6.2), (see also Example 1.1):

$$Re = \frac{VD}{\nu} = 40*(0.625/12)/6.0\text{E-}6 = 3.47\text{E+}5$$

This is the same in both the US and SI unit systems and is given in the 7th row of Table 10.2. Other parameters we use are:

K=5 for industrial turbulent flow
α=3.1416 for simply-supported tube

For an isolated tube, the hydraulic diameter is the same as the tube diameter. Therefore,

$$D_H / D = 1.0$$

The cross-sectional area, moment of inertia and total linear mass density of the tube have been calculated before in Example 9.1. They are reproduced in the first five rows in Table 10.2, from which u can be calculated. In the US unit system, for example,

$$u = VL\sqrt{\frac{\rho A}{EI}} = 480*50\sqrt{\frac{2.734E-6*0.370}{29.22E+6*0.00283}} = 0.07627$$

This is the same in both unit systems. Substituting the computed parameters in Table 10.2 into Equation (10.2), we get,

$$y_{max} = 8.53E\text{-}4 \text{ in } (2.17E\text{-}5 \text{ m}) \quad \text{zero-peak}$$

Besides being very simple, another advantage of Paidoussis' Equation (10.2) is that each factor in the equation is dimensionless. This eliminates a lot of ambiguities one often encounters in the published literature.

Method 2: Using Equation of Wambsganss and Chen

The first step in using Equation (10.1) is to calculate the natural frequency of the tube. This has been carried out in Example 9.1, where it is found that,

$$f_I = 21.9 \text{ Hz}$$

Equation (10.1) also requires that the mode-shape functions be normalized to maximum=1.0. For a tube simply-supported at both ends,

$$\psi_1(x) = \sin\frac{\pi x}{L}$$

The rest of the calculation must be done in ft and slug mass units. At the mid point of the tube, where the mode-shape function (and therefore the vibration amplitude) is maximum,

$$\psi_1(x = L/2) = 1.0$$

From Example 9.1:

$$m_I = 6.81E\text{-}5 \text{ (lb-s}^2/\text{in})/\text{in} = 6.81E\text{-}5*12*12 = 9.81E\text{-}3 \text{ slug/ft}$$

$$V = 40 \text{ ft/s}$$

$$D = D_H = 0.625/12 = 0.0521 \text{ ft}$$

$L=50/12=4.167$ ft

$\gamma=1.82/62.4=0.0292$

Substituting into Equation (10.1),

$$y_{rms}(x=L/2) = \frac{0.025*2.56E-3*0.0292*0.0521^{\wedge}1.5*0.0521^{\wedge}1.5*40^{\wedge}2*1.0}{4.167^{\wedge}0.5*21.9^{\wedge}1.5*9.81E-3*0.01}$$

$$=2.11\text{E-5 ft} = 2.54\text{E-4 in} = 6.45\text{E-6 m}$$

If we carry out the calculation in SI units:

$$y_{rms}(x=L/2) = \frac{0.025*7.84*0.0292*0.01588^{\wedge}1.5*0.01588^{\wedge}1.5*12.19^{\wedge}2*1.0}{1.27^{\wedge}0.5*21.9^{\wedge}1.5*0.469*0.01}$$

$$= 6.44\text{E-6 m}$$

By comparison, the response of the same tube to cross-flow of the same velocity and fluid density is, as computed in Example 9.1, equal to 0.035 in. rms. Thus, vibration due to axial flow is much less of a concern than that due to cross-flow. Since Equation (10.1) calculates the rms while Equation (10.2) calculates the zero-peak amplitudes, direct comparison of the two results is difficult. Assuming the response follows a Gaussian distribution, the above result indicates that 99.7% of the time, the vibration amplitude at the mid-span of the tube is less than $3*y_{rms}$, or about 7.62E-4 in. This is consistent with the result obtained from Equation (10.2).

Example 10.2

Return to the above example, assume the flow is inside the tube, with a velocity and density equal to those outside of the tube in Example 10.1. For the purpose of comparing the results, also assume that the *inside* diameter of the tube is the same as the outside diameter of the tube in Example 10.1, while the wall thickness, material density and outside fluid density is such that the fundamental frequency remains at 21.9 Hz and the total linear mass density remains the same as that in Example 10.1. Estimate the mid-span vibration amplitude of the tube using the acceptance integral method developed in Chapter 8.

Solution

We first summarize the known parameters in Table 10.3:

Table 10.3: Known Parameters for Example 10.2

	US Unit	SI Unit
Length of tube, L	50 in	1.27 m
ID, $D_i = D_H$	0.625 in	0.01588 m
Damping ratio, ζ	0.01	0.01
Flow velocity in tube, V	40 ft/s	12.19 m/s
	(480 in/s)	
Fluid density, ρ	1.82 lb/ft^3	
	(2.734E-6 (lb-s^2/in)/in^3)	29.15 Kg/m^3
Total linear density of tube, m_t	6.81E-5 (lb-s^2/in)/in	0.469 Kg/m
Fundamental modal frequency, f_l	21.9 Hz	21.9 Hz

Table 10.4: Computed Parameters for Example 10.2

	US Unit	SI Unit
$\Omega = 2\pi f_1 \delta * / V$	0.09	0.09
$G_p(f)/\rho^2 V^3 D_H$ (from Fig. 8.15 at Ω=.09)	5.0E-5	5.0E-5
G_p	2.58E-8 psi^2/Hz	1.22 Pa^2/Hz
Δ_1	6.25E-3	6.25E-3
$4f_lL/V$	9.125	9.125
J_{ll} (from Figure 8.21)	0.04	0.04
$\psi_1(x = L/2)$	0.2 in$^{-1/2}$	1.255 m$^{-1/2}$

We shall use Equation (8.50) to solve this problem. In the notation of this problem and restricted to one dimension and the fundamental mode, this equation takes the form (also compare with Equation (9.4)),

$$< y^2(x = L/2) >= \frac{LD_i^2 G_p(f_\alpha)\psi_1^2(x = L/2)J_{11}(f_1)}{64\pi^3 m_1^2 f_1^3 \zeta_1} \tag{10.3}$$

With a uniform density and the mode shape normalized to unity, the modal generalized mass $m_1=m_t$. The only variables we need to calculate are the turbulence random pressure PSD, G_p and the joint acceptance J_{11}. The following outlines the steps to calculate these two parameters.

The Random Pressure PSD

Here, some judgment must be made. Three different random pressure PSD plots are given in Chapter 8—Figures 8.15, 8.16 and 8.17. Of these, Figure 8.17 is for larger piping systems with valves and elbows and turns. Figure 8.15 is based on data obtained from tests on straight pipes and is closest to the present problem; it will be used to calculate the random pressure PSD. We shall return to Figure 8.16 at the end of the problem solution. To read off the dimensionless PSD from Figure 8.15, we must first calculate

$$\Omega = 2\pi f_1 \delta * / V$$

For small-diameter pipes and tubes, the displacement boundary layer thickness is equal to one half the internal diameter. This, together with the known first modal frequency and the velocity, enable us to calculate Ω=0.09. From Figure 8.15,

$$G_p(f)/\rho^2 V^3 D_H = 5.0E-5$$

from which the random pressure PSD can be computed. The values in both US and SI units are given in Table 10.4, together with the made-shape value at the mid-point of the tube, where the vibration amplitude is maximum.

The Joint Acceptance J_{11}

The longitudinal joint acceptance can be read off from Figure 8.21. We must first calculate the two parameters,

$$\Delta_1 = \delta * / L \text{ and } 4f_1 L/V$$

From the given values in Table 10.3, we can readily calculate

$$\Delta_1 = 6.25E-3 \quad \text{and} \quad 4f_1 L/V = 9.125$$

From Figure 8.21, we can see that the joint acceptance J_{11} is very small,

$$J_{11} \approx 0.03$$

Based on discussions in Section 10.1, the lateral joint acceptance is assumed to be unity:

$$J'_{11} = 1.0$$

Using these values, the rms response at the mid-span of the tube can be computed. For example, in the US unit system:

$$<y^2> = \frac{50*0.625^\wedge 2*2.58E-8*0.2^\wedge 2*0.03}{64*3.1416^\wedge 3*6.81E-5^\wedge 2*21.9^\wedge 3*0.01} = 6.26E-7$$

From which,

$$y_{rms} = 7.91E\text{-}4 \text{ in} \quad (2.01E\text{-}5 \text{ m})$$

At this point it should be emphasized that since the inside diameter of the tube in Example 10.2 is equal to the outside diameter of the tube in Example 10.1, and we purposely make the fundamental modal frequency and the linear mass densities of the two tubes the same, if Equation (10.3) is used to find the mid-span vibration amplitude in Example 10.1, we would get exactly the same response as above, compared with y_{rms}=2.54E-4 in. From Equation (10.1) and $y_{0\text{-}peak}$=8.53E-4 in. from Equation (10.2). Thus both equations (10.1) and (10.2) give results that are consistent with those obtained from the more rigorous Equation (10.3).

Now let us return to Figure 8.16, which is approximated by Equation (8.68),

$$\frac{G_p(f)}{\rho^2 V^3 D_H} = 0.272E-5/S^{0.25}, \quad S<5;$$
$$= 22.75E-5/S^3, \quad S>5$$

(8.68)

where

$$S = 2\pi f D_H / V = 2\Omega = 0.18$$

Substituting into Equation (8.68), we get in US units:

$$G_p \approx 0.42E-5 \text{ psi}^2/\text{Hz}$$

If we use either Figure 8.16 or Equation (8.68) to estimate the turbulence pressure PSD in the above calculations, we would get,

$$y_{rms} = 2.3E\text{-}4 \text{ in}$$

compared with 2.54E-4 in. calculated with Wambsganss and Chen's Equation (10.1).

10.4   Stability of Pipes Conveying Fluid

The stability of pipes containing flowing fluid is discussed extensively by Paidoussis (1998), Chen (1985) and, to a smaller extent, by Blevins (1990). Readers who are interested in the detailed mathematical treatment of this topic should refer to the above references. The following offers simplified derivation of the most commonly used equations based largely on physical reasoning.

Buckling Instability

Figure 10.3 shows a straight pipe of inside cross-sectional area A_i supported at the two ends and carrying fluid with density ρ flowing at a velocity V. Suppose a slight lateral displacement $y(x)$ is given to the pipe. From our introductory strength of material course, the elastic force due to the stiffness of the pipe is given by

$$EI\frac{\partial^4 y}{\partial x^4}$$

This acts in the direction of the displacement (thus, the reaction force acts in the direction opposite to y). Since the pipe is now curved, the flowing fluid will exert a centrifugal force on the pipe. From fundamental dynamics, this centrifugal force is equal to

$$\rho A_i V^2 \frac{\partial^2 y}{\partial x^2}$$

and acts in the direction opposite to displacement (thus, the reaction force acts in the direction of y). If this centrifugal force is smaller than the stiffness force, then the stiffness force will restore the pipe to its original equilibrium position. The fluid conveying pipe is then stable. However, if the velocity and density of the flowing fluid are high enough, the centrifugal force will exceed the restoring stiffness force. The pipe will deflect even more. A condition of instability is thus established. The velocity at which the centrifugal force equals the stiffness force of the pipe is call the critical velocity for static divergence, or buckling instability:

$$\rho A_i V^2 \frac{\partial^2 y}{\partial x^2} = -EI\frac{\partial^4 y}{\partial x^4}$$

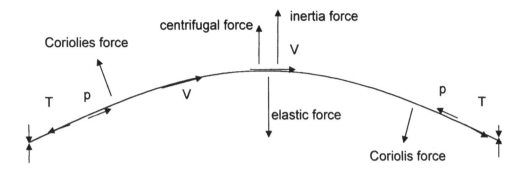

Figure 10.3 A simply-supported pipe conveying fluid

Using Equation (3.15),

$$\{y\} = \sum_n a_n(t)\psi_n(x)$$

with $\psi_n(x) = \sqrt{\dfrac{2}{L}}\sin\dfrac{n\pi x}{L}$, one can readily show that the critical velocity for buckling or static divergence instability is given by

$$V_c = \frac{\pi}{L}\sqrt{\frac{EI}{\rho A_i}} \qquad\qquad (10.4)$$

for a pipe simply-supported at both ends. The critical velocity for a pipe clamped at both ends can be derived similarly using the expressions for the corresponding mode shapes given in Table 3.1. The term "static" arises from the fact that the inertia force term, $m_i\partial^2 y/\partial t^2$, due to the vibration of the pipe has not been included in the above derivation. This, however, is not the only dynamic force acting on the pipe. Since the pipe is hinged at the two ends, as the pipe deflects at a velocity of $\partial y/\partial t$, a pipe element rotates about a fixed point in space (with the axis of rotation perpendicular to the plane of vibration, see Figure 10.3), and with the fluid flowing along its length at velocity V. From our dynamics course we know there exists yet another force, called the *Coriolis* force, acting on the pipe element. The magnitude of this force is equal to $2\rho A_i V \partial^2 y/\partial x \partial t$. If, in addition, the fluid inside the pipe has a pressure p and the pipe has an initial axial load T (positive is tension), then the equation of motion of a pipe conveying fluid can be written as

$$EI\frac{\partial^4 y}{\partial x^4} + (pA_i - T)\frac{\partial^2 y}{\partial x^2} + \rho A_i V^2 \frac{\partial^2 y}{\partial x^2} + 2\rho A_i V\frac{\partial^2 y}{\partial x\partial t} + (m + \rho A_i)\frac{\partial^2 y}{\partial t^2} = 0 \qquad (10.5)$$

inertia force

Coriolis force

centrifugal force

pre-loads

elastic force

The condition for buckling or static divergence in this case is given by,

$$EI\frac{\partial^4 y}{\partial x^4} + (\rho A_i V_c^2 + pA_i - T)\frac{\partial^2 y}{\partial x^2} = 0 \qquad (10.6)$$

Equation (10.5) can be solved by the modal decomposition method using one of the mode-shape functions given in Table 3.1. For a pipe simply-supported at both ends, the critical velocity for the fundamental mode is given by,

$$V_c = \sqrt{\frac{\pi^2 EI}{L^2 \rho A_i} + (\frac{T}{\rho A_i} - \frac{p}{\rho})} \qquad (10.6)$$

Compared with Equation (10.4), it can be seen that an initial tensile load will increase the critical velocity, while an internal pressure will decrease the critical velocity.

The presence of the Coriolis force term in Equation (10.5) greatly complicates its solution process. Unlike any other terms, the Coriolis force is anti-symmetric. That is, if we replace x with $-x$, the Coriolis force term will change its sign while all the other force terms will remain unchanged. In physics terms, the Coriolis force has odd parity and thus will couple symmetric modes to anti-symmetric modes, and vice versa. In addition, the Coriolis force, being proportional to the velocity of transversal vibration, is out of phase with the inertia force that is proportional to the acceleration of transversal vibration. This means that the solution to Equation (10.5) must be consist of both $\sin 2\pi f_n t$ and $\cos 2\pi f_n t$ terms. For a pipe simply-supported at both ends, exact solution of Equation (10.5) (see Blevins, 1990) leads to the equation to the two natural frequencies for each classical mode n for a pipe containing flowing fluid:

$$\left(\frac{\bar{f}_{n1,n2}}{f_n}\right)^2 = \alpha \pm \left\{\alpha^2 - 4\left[1 - \left(\frac{V}{V_c}\right)^2\right]\left[4 - \left(\frac{V}{V_c}\right)^2\right]\right\}^{1/2} \qquad (10.7)$$

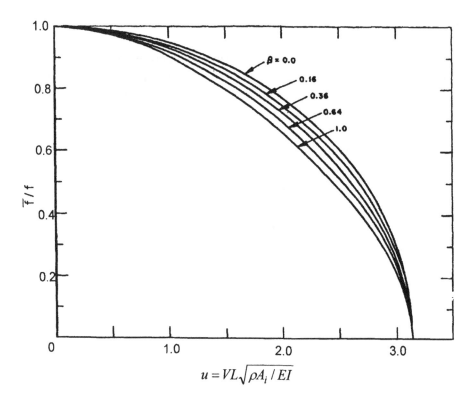

Figure 10.4 Frequency reduction due to flowing fluid (Chen and Rosenberg, 1971)

where

$$\alpha = \frac{17}{2} - \left(\frac{V}{V_c}\right)^2\left[\frac{5}{2} - \left(\frac{128}{9\pi^2}\right)\left(\frac{\rho A_i}{m + \rho A_i}\right)\right]$$

For the lower of the fundamental modes, the following approximate equation can be used:

$$\bar{f}_1 / f_1 = \sqrt{1 - (V/V_c)^2} \qquad\qquad (10.8)$$

Figure 10.4, reproduced from Chen and Rosenberg (1971), plots the frequency reduction ratio $\bar{f}_1 / f_1$ for a simply-supported pipe carrying flowing fluid. Here f_l is the fundamental frequency of the pipe when the velocity of the fluid flow is zero. The added mass due to the fluid, as well as the pre-loads T and p, should be included when calculating f_l. When the velocity reaches the critical value for static divergence, the fundamental frequency decreases to zero.

Example 10.3

A large pipe containing pressurized water is simply-supported at the two ends. The dimensions of the pipe, the material properties of the pipe, and the pressure, temperature of the water in the pipe are given in Table 10.5 in both the US and SI unit systems. The pipe is not axially constrained at the two ends. (i) Calculate the fundamental modal frequency of the pipe when the water is not flowing and the internal pressure p is zero. (ii) Calculate the critical velocity of the flow at which the pipe will suffer buckling instability. Suppose the velocity of water flow is 50 ft/s (15.24 m/s), (iii) What would be the fundamental frequencies of the pipe? (iv) Assuming the maximum vibration amplitude of the pipe at mid-span is 0.1 in (0.000254 m), what are the maximum values of the elastic, pressure, centrifugal, Coriolis, and inertia forces on the pipe and at what location do these maxima act?

Table 10.5: Input Parameters for Example 10.3

	US Units	SI Units
Water pressure, p	2500 psi	1.724E+7 Pa
Water temperature	650 F	343 C
Flow velocity, V	50 ft/s	
	(600 in/s)	15.24 m/s
OD, D_o	42.5 in	1.08 m
ID, D_i	38.0 in	0.965 m
Length, L	60 ft	18.29 m
	(720 in)	
Tube material density, ρ_m	0.306 lb/in^3	
	(7.919E-4 (lb-s^2/in)/in^3)	8470 Kg/m^3
Young's modulus, E	25.0E+6 psi	1.72E+11 Pa
Mid-span p-p amplitude	0.1 in	0.00254 m

Solution

Following our standard practice, we first convert all the units in the US system into a consistent unit set (see Chapter 1). These values are given in parenthesis in Table 10.5. From the ASME Steam Table (1979), the density of water at the given temperature and pressure is ρ=38.23 lb/ft^3 (612.4 Kg/m^3). Since the former is *not* a consistent unit, care must be exercised to convert this density (as is done for ρ_m in Table 10.5) into (lb-s^2/in)/in^3 so that it is consistent with other units used in the US unit system (see Chapter 1). This value, as well as other computed values, are given in Table 10.6 in both unit systems. Since the velocity of water flow, V, is far below the computed critical velocity V_c, the frequency reduction due to water flow is negligible either from Equation (10.8) or from Figure 10.4. Any appreciable frequency reduction is from the static pressure. The force components can be obtained from the mode-shape function, assuming the fundamental mode dominates the response. For a maximum vibration amplitude of δ at the mid-span,

$$y = \delta \sin \frac{\pi x}{L}$$

$$\frac{\partial y}{\partial x} = \frac{\pi \delta}{L} \cos \frac{\pi x}{L}, \qquad \frac{\partial^2 y}{\partial x^2} = -\left(\frac{\pi}{L}\right)^2 \delta \sin \frac{\pi x}{L},$$

$$\frac{\partial^3 y}{\partial x^3} = -\left(\frac{\pi}{L}\right)^3 \delta \cos \frac{\pi x}{L}, \qquad \frac{\partial^4 y}{\partial x^4} = \left(\frac{\pi}{L}\right)^4 \delta \sin \frac{\pi x}{L}.$$

Table 10.6: Computed Parameters for Example 10.3

	US Units	SI Units
Water density, ρ (steam table)	38.2 lb/ft^3	
	5.736E-5 (lb-s^2/in)/in^3	614.2 Kg/m^3
Outside area, $A_o = \pi D_o^2 / 4$	1419 in^2	0.9152 m^2
Inside area, $A_i = \pi D_i^2 / 4$	1134 in^2	0.7317 m^2
Metal area, $A_m = A_o - A_i$	284.5 in^2	0.1836 m^2
Linear mass density, $m = \rho_m A_m$	0.2253 (lb-s^2/in)/in	1555 Kg/m^3
Water mass per unit length, ρA_i	0.0649 (lb-s^2/in)/in	449.4 Kg/m^3
$m + \rho A_i$	0.2902 (lb-s^2/in)/in	2004 Kg/m^3
Moment of inertia, I	57800 in^4	0.0241 m^4
$f_1 = \dfrac{\pi}{2L^2} \sqrt{\dfrac{EI}{m + \rho A_i}}$	6.76 Hz	6.76 Hz
V_c (Equation (10.6))	19490 in/s	494.6 m/s
V_c, if $p=0$ (Equation (10.4))	20580 in/s (1715 ft/s)	521.3 m/s
Maximum elastic force	52.4 lb/in	9152 N/m
Maximum pressure force	-5.40 lb/in	-945.5 N/m
Maximum centrifugal force	-0.0445 lb/in	-7.82 N/m
Maximum Coriolis force	1.44 lb/in	253.4 N/m
Maximum inertia force	-52.3 lb/in	-9144 N/m

From Equation (10.5), we get:

Maximum elastic force = $EI(\pi/L)^4\delta$ at mid-span

Maximum static pressure force = $-pA_i(\pi/L)^2\delta$ at mid-span

Maximum centrifugal force = $-\rho A_i V^2(\pi/L)^2\delta$ at mid-span

Maximum Coriolis force = $\pm 4\pi^2\rho A_i V\bar{f_1}\delta/L$ at $x=0$ and $x=L$

Maximum inertia force = $-(2\pi\bar{f_1})^2(m+\rho A_i)\delta$ at mid-span

These values are given in the last five rows of Table 10.6. Because of the approximations used in the calculations, the forces do not exactly balance one another. The purpose of this example is to illustrate the relative importance of the forces in a normal industrial piping system. It is obvious that the forces due to the fluid flow—the centrifugal and Coriolis forces—are negligible compared with the elastic and inertia forces. This example shows that when properly supported, instability is not a concern in industrial piping systems under normal operational conditions, when the velocity of fluid flow in the pipe is usually far below the critical velocity for buckling instability. The situation is different when a pipe carrying high-pressure fluid is severed. From our experience with a runaway garden hose, all of us are well aware that a severed pipe carrying high-pressure fluid may be unstable, resulting in dangerous "pipe whips."

Industrial piping systems must be properly restrained against accident conditions, such as seismic events. Provisions must be made for high-energy piping systems to minimize "secondary damages" to adjacent equipment and injuries to nearby personnel in the case of pipe sever. Piping systems adequately designed against these accident conditions usually operate far below the instability threshold under normal conditions.

10.5 Leakage-Flow-Induced Vibration

Examples 10.1 to 10.3 show that under normal conditions, vibration due to axial flow is small, much smaller than that induced by cross-flow with comparable fluid velocities and densities. These two examples involve either flow along or inside a tube, with the cross-section of the flow channel either larger than or at least comparable to the cross-sections of the structures. As mentioned in the introduction, the situation can be dramatically different when narrow flow passages are surrounded by structures of much larger dimensions—a situation often encountered in the power and process industries. Under such condition, a phenomenon called leakage-flow-induced instability can occur.

So far all the flow-induced vibration phenomena we have discussed—vortex-induced vibration, turbulence-induced vibration or fluid-elastic instability—involve flow channels of fairly large cross-sectional areas and, as a result, large volumetric flow rates. By contrast, leakage-flow-induced instability usually occurs in very narrow flow channels, the most common of which are flow channels of annular cross-sections bounded by at least one flexible structural member. The volumetric flow rate can be innocently small, hence the name "leakage-flow." The system draws energy from the potential energy in the static pressure of the fluid and set into spontaneous vibration.

Leakage-flow-induced instability can be studied by solving numerically the coupled equations of fluid and structural dynamics. The readers are referred to the treatise by Paidoussis (1998). Collected technical papers, most of them dealing with numerical analysis or testing of specific components, can be found in the Proceedings of the International Symposia on Flow-Induced Vibration and Noise (Paidoussis and Au-Yang, 1984, 1992; Paidoussis et al, 1988). A review article on guidelines to avoid leakage-flow-induced vibration was published by Mulcahy (1983). Most of the analytical papers involve complex mathematical equations and tedious computational algorithms. Those who want to accurately estimate the instability threshold of their fluid-structure system due to leakage-flows have no alternative but to carry out the numerical analysis. Very often, however, we try to avoid the condition under which leakage-flow-induced instability can occur. For this, it is much more important to understand the physics, rather than the mathematical details, of the phenomenon. The following discussion is based on Miller (1970).

Figure 10.5 shows a two-dimensional flow channel with a center rod that is laterally supported by springs. The flow is from the left to the right, with the upstream ambient static pressure equal to p_1 and the downstream ambient static pressure equal to p_2. At the upstream end of the rod is a flow restrictor. Suppose we give the rod a small upward displacement. The small gap between the flow restrictor and the upper wall will start to close, resulting in a smaller flow cross-sectional area and reducing volumetric flow for the upper channel. As a result, the flow in the upper channel starts to decelerate. Since the ambient pressures p_1 and p_2 are fixed, the only way the fluid can decelerate is to have the static pressure p in the upper channel dip below the downstream pressure p_2, so there is a negative net force, which pushes the fluid mass to the left (i.e., the flow is against a negative pressure gradient), as shown in the figure. Exactly the opposite happens in the lower flow channel. As the small gap between the flow restrictor and the lower channel wall opens up, the volumetric flow rate in the lower channel begins to increase. The static pressure in the lower channel must be higher than the downstream ambient pressure p_2, so that a net force is formed to push fluid mass downstream (flow in the direction of a positive pressure gradient). This static pressure profile is shown below the flow channel in Figure 10.5. Thus, a net upward force is formed due to the pressure difference between the upper and lower channels, which tends to push the rod further up. A condition of instability is established.

Now suppose the flow is from right to left, as shown in Figure 10.6, and we again give the rod a small upward displacement so that the gap at the *downstream* end of the upper channel starts to close. As the flow cross-sectional area of the upper channel decreases, the volumetric flow decreases and the flow starts to decelerate. Since the ambient pressures p_1 and p_2 are fixed, the only way the flow in the upper channel can decelerate is when the static pressure in the upper channel rises above p_1, so that the flow is against a negative pressure gradient. Exactly the opposite happens in the lower channel. As the gap in the lower channel opens up, the volumetric flow increases, i.e.; the flow accelerates from right to left in the lower channel. The only way this can happen

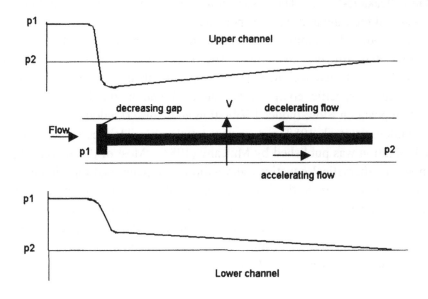

Figure 10.5 Flow from restricted end to open end of a flow channel
generates negative damping

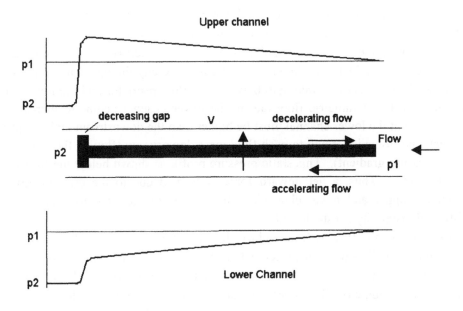

Figure 10.6 Flow from open end to restricted end of a flow channel
generates positive damping

is when the pressure gradient (in the direction of the flow) is positive. Thus, a downward net force is formed, which tends to restore the rod to its equilibrium position. A condition of stability is established.

One can easily generalize the above discussion and find that the free end of a cantilever has the same effect as the restricted end of the rod in the above example.

Rules to Avoid Leakage-Flow-Induced Vibration

From the above discussions, it is obvious that if we can calculate the "negative damping ratio" generated by the flow, we can predict the critical velocity at which a system would become unstable. Unlike the cases of vortex shedding and fluid-elastic instability, where relatively simple, although not very rigorous, empirical equations exist for predicting the critical velocities, no equivalent simple equation exists for predicting the "critical velocity" for leakage-flow-induced instability. It is this lack of quantitative assessment of the critical velocity and, indeed, general lack of interest, that made leakage-flow-induced vibration among the top most costly flow-induced vibration problems in the first 40 years of commercial nuclear power. It is only very recently that analytical methods, usually based on complicated numerical solution of the simultaneous equations of fluid dynamics and structural dynamics, have been developed to predict the onset of instability (see e.g., Fujita and Shintani, 1999; Inada and Hayama, 2000; and the references cited there). These analytical results generally agree with the phenomenological explanations given above.

For the majority of engineers, the present approach to prevent leakage-flow-induced vibration is to design around it. Based on the above discussion, leakage-flow-induced instability can be eliminated if the following criteria are met (Mulcahy, 1983):

- Avoid, if at all possible, flow in very narrow gaps bounded by flexible structural boundaries.
- In flow channels bounded by flexible structural boundaries, avoid the situation when higher-pressure drop occurs in the upstream end of the channel compared with that in the downstream end (Figure 10.7 (A) and (C)).
- In the case of a cantilever rod surrounded by a narrow fluid gap, avoid the condition when the direction of the flow is from the free end to the fixed end of the cantilever.
- Avoid massive end pieces in narrow flow channels (Figure 10.7 (B)).

10.6 Components Prone to Axial and Leakage-Flow-Induced Vibration

Axial and particularly leakage-flow-induced vibration problems are far more common in power and process plant components, and indeed in our everyday lives, than most engineers realize. Figures 10.8 and 10.9 show some common components in which axial and leakage-flow-induced vibrations are known to have occurred. Some of these will be discussed briefly in this section. Other components will be discussed in more detail in Section 10.7.

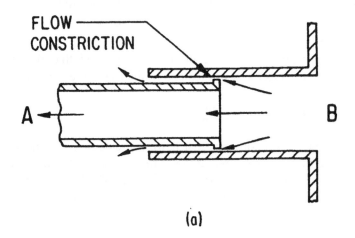

(a)

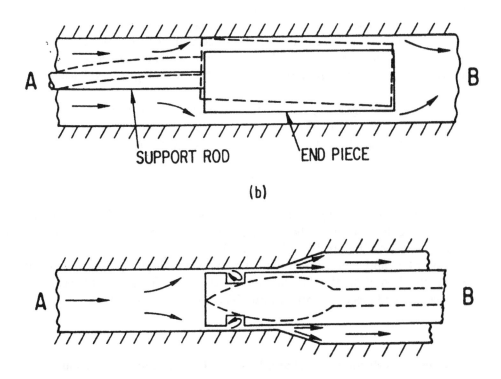

(b)

(c)

Figure 10.7   Geometries that are prone to leakage-flow-induced vibration: (a) Flow restrictor at upstream end of narrow flow channel; (b) Massive end piece in narrow flow channel; (c) Flow from free end to supported end of a rod and in a convergent-divergent flow channel (Mulcahy, 1983)

The Cavitating Venturi (Figure 10.8)

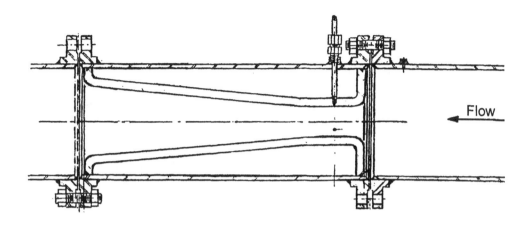

Figure 10.8 A typical cavitating venturi

This device is installed in piping systems to limit the water mass flow by "choking" it. The bubble collapses combined with the exit turbulent shear flow generate intensive acoustic energy that accounts for the high noise levels, usually in excess of 100 dB, associated with cavitating venturis. While this high noise level would require hearing protection for personnel working near the device and may even violate certain regulations, most of the energy associated with this noise is of high-frequency content. As a result, cavitating venturis do not always cause vibration problems in piping systems, the natural frequencies of which are usually below 100 Hz. However, this high-frequency noise may shake loose connectors and threaded joints, or excite the high-frequency shell-modes that can fatigue welded joints in hours. The choking flow may also set up low-frequency pulsation in the water upstream of the cavitating venturi, thereby exciting the piping system. In one application, the noise level generated by the cavitating venturis exceeded 105 dB. Hand wheels attached to the piping system were backed off and fell onto the floor, and the large vibration amplitude damaged nozzles attached to the main piping system.

Cavitating venturis can and have been used in plants without encountering any unacceptable vibration problems. The following general rules should be followed:

- Personnel near the device will have to wear hearing protections, as the noise level will be high. Obviously, cavitating venturis should not be used near populated areas, unless they are enclosed with acoustically well insulated structures.

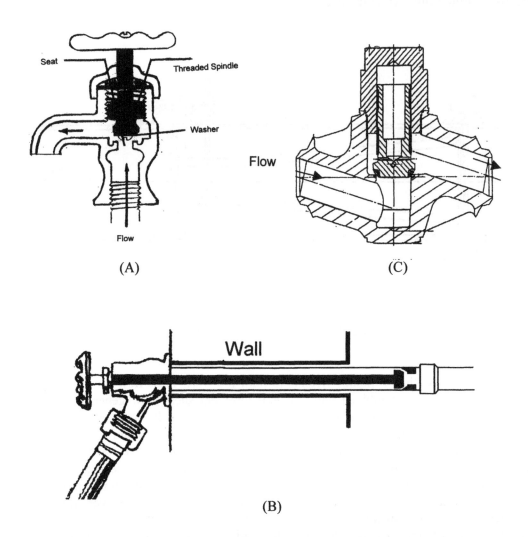

Figure 10.9 Components that have experience leakage-flow-induced instability
(A) Common faucet; (B) Frost-proof, outdoor faucet; (C) Lift check valve;
(to be continued)

- All threaded joints, particularly hand wheels, in the attached piping system should be properly locked. Welded joints should be designed for shell-mode deformation of the pipes.
- Do not attach nozzle-mounted heavy components, such as valves, to the piping system.
- The affected piping system should be properly supported.

We shall return to the cavitation venturi in Chapter 12, which is on acoustically induced vibration and noise.

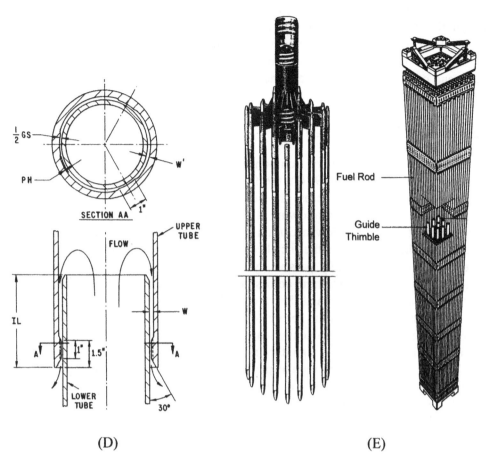

(D)   (E)

Figure 10.9 (continued)   Components that have experienced leakage-flow-induced instability: (D) Tube-in-tube slip joint (unstable if flow is reversed; Mulcahy, 1984); (E) Nuclear reactor control rods (which are inserted into the guide thimbles.)

The Common Faucet (Figure 10.9 (A))

In most common faucets, a threaded spindle with a flat or beveled washer at its end controls the flow. The gap between the washer and the seat determines the volumetric flow rate. When the gap is small, as when the faucet is first turned on or near the end of its closing stroke, tremendous pressure drop occurs through this flow restrictor, which is at the upstream end of the flow channel formed by the spindle and the faucet body. As long as the threads are new and engage the faucet body tightly, the structure is essentially rigid and no leakage-flow-induced vibration can occur. However, if the threads are worn, the spindle may experience violent vibration that can shake the attached water pipe. Those who have experienced this probably notice that the vibration would abruptly start when the water is first turned on, may abruptly stop when the faucet is fully open, and may recur when the faucet is almost closed. This is because leakage-flow-induced instability occurs only when there is enough pressure drop at the upstream end of the flow channel to establish a negative damping that exceeds the system damping of the faucet spindle. This pressure drop diminishes as the faucet is opened.

Some "frost-proof" outdoor faucets have a long spindle extending all the way from the outside wall to the inside of the house to avoid freezing. With the seat at the upstream end of long, narrow flow channels, this type of faucets is particularly prone to leakage-flow-induced vibration.

Lift Check Valves (Figure 10.9 (C))

Lift check valves are very common components in the power and process industries. In one design, shown in Figure 10.9 (C), the flow is controlled by a spring-loaded piston. When the pressure differential across the piston is high enough to overcome the spring force, the piston will lift, thus allowing the fluid flow through the valve. After the pressure differential drops below a certain threshold value, the spring would seat the piston and stop the flow. Thus the lift check valve works very much like the common faucet except it is a completely passive system with opening and closing completely controlled by the pressure differential across the valve. As designed, the spring force and the tightness between the piston and the valve body should prevent the piston from flow-induced vibration. However, if the spring is too soft or is broken and the piston is worn, then like the common faucet, it can experience leakage-flow-induced instability.

Figure 10.10 is the vibration signature measured by an accelerometer mounted on a piston lift check value. The very regular period, as well as the very high amplitude of the signal, clearly indicates that this piston check valve is experiencing leakage-flow-induced instability. Leakage-flow-induced vibration of such magnitude usually will damage the valve seat and even welded joints in the attached piping in a very short time.

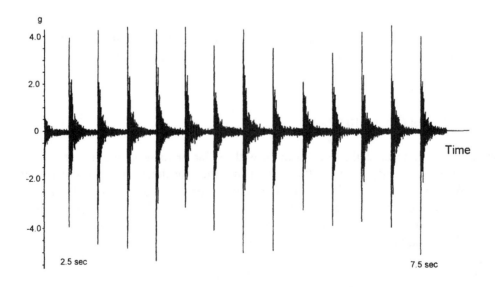

Figure 10.10 Vibration signature from an accelerometer mounted on a
piston check valve showing leakage-flow-induced instability

Tube-in-Tube Slip Joint (Figure 10.9 (D))

Tube-in-tube slip joints are common power and process plant components and were studied extensively by Mulcahy (1984). To maintain a tight fit, very often one end of the slip joint has a tighter clearance. Since the static pressures inside and outside of the tube are usually not the same, leakage-flow through the slip joint may result. As long as the flow is in a direction such that the restrictor is at the downstream end of the flow path, as shown in Figure 10.9 (C), the tubes will be stable. However, if, due to a change in the ambient pressures, the flow reverses its direction and goes from the restricted end to the open end, leakage-flow-induced instability may occur.

Reactor Control Rods (Figure 10.9(E))

The power output of a nuclear reactor is controlled by rod clusters that contain neutron-absorbing materials, which are inserted into guide thimbles inside the fuel assemblies (Figure 10.9(E)). During normal operations, the control rods are withdrawn. The further the rods are drawn up, the less neutron-absorbing materials are present in the fuel assemblies and more power is generated. In an emergency, such as, for example, loss of station electric power, the control rod drive mechanism would disengage, allowing the control rods to drop into the guide thimbles and thus shut down the reactor immediately. The time required for the control rods to drop has been carefully calculated for the safe shutdown of the reactor. Even though a back-up system (borated water released from a tank to flood the reactor core) would also be activated to shut down the reactor in such emergencies; both systems must be in perfect working order before a nuclear reactor can be operated.

   The control rods are extremely long and flexible "cantilevers" supported at the tops by the "spider." The natural frequencies of the individual rods are typically less than 1.0 Hz. The clearance between these rods and the guide thimble through which these rods are inserted form leakage-flow paths through which coolant flows from the bottom to the top of each guide thimble. This is a condition that favors leakage-flow-induced instability.

   It is no secret in the nuclear industry that wear marks between the control rods and the guide thimbles have been observed in operational nuclear plants. Excessive wear would compromise the drop time of these rods. To overcome this problem, the industry resorts to "park" the rods at different elevations, so that the tips do not wear against the same spot in the guide thimble, or plate the guide thimbles and the rods with long wearing material, such as chromium carbide. In the late 1980s, the industry started to look into the control rod wear problem more closely by both testing and analytical studies on the mechanics of leakage-flow-induced vibration (see, e.g., Fujita and Shintani, 1999; Inada and Hayama, 2000; and the references cited there).

10.7   Case Studies in Axial and Leakage-Flow-Induced Vibrations

In the following, some well-documented and costly axial and leakage-flow-induced vibration problems in the nuclear industry are reviewed. Similar problems can also occur in other power or process industries. There are two reasons why the nuclear industry is mentioned in particular: First, it is the author's own experience with this industry. Second and more important, is that, being a very highly regulated industry, nuclear plants are required by law to report every flow-induced vibration problem to the regulators. These incidents are well documented, their root cause thoroughly analyzed and the information is available to the general public. No equivalent systematic documentation is available in any other industries.

The following case studies summarize the components, the symptoms, the causes and the fixes of these incidents. More details are given in the cited references.

Case Study 10.1: Boiling Water Reactor Thermal Shield (Corr, 1970)

One of the earliest documented flow-induced vibration problems in commercial nuclear power happened at the Big Rock Point Nuclear Plant boiling water reactor (BWR), the thermal shield (Figure 10.11) of which was a large shell 92 in. (2.34 m) long by 103 in. (2.62 m) OD, with a 1.5 in. (0.038 m) annular gap surrounding it. At the bottom of the gap was a seal with water flowing from the bottom to the top of the annulus. The shell was bottom-supported. In 1964, several thermal shield-retaining bolts were found to be broken. The support structure was stiffened by a factor of 100. In subsequent tests, sudden onset of large amplitude vibration of the thermal shield at 4 Hz was observed. The following vibration characteristics were observed:

- The thermal shield was vibrating in a rigid body, transverse translation mode with elliptical orbits.
- Sudden onset of obviously unacceptable large amplitude vibration.
- Vibration amplitude varied non-linearly with the flow rate, with an apparent critical flow rate for the onset of large amplitude vibration.
- Initial low level vibration indicated decrease of total system damping as the flow rate was increased.

All of the above symptoms were consistent with leakage-flow-induced instability. The annular gap between the reactor vessel and the thermal shield formed a very narrow flow path compared with the size of the surrounding structure. Even after stiffening by a factor of 100, the thermal shield was still a relatively flexible structure. The seal at the bottom of the annular gap caused a much larger pressure drop at the upstream end of this narrow flow path compared with the open end at the top. A situation that favors leakage-flow-induced instability was formed (compared with Figure 10.5). Stiffening the supporting structure would only delay the onset of instability, but not enough to completely eliminate it. Three more modifications were implemented:

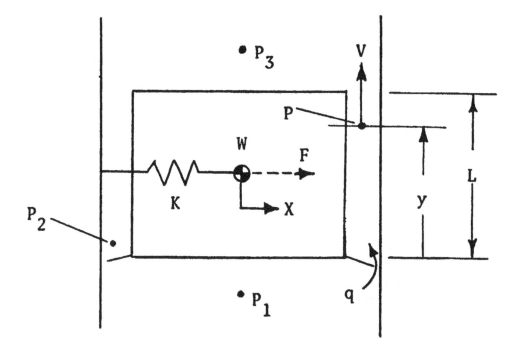

Figure 10.11 Thermal shield in a boiling water reactor (Corr, 1970)

- The bottom seal was immobilized with wedges.
- A top seal was added to the top of the annular gap to increase the pressure drop at the downstream end of the leakage-flow path.
- Hold-down weights were added to the thermal shield to immobilize it.

The Big Rock Point Plant has since been de-commissioned as it reached the end of its design life. No large amplitude vibration problem was observed after the fix. Of the above three modifications, the most significant was probably the second, which alone might be enough to correct the problem.

Case Study 10.2: Feedwater Sparger Thermal Sleeve (Torres, 1980)

There are from two to six feedwater spargers in each boiling water nuclear reactor (BWR), each comprising an arc-shaped closed-end pipe with holes along its length and a "T" at its center (Figure 10.12 (A)). The function of the sparger is to evenly distribute the feedwater, which enters from the feedwater nozzles into the reactor. Feedwater nozzles are common components in the power industry and it is common practice to use a thermal sleeve (Figure 10.12 (B)) inside the nozzle to minimize the effect of thermal cycling. In this particular case, the thermal sleeve was welded to the sparger in the form of a "T" junction and was slip-fit into the nozzle. This is also a common practice in the power industry. The purpose was to accommodate the differential thermal expansion between the sparger and the reactor vessel, which were made of different materials. To avoid trapped "dead" water between the thermal sleeve and the nozzle, which may be harmful to the materials, it is also standard industry practice to allow some small flow through this gap. Unfortunately, this was exactly the condition that favored leakage-flow-induced instability.

In the 1970s, feedwater spargers in a BWR experienced extensive cracking at the weld junction shortly after it went into operation. A replacement set was installed. After six weeks of normal operation, inspection revealed the replacement set cracked at the same place. An extensive flow-induced vibration test was carried out to determine the root cause of this cracking. It was found that with a 1.14 in. (29 mm) ID thermal sleeve and a radial gap between the OD of the thermal sleeve and the ID of the nozzle less than 0.02 in. (0.5 mm), the thermal sleeve started to experience excessive vibration at a flow rate of 66 gal/s (0.25 cu.m./s). The response was not linear with the flow rate—large-amplitude vibration persisted until the flow rate was reduced to less than 42 gal/s (0.16 cu.m./s). This result was quite repeatable.

The test was repeated with the annular gap between the thermal sleeve and the nozzle plugged. With flow rate increased to 140% of the above threshold value, the thermal sleeve would not vibrate, showing that the flow through the nozzle was not the cause of the large amplitude vibration. The test was then repeated with the main nozzle flow path blocked with a plate so that the only flow was through the small gap between the thermal sleeve and the nozzle. The water source was taken from the city water supply hooked directly to the nozzle. At a flow rate of only 1.6 gal./s (0.006 cu.m./s) and a pressure differential of only 14 psi (96.5 KPa) across the thermal sleeve, the entire schedule 80 (15.24 cm diameter) stainless steel pipe to which the test components were attached was set into large-amplitude, spontaneous vibration. Tests with exceptionally small clearances between the thermal sleeve and the nozzle, which almost eliminated the leakage-flow, showed no large-amplitude vibrations.

This test dramatically demonstrated the potentially very detrimental effect of tiny amount of leakage-flow through very narrow flow paths. The feedwater spargers were then re-designed with either the leakage-flow path blocked off or with the slip joint completely eliminated. No cracking was experienced in the subsequent 25 years of operation.

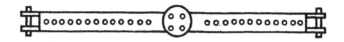

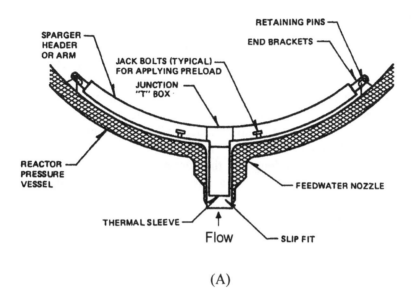

(A)

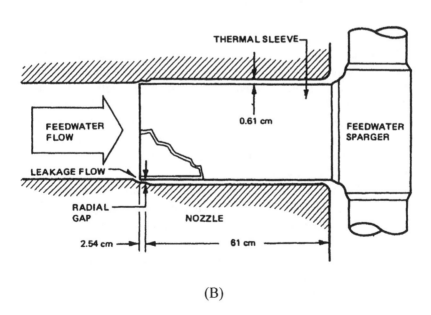

(B)

Figure 10.12 (A) Boiling water reactor feedwater sparger and
(B) Thermal sleeve inside feedwater nozzle (Torres, 1980)

Case Study 10.3: Surveillance Specimen Holder Tube (Figure 10.13)

As its name implies, this component holds material samples in the annular gap between the thermal and the reactor vessel of some nuclear reactors so that the long-term radiation effects on materials can be evaluated. The assembly consisted of two components: an outer tube and an inner tube containing the material samples, with reactor coolant flowing through the annular region between the tubes. In one original design, the inner tube had spacer disks for centering and separating the specimen capsules. After less than two years of operation, the disk cut through the 3/8-inch (9.5 mm) stainless steel wall of the outer tube, dropping the bottom portion of this tube onto the bottom of the reactor vessel.

The component was designed to withstand the high cross-flow resulting from the direct impingement of the flow through the inlet nozzles. Where the inner tube cut off the outer protecting sheath, the flow was mainly axial. Extensive analysis and laboratory tests indicated that base motion, due to the motion of the thermal shield on which the component was mounted, could not have induced vibration amplitude large enough to wear through the wall of the outer tube. Because of the small diameter of the tube compared with the acoustic wavelength at the dominant blade passing frequencies (100, 200, 300 Hz) of the reactor coolant pump, acoustically induced vibration was determined not to be the cause of the problem (see Chapter 12, Appendix). This leaves annular-flow-induced vibration as the only possible cause of such rapid wear, which could only be a result of some type of instability mechanism, such as leakage-flow-induced instability.

The assembly was subsequently re-designed to be stiffer. The flow holes on the outer jacket were eliminated. The problem did not recur after another 20 years of operations.

Case Study 10.4: Kiwi Nuclear Rocket (Koenig, 1986, Zeigner, 1965)

From the 1950s to 1973, the United States experimented with nuclear rockets. The Kiwi series of nuclear rocket prototypes, all of which were graphite reactors, were not designed to fly. They were merely ground test beds for the nuclear fuels. The very first Kiwi nuclear rockets had disc-like fuel elements loosely stacked together. It lasted only a few seconds, when severe flow-induced vibration was encountered, followed shortly by ejection of the fuel elements from the core. Without any positive restraint, the fuel elements behaved like loose parts in the hydrogen coolant flow. The fuel discs rattled and knocked against one another until they disintegrated.

Starting with Kiwi A3, cylindrical fuel elements stacked six end-to-end (Figure 10.14) were used. Each cylinder had four flow holes through it. These flow holes were supposed to line up, virtually on their own, so that cooling hydrogen could pass through. In practice they did not. Leakage-flow between the fuel element and the outer container soon started to excite the fuel elements, which again were loosely stacked together without any positive restraint.   Within minutes the fuel elements were completely damaged and were ejected from the core.

Later Kiwis had continuous beam fuel elements and seemed to have encountered no significant vibration problems.

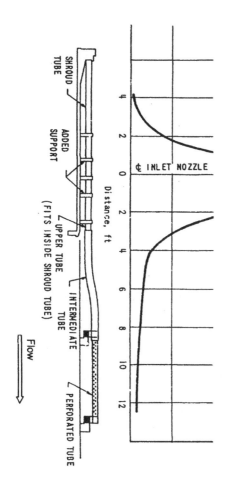

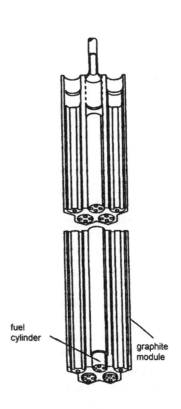

Figure 10.13 Original design of the
surveillance specimen holder tube that
encountered axial flow-induced
vibration problem.

Figure 10.14 Kiwi A3 nuclear rocket

References

ASME, 1979, Steam Tables, Fourth Edition.

Au-Yang, M. K. and Jordan, K. B., 1980, "Dynamic Pressure Inside a PWR— A Study Based on Laboratory and Field Test Data," Journal Nuclear Engineering and Design, Vol. 58, pp. 113-125.

Au-Yang, M. K.; Brenneman, B. and Raj, D., 1995, "Flow-Induced Vibration Test of an Advanced Water Reactor Model, Part I—Turbulence Induced Forcing Function," Journal Nuclear Engineering and Design, Vol.157, pp. 93-109.

Blevins, R. D., 1990, Flow-Induced Vibration, Second Edition, Van Nostrand Reinhold, New York.

Clinch, J. M., 1969, "Measurement of Wall Pressure at the Surface of a Smooth-Walled Pipe Containing Turbulent Water Flow," Journal Sound and Vibration, Vol. 9, pp. 398-419.

Chen, S. S., Rosenberg, G. S., 1971, Vibration and Stability of a Tube Conveying Fluid, USAEC Report, ANL-7762.

Chen, S. S.,1985, Flow-Induced Vibration of Cylindrical Structures, Report ANL-85-51, Argonne National Laboratory.

Corr, J. E., 1970, "Big Rock Point Vibration Analysis," in Proceedings of the Conference on Flow-Induced Vibration in Reactor Components, 1970, Report ANL-7685, Argonne National Laboratory, pp. 272-289.

Fujita, K. and Shintani, A., 1999, "Flow-Induced Vibration of An Elastic Rod due to Axial Flow," in Flow-Induced Vibration, ASME Special Publication PVP-389, edited by M. J. Pettigrew, ASME Press, New York, pp. 199-215.

Inada, F., Hayama, S., 2000, "Mechanism of Leakage-Flow-Induced Vibrations—Single-Degree-of-Freedom and Continuous Systems," Proceedings of the 7th International Conference on Flow-Induced Vibration, Lucerne, Switzerland, edited by S. Ziada and T. Staubli, Balkema, Rotterdam, pp. 837-844.

Koenig, D. R., 1986, Experience Gained from the Space Nuclear Rocket Program (ROVER), Report No. LA-10062-H, Los Alamos National Laboratory.

Miller, D. R., 1970, "Generation of Positive and Negative Damping with a Flow Restrictor in Axial Flow," Proceedings Conference on Flow-Induced Vibration in Reactor Components, Report ANL-7685, Argonne National Laboratory, pp. 304-307.

Mulcahy, T. M., 1983, "Leakage-Flow-Induced Vibration of Reactor Components," The Shock and Vibration Digest, Vol. 15, No. 9, pp. 11-18.

Mulcahy, T. M., 1984, "Leakage-Flow-Induced Vibration of a Tube-in-Tube Slip Joint," in Proceedings Symposium on Flow-induced vibration, Vol. 4, edited by M. P. Paidoussis and M. K. Au-Yang, ASME Press, New York, pp. 15-24.

Paidoussis, M.P., 1981, "Fluidelastic Vibration of Cylinder Arrays in Axial and Cross-Flow—State-of-the-Art," in Flow-Induced Vibration Design Guidelines, ASME Special Publication PVP-Vol. 52, edited by P. Y. Chen, ASME Press, pp. 11-46

Paidoussis, M. P. and Au-Yang, M. K., 1984, editors, Proceeding: Symposium on Flow-Induced Vibration, Vol. 4, Vibration Induced by Axial and Annular Flows. ASME Press, New York.

Paidoussis, M. P.; Au-Yang, M. K. and Chen, S. S., 1988, editors, Proceeding: International Symposium on Flow-Induced Vibration and Noise, Vol. 4, Flow-Induced Vibrations due to Internal and Annular Flows, ASME Press, New York.

Paidoussis, M. P. and Au-Yang, M. K., 1992, editors, Proceeding: International Symposium on Flow-Induced Vibration and Noise, Vol. 5, Axial and Annular Flow-Induced Vibration and Instabilities, ASME Press, New York.

Paidoussis, M. P., 1998, Fluid-Structure Interaction, Vol. I, Academic Press, New York.

Torres, M.R., 1980, "Flow-Induced Vibration Testing of BWR Feedwater Spargers" in Flow-Induced Vibration of Power Plant Components, edited by M. K. Au-Yang, ASME Press, New York, pp. 159-176.

Wambsganss, M. W. and Chen, S. S., 1971, Tentative Design Guide for Calculating the Vibration Response of Flexible Cylindrical Elements in Axial Flow, Report ANL-ETD-71-07, Argonne National Laboratory.

Zeigner, V. L., 1965, Survey Description of the Design and Testing of Kiwi B4E 301 Propulsion Reactor, Report LA-3311-MS, Los Alamos Scientific Laboratory.

CHAPTER 11

IMPACT, FATIGUE AND WEAR

Summary

The three principal damage mechanisms resulting from flow-induced vibrations are impact (which may result in fatigue wear), fatigue and wear. Because turbulence-induced vibration is random, the zero-to-peak vibration amplitudes can occasionally exceed several times the computed rms response. Thus, the fact that the nearest component is three times the rms vibration amplitude away does not guarantee that impacting between the two components will not occur. The number of impacts between two vibrating components over a given time period can be estimated based on the theory of probability, assuming that the vibration amplitudes follow a Gaussian distribution.

Likewise, since the zero-to-peak amplitude of vibration of a structure excited by flow turbulence can, over a long enough period of time, exceed arbitrarily large values, there is no endurance limit in random vibration. Given a long enough time, any structure excited by any random force will theoretically fail by fatigue. The cumulative fatigue usage can again be calculated based on the probabilistic theory. Since in cumulative fatigue analysis, the usage factor is computed based on the absolute value of the zero-to-peak vibration amplitudes, which follow the Rayleigh distribution function if the ± vibration amplitudes follow the Gaussian distribution, the Rayleigh probability distribution function must be used in computing the cumulative fatigue usage factor of a component excited by turbulent flow. From the ASME fatigue curves (which are based on the 0-to-peak vibration amplitudes), corresponding fatigue curves based on rms vibration amplitudes had been derived for several types of materials. These are given in Figures 11.6 to 11.10.

Compare with fatigue usage calculations, wear analysis due to flow-induced vibration is orders of magnitude more complex. This is because the wear mechanisms are not only dependent on the dynamics of the structures, but also on the material and the ambient conditions. Generally, there are three major types of wear mechanisms: Impact wear is that caused by moderate to fairly large vibration amplitudes, with resulting high impact forces that can cause surface fatigue and rapid failure of the structure. Blevins (1984) proposed the simple equation

$$s_{rms} = c\left(\frac{E^4 M_e f_n^2 <y^2>_{max}}{D^3}\right)^{1/5}$$

(11.16)

to estimate the rms surface stress of a heat exchanger tube impacting its support, with the contact stress parameter c obtained from tests (These are given in Figure 11.13). Blevins postulated that if the computed stress is below the endurance limit, then impact wear is not a concern. However, if it exceeds the endurance limit, then rapid wear of the material can be expected.

The second type of wear mechanism is sliding wear. The volumetric wear due to sliding wear is usually expressed in terms of Archard's (1956) equation,

$$Q = KF_nS \tag{11.17}$$

In spite of its simplicity, the above equation is difficult to use because none of the three factors on its right-hand side can be easily computed. Usually, the sliding wear coefficient K is obtained from long-duration tests. For a tube whirling inside an oversize support hole at a frequency f_n, such as when the tube becomes fluid-elastically unstable, Connors (1981) proposed the following simple equation to estimate the sliding distances and the normal contact force:

$$S = \pi f_n g t \tag{11.18}$$

$$F_n = \frac{\pi D y_{0-p}}{\mu\left(\dfrac{L^2}{A_m E} + \dfrac{D^2 L^2}{4EI}\right)} \tag{11.19}$$

However, for a fluid-elastically unstable tube, tube-to-tube impacts will cause tube failure much sooner than that due to wear at the supports.

Based on operational experience, it appears that the wear mechanism responsible for the observed wear marks in most of today's operating nuclear steam generator tubes is fretting wear, which is a special terminology for small amplitude impact/sliding wear. Fretting wear volume is more commonly expressed in terms of the "wear work rate," defined as:

$$\frac{dW}{dt} = F_n \frac{dS}{dt} \tag{11.23}$$

In terms of the wear work rate, the volumetric fretting wear is given by the equation,

$$\dot{Q} = K\dot{W} \tag{11.25}$$

again with the coefficient of fretting wear, K, obtained by tests. Limited data on the fretting wear coefficient, mainly on material pairs used in nuclear steam generators, are available in the open literature and summarized in Table 11.3.

Even with the fretting wear coefficient known, the wear work rate must be computed by solving the non-linear equation of tube dynamics in the time domain. The wear work rates are usually obtained by numerical summation of the product of the instantaneous contact force and sliding distance. For a tube vibrating inside an oversize support hole, Connors (1981) suggested the following simplified equation for the total sliding distance:

$$S = 2\pi D y_{0-peak} f_n t / L \tag{11.26}$$

With this and the equation for the contact force given above, the wear work rate, and hence the wear volume and wear depth on the heat exchanger tube, can be computed. The results generally are consistent with those obtained by non-linear structural dynamic analysis.

Based on the fact that the apparent damping in a loosely-supported heat exchanger tube comes mainly from the tube-to-support interaction, i.e., wear between the tube and its supports, the wear work rate of a heat exchanger tube vibrating in oversize support holes can be shown to be,

$$\dot{W} = 8\pi^3 (\zeta_n f_n^3 m_n / \mu)(y_{0i}^2 / \psi_{ni}^2) \tag{11.28}$$

Thus, the volumetric wear of a tube can also be assessed by comparison with that of another tube in which the volumetric wear rate is known either from operational experience, or from a separate non-linear time domain analysis. The damping ratios and the vibration amplitudes of both tubes must be known either by measurement or by linear structural dynamic analysis.

Nomenclature

A_m	Metal cross-sectional area of tube
c	Contact stress parameter (in Blevins' equation for impact wear)
C	Damping coefficient
D	OD of tube
E	Young's modulus
f	Vibration frequency, in Hz
f_n, f_0	Natural frequency, in Hz
F_c	Contact force between components
F_{inc}	Incident (external) force
F_n	Normal contact force
g	Diametral (or two-sided) clearance between a tube and its support
h	Thickness of tube wall
I	Area moment of inertia
k	Stiffness
k_c	Contact stiffness between two components
K	Wear coefficient
L	Total length of spans on both sides of the loose support
m_n	Generalized mass

M	Mass
M_e	Effective mass of the tube spans
n	Number of vibration cycles
n_a	Number of vibration cycles with stress levels between s_a and $s_a + \delta s_a$
N	Total number of vibration cycles over a time period
N_a	Number of allowable cycles at stress level s_a
$p(z)$	Probability distribution that a response has the value between $p(z)$ and $p(z + \delta z)$
P	The total probability the vibration amplitudes lie within $\pm z y_{rms}$
Q	Wear volume
S	Total sliding distance between contact surfaces
t	Time
U_a	Fatigue usage at stress level s_a
$\dot{W}$	Wear work rate
y_{0-p}	0-peak response
y_{rms}	Root mean square response
$<y>$	Mean value of y
z	Number (integer or fractional) of rms response
ζ	Damping ratio
μ	Coefficient of friction between contact surfaces
σ_y	Standard deviation of y
ψ	Mode-shape function

11.1   Introduction

The ultimate objectives of flow-induced vibration analysis are two: (1) to assess the environmental impact of the affected components; and (2) to assess the concern of failure, or to assess the residual life, of the affected component or systems.   The most common example of the first is noise generated by vibration. The cavitating venturi, discussed in the previous chapter, is one such example. More examples will be given in the following chapter on acoustically induced vibrations. This chapter focuses on the second objective of flow-induced vibration analysis—failure caused by flow-induced vibrations.

There are three principal damage mechanisms resulting from flow-induced vibration: impact, fatigue and wear. These will be discussed separately in the following sections.

11.2   Impacts due to Turbulence-Induced Vibration

As discussed in Chapter 8, because of the random behavior of the forcing function and the resulting structural responses, we have up to this point decided to give up any attempt to calculate the detailed time history responses of the structures excited by turbulence.

The acceptance integral method discussed extensively in Chapters 8, 9 and 10 was developed with a clear understanding of this limitation. The result is that this method, which was virtually the only method of estimating the vibration responses due to a random, spatially distributed forcing function such as flow turbulence, up to the early 1990s, enables us to calculate only the rms responses (displacement, stress, reaction force at the supports and their respective crossing frequencies), not the detailed time histories of the responses. Unfortunately, a structure experiencing random vibration (due to turbulence or other random forces) with a rms amplitude of y_{rms} does not always have an amplitude of $\pm y_{rms}$, or $\pm \sqrt{2}\, y_{rms}$ as in the case of sinusoidal excitation. Instead, as discussed in Section 8.2, in the majority of engineering problems, as long as the structure is linear and the damping is light, the probability density distribution of the vibration amplitudes follow the "bell curve," that is, the Gaussian distribution. Thus the probability density for any vibration amplitude to assume the value

$$y = zy_{rms} \tag{11.1}$$

where z is any fractional or integer real number, is given by,

$$p(z) = \frac{1}{\sqrt{2\pi}} e^{-z^2/2} \tag{11.2}$$

Equation (11.2) is the normalized Gaussian distribution. The above representation of the actual vibration amplitudes in terms of the rms amplitude is strictly empirical based on our everyday observations.

Based on Equations (11.1) and (11.2), theoretically a structure vibrating randomly with a rms amplitude of 0.001 inch can occasionally vibrate with a peak-peak amplitude of ± 1 inch or more. For many applications, such as impact and fatigue analyses, it is necessary to know the zero-peak amplitudes of vibration. For a quick estimate, some engineers often use the "3σ" approximation by assuming that the zero-to-peak amplitude is equal to three times the rms amplitude. As long as the component is more than 3 times its rms vibration amplitude away from a structural boundary, one assumes that the vibrating component will not impact the structural boundary. Unfortunately this is not rigorously true. From Equation (11.2), the cumulative probability that the vibration amplitude is less than zy_{rms} is,

$$P = \int_{-z}^{+z} p(z)dz \tag{11.3}$$

The probability that the excursion is *more* than $+zy_{rms}$ and *less* than $-zy_{rms}$ is *1-P*. The integral in Equation (11.3) has been evaluated and tabulated by statisticians and is often referred to as the area under the normalized or standardized Gaussian curve. For $z=3$, or

$x < -3y_{rms}$ and $x > 3y_{rms}$, $P=0.9973$. That is, 0.27% of the time, the vibration amplitudes will exceed the $-3y_{rms}$ and $+3y_{rms}$ bounds. Over a long period of time, the total number of times that the vibration amplitude exceeds the $\pm 3y_{rms}$ limits may not be negligible. Figure 11.1 shows 4.0 second of the time history response of a heat exchange tube subject to cross-flow turbulence-induced vibration, using a time history response calculation technique to be discussed in Section 11.5. Already there are a few excursions beyond the $\pm 3y_{rms}$ limits.

The number of zero-to-peak vibration amplitudes that exceeds the $\pm zy_{rms}$ limits can be estimated from the normalized Gaussian distribution table (or statistical analysis software) usually found in books on statistics. First, let us refresh our memory on a few essentials of the Gaussian statistics discussed in Section. 8.2. The *standard deviation* σ for a random process, such as the response y at a point of a structure, is defined in Equation (8.7). With the general function replaced by the specific Gaussian distribution function p, this takes the form:

$$\sigma_y^2 = \int_{-\infty}^{\infty}(y-<y>)^2 p(z)dz \tag{11.4}$$

In turbulence-induced vibration, the mean value $<y>$ is either zero or, if it is not zero, the static mean value will be calculated separately and added to the fluctuating component at the end of the dynamic analysis. Assuming we are considering only the fluctuating component from now on, we can assume $<y>=0$ and, as a result,

$$\sigma_y^2 = \int_{-\infty}^{+\infty} yp(z)dz = y_{rms}^2 \tag{11.5}$$

That is, the standard deviation is the rms response. This is an important step as in the following discussion, we shall use y_{rms} instead of σ. This is necessary because statisticians universally use the terminology standard deviation with the notation σ, whereas engineers tend to use the terminology rms values.

Figure 11.2 shows the Gaussian distribution with zero mean. Suppose over a period of time, we expect a total of N cycles will be accumulated. From Figure 11.2, $Np(z)dz$ of these N cycles will have the negative amplitudes between $-zy_{rms}$ and $-(z+dz)\,y_{rms}$ and positive amplitudes between $+zy_{rms}$ and $+(z+dz)\,y_{rms}$, where p is the *normalized* Gaussian distribution,

$$p(z) = \frac{1}{\sqrt{2\pi}} e^{-z^2/2} \tag{11.6}$$

The number of cycles with amplitudes lying between $-zy_{rms}$ and $+zy_{rms}$ is therefore,

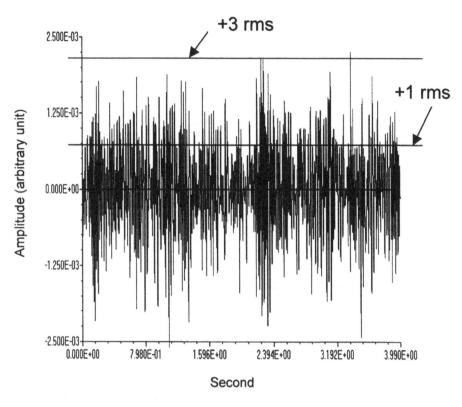

Figure 11.1 Time history response of a point in a heat exchanger tube
due to cross-flow turbulence

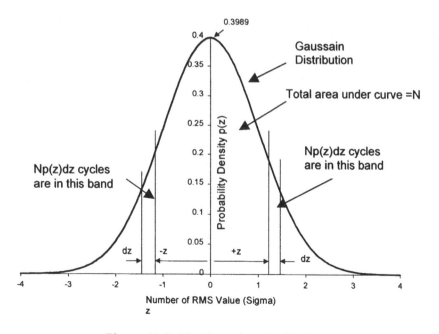

Figure 11.2 The Gaussian distribution

Table 11.1: Areas under the normalized Gaussian Curve

(Source: Pearson and Hartley, "Biometrika Tables for Statisticians",
Vol. 1, Cambridge University Press, re-compiled in engineers' language.)

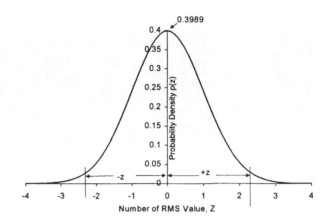

z	P(z)	Probability Excursion outside $\pm z$ times rms value $(1-P(z))*2$	Probability Excursion within $\pm z$ times rms value $1-(1-P(z))*2$
1	0.8413447	3.1731E-01	0.68268940000
2	0.9772499	4.5500E-02	0.95449980000
3	0.9986501	2.6998E-03	0.99730020000
3.2	0.9993129	1.3742E-03	0.99862580000
3.4	0.9996631	6.7380E-04	0.99932620000
3.6	0.9998409	3.1820E-04	0.99968180000
3.8	0.9999277	1.4460E-04	0.99985540000
4	0.9999683	6.3400E-05	0.99993660000
4.2	0.9999807	3.8600E-05	0.99996140000
4.4	0.9999946	1.0800E-05	0.99998920000
4.6	0.9999978875	4.2250E-06	0.99999577500
4.8	0.9999992067	1.5866E-06	0.99999841340
5	0.9999997133	5.734E-07	0.99999942660

$$P = N \int_{-z}^{+z} p(z)dz \qquad (11.7)$$

Table 11.1 gives the area under the normalized Gaussian curve, Equation (11.6), the
probability the vibration amplitudes exceeds $\pm z * rms\ value$ and the probability that the

vibration amplitudes remain within $\pm z * rms\ value$. Most tables in statistics books contain these values only up to $z=3.0$.

Example 11.1

In a heat exchanger, the gap clearance between adjacent tubes is 0.21 inch (5.33 mm). The tubes are subject to cross-flow-induced random vibration with mid-span rms amplitudes of 0.035 inch (0.89 mm), or 1/6 of the clearance between the two tubes, with a frequency of 50 Hz. Estimate how many times one tube will hit its neighbors in one year.

Solution

Assuming the vibration amplitudes follow the Gaussian distribution, the probability that the mid-span vibration amplitudes exceed ± 3 times the rms vibration value, or 0.105 inch (2.67 mm), is, from Table 11.1, equal to 2.7E-3. The probability that the mid-span excursion of its neighbor exceeds 0.105 inch is also 2.7E-3. The probability that the mid-span excursion of both neighboring tubes exceeds an amplitude of 0.105 inch is thus $(2.7E-3)^2 = 7.29E-6$. Since this tube is surrounded by it neighbors, the probability that it would hit a neighbor is 7.29E-6.

In one year, this tube and its neighbors each vibrate 50*60*60*24*365 =1.577E+9 times. Therefore, it will hit its neighbors 11,500 times. This is clearly not acceptable.

The above calculation is approximate. It clearly shows that using the "3σ" method to assess long term effects in random vibration is not conservative.

11.3   Cumulative Fatigue Usage due to Turbulence-Induced Vibration

Fatigue and usage factor analyses pose problems that are one level higher than impact. This is because fatigue curves were generated based on tests with deterministic (sinusoidal) loads. The fatigue curves given by the ASME Boiler Code (1998), for example, are based on the zero-to-peak amplitudes. For deterministic vibration, if s_a is the amplitude of the stress and the number of allowable cycles at this stress level, as given by the ASME fatigue curve, is N_a, then the usage factor after n_a cycles of vibration at this stress level is (see Figure 11.3):

$$U_a = n_a / N_a \tag{11.8}$$

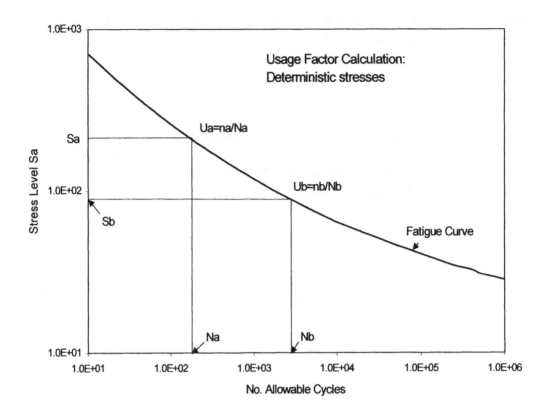

Figure 11.3 Cumulative fatigue usage

After the structure has vibrated n_a cycles with a peak stress of s_a at which the allowable number of cycles is N_a, and n_b cycles with a peak stress of s_b at which the allowable number of cycles is N_b, and so on, the cumulative usage factor will be,

$$U = n_a / N_a + n_b / N_b + n_c / N_c + ...$$ (11.9)

When U=1.0, fatigue failure is assumed to occur.

The above analysis poses a problem when the vibration is random in nature. Since we can only calculate the rms stress and the modal frequency, we cannot easily calculate the fatigue usage over a period of time. For a crude estimate, engineers again often use the "3σ" approach by assuming that the peak vibration amplitude is three times the rms value. Like impact analysis discussed above, this often leads to unconservative results over a long time period.

Crandall's Method

In the 1950s before computers were generally available, Crandall (1963) solved the above problem by fitting a published fatigue curve with a straight line in the log-log plot

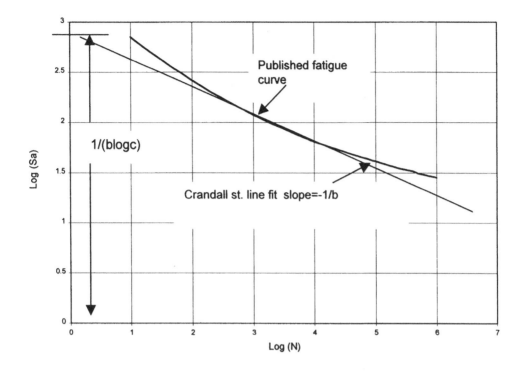

Figure 11.4 Crandall's method

(Figure 11.4), and found the cumulative fatigue usage by analytically integrating the usage due to each finite value of the stress. Following this, Crandall derived a closed-form solution to the usage factor for a structure undergoing random vibration,

$$U = \frac{n}{c}(\sqrt{2}s_{rms})b\Gamma(1+b/2)$$ (11.10)

where Γ is the Gamma function, -1/b is the slope and 1/$blogc$ is the intercept of Crandall's straight line approximation to the fatigue curve in the log-log plot (Figure 11.4). Crandall's method is neither simple nor reliable. Considerable variations in the computed usage factor can result depending on how the engineer approximates the fatigue curve with a straight line. In spite of this, Crandall's method is still being used today as a more accurate method to calculate the usage factor than the "3σ" method. This may be true. However, in view of the wide availability of inexpensive personal computers, Crandall's method is no longer necessary. Much more accurate results for the cumulative usage can be obtained by direct numerical integration using the probability distribution functions of the stress levels. This will be discussed in the following sub-section.

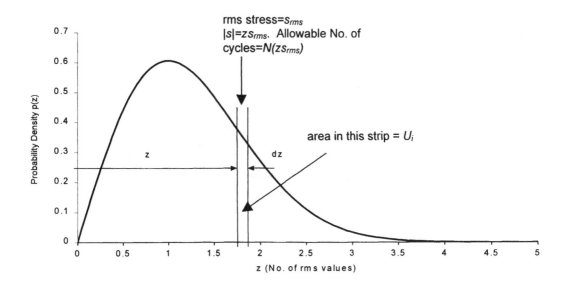

Figure 11.5 Rayleigh distribution and cumulative fatigue usage

Cumulative Fatigue Usage by Numerical Integration

As was shown by Crandall and Mark (1963), if we assume that the vibration amplitude (either displacement or stress) $\pm y$ follows a Gaussian distribution, then its zero-peak level (which is an absolute number) will follow a Rayleigh distribution (Figure 11.5),

$$p(z) = ze^{-z^2/2} \tag{11.11}$$

The zero-peak stress level is given by

$$|s(z)| = zs_{rms} \tag{11.12}$$

It is this zero to peak stress values that are plotted in the ASME fatigue curve and fatigue usage is based on $|s|$, not s.

Based on the Rayleigh distribution, one can easily calculate the cumulative fatigue usage by direct numerical integration. If the computed rms stress is s_{rms} and the total number of cycles at this rms stress is N, the probability density distribution for the number of cycles $n(|s|)$ with stress levels $|zs_{rms}|$ is:

$$n(|s|) = Np(z) \tag{11.13}$$

where

$$p(z) = ze^{-z^2/2} \quad z \geq 0 \tag{11.14}$$

is the Rayleigh Distribution. The number of cycles with stress amplitudes between $|s|$ and $|s|+|\delta s|$ is,

$$\delta n = Np(z)\frac{\partial p(z)}{\partial z}\delta z$$

This number of cycles imparts a usage of

$$\delta U = \delta n / N(s = zs_{rms})$$

where $N(s = zs_{rms})$ is the number of allowable cycles at the stress level $s=zs_{rms}$, obtained from the ASME Boiler Code (1998) or other sources. From the Rayleigh distribution function, Equation (11.4), we get,

$$\delta U = \frac{N}{N(zs_{rms})}e^{-z^2}(1-z^2)\delta z$$

The cumulative fatigue usage can then be computed by numerical integration,

$$U = \sum_i U_i = N\sum_i \frac{e^{-z_i^2}(1-z_i^2)\Delta z_i}{N(z_i s_{rms})} \tag{11.15}$$

where

N = total number of cycles = crossing frequency*time (see Chapter 7 on the crossing frequency)

z_i = abscissa in the Rayleigh distribution

Δz_i = integration step sizes

$N(z_{srms})$ = allowable number of cycles at stress level $s=zs_{rms}$

Fatigue Curves based on RMS Stress

We can find the reverse solution, given the zero-to-peak stress levels and the corresponding allowable number of cycles (example: from the ASME fatigue curve), what are the corresponding allowable number of cycles in term of rms stresses?

This reverse solution needs more sophisticated programming, but it has been done. Figures 11.6 to 11.10, reproduced from Brenneman (1986), are rms fatigue curves for some materials. These curves were derived from the corresponding ASME zero-peak

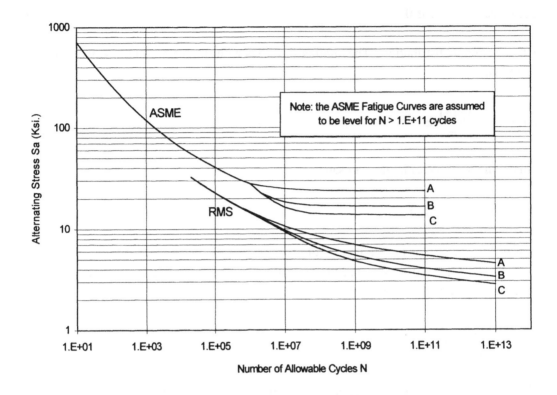

Figure 11.6   Fatigue curves for Austenitic steel based on rms stress derived
from the corresponding fatigue curves given in the ASME Boiler code (1998)
(E=28.3E+6 psi) (Brenneman, 1986)

fatigue curves (ASME, 1998). To find the usage factor using these curves, one just reads off the allowable number of cycles N_{rms} corresponding to the computed rms stress s_{rms} for the Gaussian random vibration, and the total number of cycles n at this stress, the usage factor is:

$$U = n / N_{rms}$$

Because of the Rayleigh distribution of the stress level, theoretically given long enough time, any random vibration can reach arbitrarily large amplitudes. Thus, there are no "endurance limits" for the rms fatigue curves. Even after the corresponding deterministic fatigue curve has leveled off, the rms curve will continue to drop, although at a much slower rate.

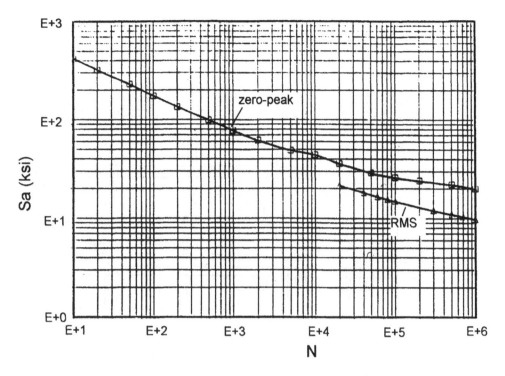

Figure 11.7 Fatigue curves for carbon, low alloy and high tensile steels
(Ultimate tensile strength=115.0 to 130.0 ksi) (Brenneman, 1986)

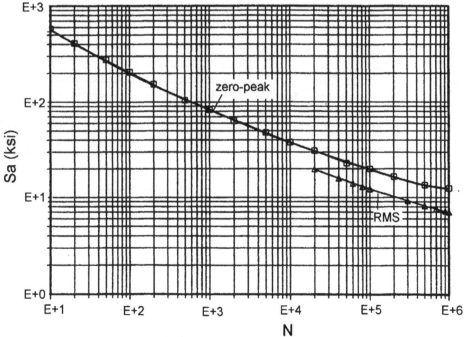

Figure 11. 8 Fatigue curves for carbon, low alloy and high tensile steels
(Ultimate tensile strength < or =80.0 ksi) (Brenneman, 1986)

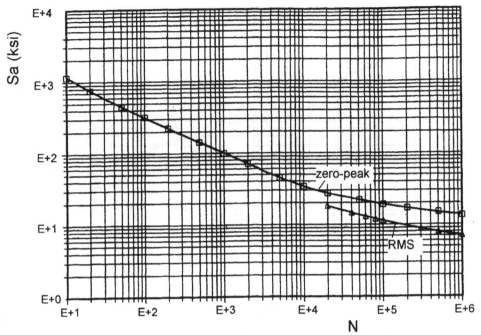

Figure 11.9 Fatigue curves for high strength steel bolting
(Maximum nominal stress $\leq 2.7 s_m$) (Brenneman, 1986)

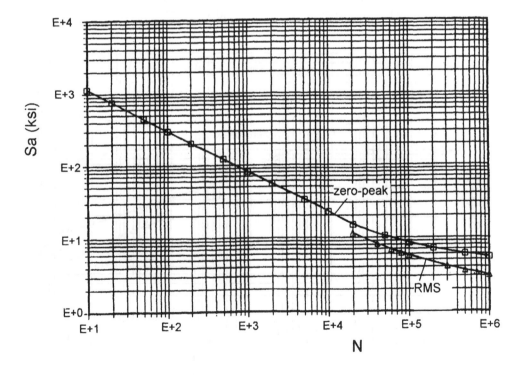

Figure 11.10 Fatigue curves for high strength steel bolting
(Maximum nominal stress $=3.0 s_m$) (Brenneman, 1986)

11.4   Wear due to Flow-Induced Vibration

When the fluid flow induces relative motion between two contacting structural surfaces, wear will result.   Very crudely, there are at least three different kinds of wear mechanisms due to flow-induced vibration (wear experts such as Ko, 1993, would distinguish many more different wear mechanisms).

Impact Wear

This term usually refers to moderate to fairly large amplitude vibrations, resulting in high-g level impacts between components. The impacting components would bounce back from each other with little relative sliding motion between the contacting surfaces. Repeated impacts between two components would rapidly cause surface fatigue of the material. Examples of this include backstop or seat tapping of disk assemblies inside check valves, and tube-to-tube impacting after a tube bundle becomes fluid-elastically unstable. Continuous seat tapping, even at small vibration amplitudes and g-levels, will ultimately damage the seat trim, resulting in seat leakage inside valves. Figure 11.11 shows the damaged disk of a check valve caused by repeated seat tapping. In this case (Dixon, 1999), the vibration amplitudes were relatively small. Damage to the seat trim became noticeable only after years of operation. Figure 11.12 shows the failure shape of an internally pressurized tube due to an axial flaw (which may or may not be caused by tube-to-tube impacting). In general, the coefficient of impact wear is much larger than those of other types of wear mechanisms.

Based on tests with simple tube models, Blevins (1984) derived a semi-empirical equation to estimate the surface stress due to impacting between heat exchanger tubes and the loose support plates,

$$s_{rms} = c \left(\frac{E^4 M_e f_n^2 <y^2>_{max}}{D^3} \right)^{1/5}$$

(11.16)

where

c	= Contact stress parameter, given in Figure 11.13
E	=Young's modulus
M_e	="Effective mass" of tube, usually taken as 2/3 the total mass of the two spans
f_n	=Natural frequency of the tube
$<y^2>_{max}$	=Maximum mean square vibration amplitude of the tube in the adjacent spans
D	= OD of the tube

Figure 11.11 Check valve disk damaged by repeated seat tapping (tiny pits in the outer
rim were caused by cavitation) (Dixon, 1999)

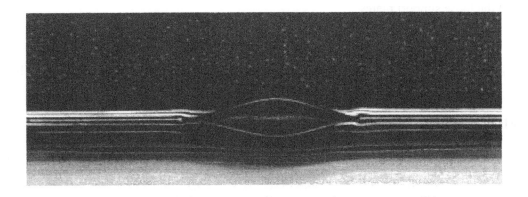

Figure 11.12 Bursting of a pressurized tube due to an axial flaw
(laboratory simulation)

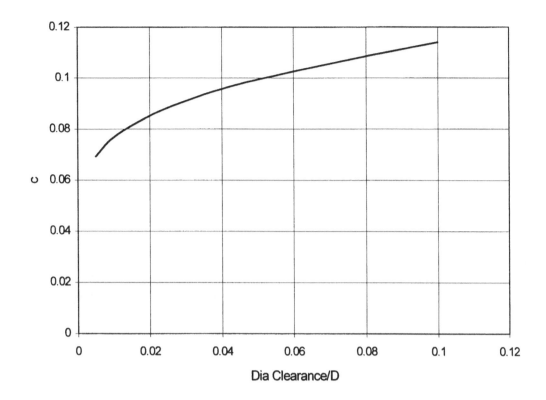

Figure 11.13 Contact stress parameter (c) as a function of tube-support clearance
(based on data from Blevins, 1984)

Note that if the tube undergoes sinusoidal vibration, as is the case when it becomes fluid-elastically unstable, the above equation still applies. On the right-hand side of the equation, the mean square vibration amplitude should then be replaced by y_{0-peak}^2. The stress computed will be the 0-peak stress.

When the surface stress is below the published endurance limit for the particular material, impact wear is usually not a problem. If the surface stress is above the endurance limit of the material, then surface damage due to impact wear may occur.

Example 11.2

Suppose in Example 9.1 (Chapter 9), the two intermediate supports each has a diametral clearance of 0.015 inch (3.81E-4 m) and assuming that all the other parameters remain the same, will impact wear be a concern?

Solution

We first reproduce in Table 11.2, the relevant given or computed parameters from Tables 9.1 and 9.2 in Chapter 9:

Table 11.2: Known Parameters for Example 11.2

	US Unit	SI Unit
Length of each span, L	50 in	1.27 m
OD, D	0.625 in	0.01588 m
ID, D_i	0.555 in	0.01410 m
Diametral clearance, g	0.015 in	3.81E-4 m
Young's modulus, E	29.22E+6 psi	2.02 E+11 Pa
Tube material density, ρ_m	0.306 lb/in^3	8470 Kg/m^3
	(7.919E-4 (lb-s^2/in)/in^3)	
Tube cross-sectional area, A_m	0.0649 in^2	4.19E-5 m^2
Moment of inertia, I	0.00283 in^4	1.18E-9 m^4
Generalized mass $m_1 = \rho_m A_m + \rho_s A_o + \rho_i A_i$	6.81E-5 (lb-s^2/in)/in	0.469 Kg/m
f_l, Hz	21.9	21.9
y_{rms}	0.035 in	8.95E-4 m

From the above table, the ratio of diametral clearance/D is g/D=0.015/0.625=0.024. From Figure 11.13, c= 0.086. The effective mass is 2/3 of the total mass of the two adjacent spans of each intermediate support,

$$M_e = (2/3)*2*50*6.81E-5 = 4.54\text{E-3 lb-s}^2/\text{in} \qquad (0.794 \text{ Kg})$$

Substituting into Equation (11.16), we get

$$s_{rms} = 3.27\text{E+4 psi} \quad (2.23\text{E+8 Pa})$$

Figure 11.6 gives the fatigue curves for austenitic steel, which is one of the most commonly used materials for nuclear steam generator tubes. At a stress level of 32,700 psi rms, the lower curves show that the allowable number of cycles is only about 10,000 cycles. At 21.9 Hz, this represents a fatigue life of less than one hour. Thus if the tube indeed rattles inside the oversize tube support hole, impact wear will be a concern.

Sliding Wear

This refers to wear between two components that are always in contact with each other. Examples include rotation of a shaft inside a sleeve bearing or the whirling of a tube inside an oversize hole after the tube bundle becomes fluid-elastically unstable. From Archard's (1956) equation, if F_n is the normal contact force between the two components and S is the total sliding distance between the two components, then the volumetric wear rate is given by the equation:

$$Q = KF_nS \qquad\qquad (11.17)$$

Unfortunately this grossly oversimplified equation contains three elusive variables that can only be either computed with monumentally complicated non-liner structural dynamic models or measured with long-duration experiments. They are:

K Wear coefficient, obtained only by measurement. Its value depends on the material pair and the ambient conditions under which the wear occurs, as well as the wear mode (impact, sliding or fretting)

F_n Normal reaction force between the two components

S Total sliding distance

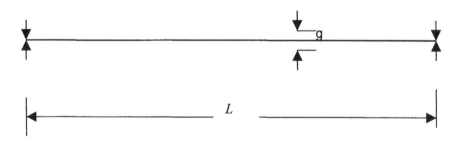

Figure 11.14 Vibration of a tube inside an oversize support

Accurate determination of F_n and S in the general case of a tube vibrating inside an oversize support hole will require non-linear structural dynamic analysis. For a tube whirling inside an oversize support hole, Connors (1981) suggested the following simplified equations: If g is the diametral gap clearance between the tube and the support (Figure 11.14), and the tube slides along the hole by an angle $d\theta$, the sliding distance is,

$$dS = gd\theta/2$$

The total sliding distance is thus given by,

$$S = \pi f_n g t \tag{11.18}$$

where f_n is the frequency of whirling and t is the total elapsed time. Connors (1981) further derived the simplified expression for the normal contact force:

$$F_n = \frac{\pi D y_{0-p}}{\mu \left(\dfrac{L^2}{A_m E} + \dfrac{D^2 L^2}{4EI} \right)} \tag{11.19}$$

where

μ	= Coefficient of friction, with 0.5 a good assumed value
A_m	= Metal cross-sectional area of the tube
I	= Area moment of inertia of the tube
L	= Total length of spans on both sides of the oversize support
E	= Young's modulus

From Equation (11.17), the wear volume after time t is,

$$Q = \pi K F_n f_n g t \tag{11.20}$$

The value for the wear coefficient K remains to be determined. Limited experimental data on wear coefficients were published in the literature, mainly for material pairs used in nuclear steam generators. Figure 11.15, reproduced from Hofmann and Schettler (1989), appears to indicate that the wear coefficients for impact/sliding and pure sliding were about the same. Unfortunately, since most other wear tests involved only impact/sliding motions, confirmation of Hofmann and Schettler's data on pure sliding is difficult.

Example 11.3

Return to Example 11.2 again and assume that the tube is made of Inconel 600 and is whirling inside an oversize support plate hole, making contact with the support plate all the time. Assume the support plate thickness is 1.5 in. (0.038 m) and is made of carbon steel. Based on Hofmann and Schettler's (1989) data, Figure 11.15 (see also Table 11.3 following sub-section on Fretting Wear), the wear coefficient is 9.7E-12 psi[-1] (1.4E-15 Pa[-1]). Estimate what would be the approximate wear depth on the tube after 10 years of continuous operation.

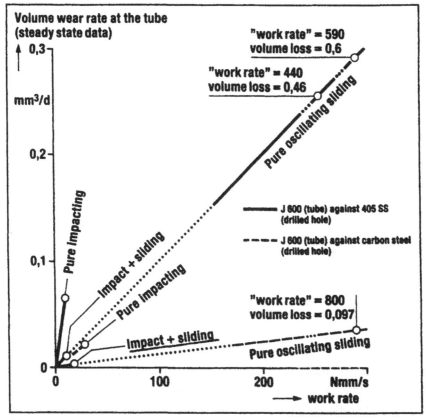

Figure 11.15 Wear coefficients for some material pairs
(Hofmann and Schettler, Copyright © 1989, EPRI report NP-5743,
Electric Power Research Institute, reproduced with permission.)

Solution

For an approximate estimate, we assume the coefficient of friction μ=0.5.   Some
engineering judgment is necessary to determine what should be the 0-peak vibration
amplitude. Since the tube is whirling inside the support plate hole, we can assume that
the motion is sinusoidal. Therefore,

$$y_{0-peak} = \sqrt{2}y_{rms} \ \ =0.05 \text{ in} \ \ (1.26\text{E-}3 \text{ m})$$

Using Equation (11.19) with L=total length of *two* spans on both sides of the oversize
support hole, together with the information in Table 11.2, one can readily calculate that,

$F_n = 11.4$ lb (50.8 N)

From Equation (11.18), at 21.9 Hz, the total sliding distance in 10 year is,

$S=3.14159*21.9*3600*24*365*10g=(2.17E+10)g=3.26E+8$ in (8.27E+6 m)

and from Equation (11.20), the volumetric wear after 10 years is,

$Q = KF_nS = 0.036$ in^3 (5.88E-7 m^3)

If the tube diameter after 10 years is D_1,

$$\frac{\pi}{4}(D^2 - D_1^2)*1.5 = 0.036$$

From which

$D_1 = 0.596$ in

The wear depth after 10 years of continuous operation will therefore be 0.029/2=0.0145 in. or 41% of the tube wall. The above calculation is approximate because, strictly speaking, the sliding velocity depends on g, which is a function of time. A rigorous solution to the total sliding distance in 10 years should therefore be evaluated by integration of Equation (11.18).

Fretting Wear

In Examples 11.2 and 11.3, the dimensions and the material properties of the tube, as well as the forcing function, are approximately the same as tubes subject to the highest cross-flows in today's operating nuclear steam generators. Thus, if impact or sliding wear were indeed responsible for the wear mechanism in these tubes, we should have observed many steam generator tube failures after 20 years of continuous operations. Field inspections using eddy current technique, followed by occasional destructive metallurgical examinations, found that while wear marks were observed in a very small number of the nuclear steam generator tubes, particularly those located in regions of high cross-flows, the wear rates are less than what are computed in Examples 11.2 and 11.3. Furthermore, the wear marks were usually confined to an arc (Figure 11.16) instead of 360-deg. around the tube. Thus, there is a third wear mechanism responsible for the observed wear marks. This is fretting wear, or wear caused by a combination of impact and sliding (Figure 11.17). However, fretting wear usually refers to a sub-class of this, with very small vibration amplitudes and impact forces typically caused by low-amplitude vibrations. A classic example of fretting wear is turbulence-induced wear in heat exchanger tubes supported by oversize holes.

Figure 11.16 Fretting wear mark on a heat exchanger tube

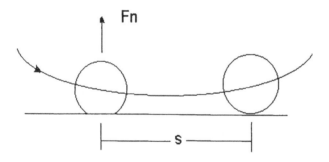

Figure 11.17 Combined impact and sliding wear

If turbulence-induced fatigue analysis is one order of magnitude more complicated than the corresponding impact analysis, then turbulence-induced wear analysis is at least two orders of magnitude more complicated than fatigue analysis. There are two reasons for this quantum jump in complication: First, wear by itself is a complicated subject involving microscopic properties of the materials. Even for the same material pair in the contacting structural components vibrating with exactly the same amplitudes and frequencies, the wear on the components can be completely different depending on the geometry of the contacting surfaces. Second, at least in the power and process industries, the most important wear problems occur in components with non-linear supports. Examples of this include heat exchanger tubes and nuclear fuel bundles. To allow for thermal expansion and also ease of fabrication, the tubes in a heat exchanger invariably passes through oversize holes in the support plates. It is this tube-to-support plate clearances, which result in relative motions between the tubes and the support plates, that cause wear in the tubes. In all the examples of turbulence-induced vibration analysis given in Chapter 9, we have assumed that the tubes are either clamped or simply-supported at the support plates, without any clearances. This enables us to solve the problem with linear structural mechanics but does not allow us to calculate the relative motion between the tube and the support plates, which is a pre-requisite for any quantitative analysis of the wear rates between these two components. Detailed treatment of non-linear structural dynamics is beyond the scope of this book. In practice, quantitative flow-induced vibration/wear analysis is almost always carried out with the aid of a finite-element, non-linear structural analysis computer program—either one specifically developed for flow-induced vibration/wear or one for general purpose analysis. Even with today's high-speed computers, clever modeling of the structure and approximation of the turbulence forcing function, which is random both in space and in time, will be necessary. In Section 11.5, the fundamental equations for flow-induced vibration with tube/support interaction are given without going into the details of non-linear structural dynamics analysis. The readers should refer to books and computer program manuals on this specific subject. In Section 11.6, some simplified methods of qualitatively assessing the relative wear rates will be given.

<u>Fretting Wear Coefficient</u>

Table 11.3: Wear Coefficients for Impact/Sliding

Material Pair[1]	Source	Units as reported	K in Pa^{-1}
Steel/Steel	Connors, 1981	505E-12 psi^{-1}	72E-15
--	Ko, 1993	28E-12 to 150E-12 psi^{-1}	4.0E-15 to 22E-15
Inconel 600/ Carbon Steel	Hofmann and Schettler, 1989	Figure 11.15	1.4E-15
Inconel 600/ SS 405	Hofmann and Schettler, 1989	Figure 11.15	12E-15
Incoloy 800/ SS 410	Fisher et al, 1994	in Pa^{-1}	40E-15
Inconel 600/ Carbon Steel	Fisher et al, 1994	in Pa^{-1}	50E-15
Inconel 600/ SS 405	Kawamura et al, 1991	Graph in vol. loss vs. watt	14E-15

[1] Wear coefficient applied to the first of the wear pairs.

As in impact and sliding wear, one of the essential parameters to calculate the fretting wear rate is the wear coefficient. From 1980 to 2000, some experimental data on the fretting wear coefficient were published in the literature, mainly for material pairs used in nuclear steam generators. They were often reported in different units, making direct comparison between them difficult. Furthermore, the exact type of motion under which the test was carried out often was not precisely described. In Table 11.3, some of these data are reproduced and converted to the common unit of Pa^{-1} (inverse Pascal), which appears to be the accepted unit in the more recent technical papers. Also in this table are the type of motion associated with each data set, as understood by the author of this book.

11.5   Fretting Wear and the Dynamics of a Loosely Supported Tube

In Chapter 7 it is shown that the equation of motion for a tube with fluid-structure interaction is given by,

$$M\ddot{y} + (C_{sys} + C_{fsi})\dot{y} + ky = F_{inc} \tag{7.14}$$

For cross-flow, the apparent damping due to fluid-structure interaction is given by,

$$C_{fsi} = 4\pi M f_o \zeta_{fsi} \quad and \quad \zeta_{fsi} = -\frac{\rho V_p^2 / 2}{\pi f_o^2 m \beta^2} \tag{7.15}$$

For an isolated tube, this term is zero. Whether this fluid-structure interaction force is present in a tube bundle even below the instability threshold is a subject of many debates. However, since Equation (7.14) is a direct consequence of Connors' (1981) equation, it is only reasonable to assume that this term is always present in a tube bundle subject to cross-flow, as long as we believe in Connors' equation.

   Equation (7.14) assumes that there is no clearance between the tube and its supports. In the presence of tube-support clearances, a non-linear, localized contact force F_c must augment Equation (7.14),

$$M\ddot{y} + (C_{sys} + C_{fsi})\dot{y} + ky = F_{inc} + F_c \tag{11.21}$$

$$F_c = -k_c(|y(t)| - g) \quad if \quad |y(t)| > g$$
$$= 0 \quad if \quad |y(t)| < g$$

where k_c is the equivalent stiffness of the tube/support combination. Since the tube is usually much less rigid than the support plate, Axisa et al (1984) suggested using the stiffness associated with the local ovalization of the tube,

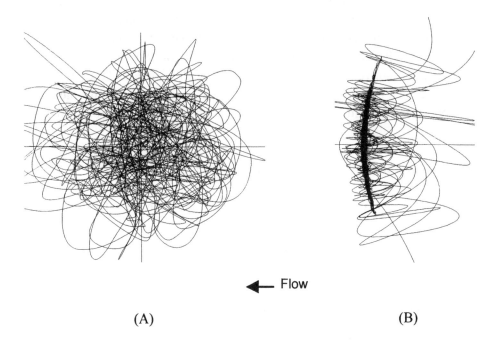

← Flow

(A)   (B)

Figure 11.18 Orbital motion of tubes due to turbulence-induced vibration in:
support plate inactive (A) and support plate active (B) modes

$$k_c = 1.9 \frac{Eh^2}{D} \sqrt{\frac{h}{D}} \tag{11.22}$$

where D, h are the tube diameter and wall thickness respectively. Work from various authors (see references below) showed that the computed tube wear rate is not very sensitive to the equivalent support stiffness. Therefore, accurate values of k_c is not necessary.

Frick et al (1984) suggested the parameter "wear work rate," defined as the product of the contact force multiplied by the sliding distance,

$$\frac{dW}{dt} = F_n \frac{dS}{dt} \tag{11.23}$$

to quantify the volumetric wear rate. From Equation (11.17), the volumetric wear rate is

$$\frac{dQ}{dt} = KF_n \frac{dS}{dt} \tag{11.24}$$

In terms of the wear work rate, the volumetric wear rate is,

$$\dot{Q} = K\dot{W} \tag{11.25}$$

Equation (11.25) has since become the preferred expression for flow-induced vibration wear analysis. The term "work rate" is, however, somewhat misleading as the true work rate is not $F_n\dot{S}$, but $\mu F_n\dot{S}$.

In practice, Equation (11.21) is solved by the finite-element together with the direct time history integration method. The non-linear tube supports are modeled by "gap elements," with specified gap clearances and equivalent support stiffness. The normal force and the sliding distances are computed and summed to find the total work rate, and hence the wear rate and the integrated volumetric loss. Since the dynamics of the tube, the work rate, the wear rate and the fluid-structure interaction force are all affected by the gap clearance, these parameters must be periodically updated as one marches forward in time. The numerical process can be extremely time-consuming. The readers are referred to the many technical papers published in the 1990s (e.g., Rao et al, 1988; Sauve', 1996) for this highly specialized topic. Figures 11.18(A) and (B) are polar plots of 4.0 seconds each of the times histories of the motion of points on two different heat exchanger tubes supported by plates with oversize circular drilled holes. These time histories were obtained by numerical solution of Equation (11.21) using a non-linear finite-element computer program, with time steps as small as 1/40,000 second. In Figure 11.18(A), the tube is vibrating in the "support plate inactive" mode. That is, the tube does not touch the support plate even during the extreme excursions of the motion. The motion is random about the static equilibrium position. In Figure 11.18(B), the tube is vibrating in the "support plate active" mode, as is evidenced by the vague profile of the arc-defining part of the drilled hole. In this case the tube is biased to the left by the flow, impacting and sliding mostly along this arc. It is this mechanism that causes the wear mark shown in Figure 11.16. Because of the finite stiffness of both the tube and the support, the excursion of the tube is seen to occasionally exceed the boundary of the support plate hole.

Connors' Approximate Method for Fretting Wear

Based on the assumption that the vibrating tube is rocking inside an oversize circular support hole without "lift-off," Connors (1981) derived an expression for the total sliding distance in time t,

$$S = 2\pi D y_{0-peak} f_n t / L \tag{11.26}$$

Note that in Equation (11.26), L is the total length of the spans on both sides of the oversize support. This, together with Equation (11.19) for the normal reaction force and

Equation (11.20) for the wear rate, enable one to calculate, although very crudely, the wear volume as a function of time.

Example 11.4

Return once again to Example 11.2 and Table 11.2: assume the 0-peak vibration amplitude is three times the rms amplitude computed in Example 9.1 and given in Table 11.2, and assume that the tube is vibrating with constant contact with the middle, oversize support holes, estimate the fretting wear depth of the tube wall after 10 years of continuous operation.

Solution

Using Equation (11.19) with L=100 in (2.54 m), and y_{0-peak}=3*0.035 in=0.105 in (0.00267 m), we can readily calculate the normal reaction force at the support to be:

F_n = 24.1 lb   (108 N)

From Equation (11.26), the total sliding distance after 10 years of continuous operation is,

S=2*3.1416*0.625*21.9*(3*0.035)*(10*365*24*3600)/100=2.85E+7 in   (7.23E+5 m)

Using the same wear coefficient of 9.7E-12 psi^{-1} as in Example 11.3, we get from Equation (11.20),

Q = 0.0062 in^3 (1.02E-7 m^3)

If we assume the wear is confined to a 90-degree arc, then following the same procedure as in Example 11.3, we find the loss of wall thickness after 10 year of continuous operation to be 0.008 in. (0.051 mm), or about 24% of the wall thickness. This is closer to what is measured in operating nuclear steam generators.

The Energy Method for Fretting Wear Estimate

Based on the fact that energy dissipation of a vibrating heat exchanger tube is mainly through its damping and that most of the damping in a heat exchange tube comes from the interaction (i.e., wear) between the tube and its support plates, Yetisir et al (1997) suggested a method of estimating the wear rate of heat exchanger tubes based on the measured damping ratios of the tubes. However, due to the large uncertainties in the measured damping ratios as well as other parameters in the turbulence forcing function, accurate determination of the quantitative wear rate based on this method is not feasible.

This method is best applied to assess the relative wear rate between two heat exchanger tubes. If the wear rate of the reference heat exchanger is known either from operational experience or from a separate non-linear vibration/wear analysis, then the wear rate of the other heat exchanger tube can be deduced.

Yetisir et al (1997) formulated the equation based on a simply-supported heat exchanger tube. The following is a revised formulation of the method. It is applicable to single or multi-span heat exchanger tubes of arbitrary boundary conditions. From Chapter 3, the power dissipated by a vibrating structure is,

$$P = 8\pi^3 \zeta_n f_n^3 m_n (y_{0i}^2 / \psi_{ni}^2) \tag{3.40}$$

If μ is the coefficient of friction, the force resisting the tube motion is μF_n. From Equation (11.23), the power dissipated by the vibrating tube is related to the wear work rate by,

$$P = \mu \dot{W} \tag{11.27}$$

or

$$\dot{W} = 8\pi^3 (\zeta_n f_n^3 m_n / \mu)(y_{0i}^2 / \psi_{ni}^2) \tag{11.28}$$

Since $y_{ni} / \psi_{ni} = a_n$ (see Chapter 3) is independent of the location, point i can be any point on the structure. If the wear rate of a reference heat exchanger tube is known from operational experience and the vibration amplitude and the mode shape at one point of this tube are known either by measurement or by analysis, the wear rate of another heat exchanger built with the same tube and support plate material pair can be estimated by ratioing with the reference heat exchanger. The uncertainties in the coefficient of friction as well as in the turbulence forcing function cancel out. The following example should clarify the method.

Example 11.5

(Since the purpose of this example is to illustrate the method, units are not explicitly mentioned except that they are given in consistent unit sets. Readers who are familiar with nuclear steam generator tube dimensions can easily figure out that US units are used.)

Figures 11.19(A) and (B) show two heat exchanger tubes of different designs, but with similar tube and support plate materials. Full-scale mock-up tests of tube A showed that the dominant mode of vibration was in the out-of-plane direction, with a modal damping ratio of 0.01. A detailed flow-induced vibration analysis based on a linear

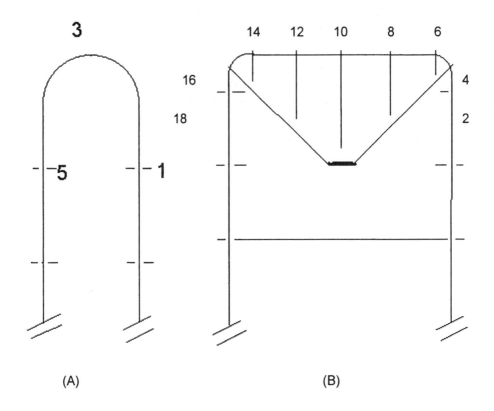

Figure 11.19 Two heat exchanger tubes

finite-element model of tube (A) had been performed (see Chapter 9). With the modal generalized mass normalized to unity, the dominant out-of-plane modal frequency, mode-shape value and vibration amplitude at point "3" were calculated. These values are given in Table 11.4.

A full-scale mock-up test of tube B has also been performed. It showed that the dominant mode of vibration was an in-plane mode, with a damping ratio of 0.05. A detailed flow-induced vibration analysis based on a linear finite-element model of tube B was also carried out. With the modal generalized mass normalized to unity, the dominant out-of-plane modal frequency, mode-shape value and vibration amplitude at point "4" were computed and were shown in Table 11.4.

Based on operational experience, it was determined that tube (B) had a combined volumetric wear rate of 0.001 units per year of operation at all the support points of the tube. Of this, most of the wear occurred at points 2, 4, 6 and 14, 16, 18, while the wear at all other locations were negligible.

Assuming most of the wear in tube A occurs at support points 1 and 5, estimate the volumetric wear rates at these two locations.

Table 11.4: Results from Linear Analysis

	Steam Generator A	Steam Generator B
Measured damping ratios	0.01	0.05
Response y (rms)	0.0125	0.0103
at	apex (Point 3)	bat wing (point 4)
Direction of response	out-of-plane	in-plane
Dominant modal freq, Hz	13.6	6.1
Generalized mass, m_A, m_B	1	1
Mode-shape value	10.6 at point 3	9.3 at point 4
Volumetric wear rate	to be determined	0.001 unit/year

Solution

Using Equation (11.28), the ratio of the wear work rates between the two heat exchanger tubes are,

$$\frac{\dot{W}_A}{\dot{W}_B} = \frac{P_A/\mu}{P_B/\mu} = \frac{\zeta_A f_A^3 y_A^2 m_A / \psi_A^2(1)}{\zeta_B f_B^3 y_B^2 m_B / \psi_B^2(2)} = \frac{0.01 * 13.6^3 * 0.0125^2 * 1/10.6^2}{0.05 * 6.1^3 * 0.0103^2 * 1/9.3^2}$$

$$= 2.52$$

Thus the total volumetric wear rate at points 1 and 5 of tube A is about 2.5 times that of heat exchanger tube B, or 0.00125 units per year at each of locations 1 and 5.

References

Archard, J. F. and Hirst, T., 1956, "The Wear of Metals Under Unlubricated Conditions," Proceedings of the Royal Society of London, Series A, Vol. 236, pp. 397.

ASME, 1998, Boiler Code Sec III Appendix N-1300 Series.

ASME, 1998, Boiler Code Sec III Appendix I.

Axisa, F., Desseaux, A., and Gilbert, R. J., 1984, "Experimental Study of Tube/Support Impact Forces in Multi-Span PWR Steam Generator Tubes," Proceeding Symposium on Flow-Induced Vibration, Vol. 3, edited by M. P. Paidoussis, ASME Press, New York, pp. 139-148.

Blevins, R. D., 1984, "A Rational Algorithm for Predicting Vibration-Induced Damage to Tube and Shell Heat Exchangers," Proceeding Symposium on Flow-Induced Vibration, Vol. 3, edited by M. P. Paidoussis, ASME Press, New York, pp. 87-101.

Brenneman, B., 1986, "RMS Fatigue Curves for Random Vibrations," ASME Journal of Pressure Vessel Technology, Vol. 108, pp. 538-541.

Connors, H. J., 1981, "Flow-Induced Vibration and Wear of Steam Generator Tubes," Nuclear Technology, Vol. 55, pp. 311-331.

Crandall, S. H. and Mark, W. D., 1963, Random Vibration in Mechanical Systems, Academic Press, New York.

Dixon, R. E., 1999, "Cavitation of Cold Leg Accumulator Check Valves," paper presented at the Nuclear Industry Check Valve (NIC) Group Conference, St. Petersburg Beach, Fl.

Fisher, N. J., Chow, A. B. and Weckwerth, M. K., 1994, "Experimental Fretting-Wear Studies of Steam Generator Materials," ASME Special Publication PVP-Vol. 273, Flow-Induced Vibration 1994, edited by M. K. Au-Yang, pp. 281-247.

Frick, T. M., Sobek, T. E. and Reavis, R. J., 1984, "Overview on the Development and Implementation of Methodologies to Compute Vibration and Wear of Steam Generator Tubes," Proceeding Symposium on Flow-Induced Vibration, edited by M. P. Paidoussis, ASME Press, New York, pp. 139-148.

Hofmann, P. J. and Schettler, T., 1989, PWR Steam Generator Tube Fretting and Fatigue Wear, EPRI Report NP-6341.

Kawamura, K., Yasuo, A. and Inada, F., 1991, "Tube-to-Support Dynamic Interaction and Wear of Heat Exchanger Tubes Caused by Turbulent Flow-Induced Vibration," ASME Special Publication PVP-Vol. 206, Flow-Induced Vibration and Wear—1991, edited by M.K. Au-Yang and F. Hara, pp. 119-128.

Ko, P. L., 1993, "Wear Due to Flow-Induced Vibration," in Technology for the 90s," edited by M. K. Au-Yang, ASME Press, New York, pp. 865-896.

Rao, M. S. M., Steininger, D. A., Eisinger, F. L., 1988, "Numerical Simulation of Fluidelastic Vibration and Wear of Multi-span Tubes with Clearance at Supports," Proceedings of the Second International Symposium on Flow-Induced Vibration and Noise, Vol. 5, edited by M. P. Paidoussis, ASME Press, New York, pp. 235-250.

Sauve' R. G., 1996, "A Computational Time Domain Approach to Fluidelastic Instability for Nonlinear Tube Dynamics" in Flow-Induced Vibration—1996, ASME Special Publication PVP-Vol. 328, edited by M. J. Pettigrew, pp. 327-335.

Yetisir, M., McKerrow, E. and Pettigrew, M. J., 1997, "Fretting Wear Damage of Heat Exchanger Tubes: A Proposed Damage Criterion Based on Tube Vibration Response," Proceeding 4th International Symposium on Fluid-Structural Interaction, Aeroelasticity, Flow-Induced Vibration and Noise, edited by M. P. Paidoussis, ASME Press, New York, pp. 291-299.

CHAPTER 12

ACOUSTICALLY INDUCED VIBRATION AND NOISE

Summary

Although not the most costly, acoustically induced vibration is probably one of the most common vibration problems in the power and process industries. Vessels, piping systems, valve cavities, heat exchanger internals, ducts and many other components are potential resonators in which standing waves can form, while fans, pumps, valves, elbows, obstruction and discontinuities in flow channels, or even the addition or removal of heat all have the potential to excite these standing waves. Once the resonant conditions are met, the resulting sound intensity in most cases will require remedial action. In some cases, acoustic excitation can cause rapid fatigue failure of piping welded points, valve internal parts and other components.

The first requirement in acoustically induced vibration analysis is to calculate the velocity of sound in the fluid media. The velocity of sound in air at 68 deg. F (20 deg. C) and one atmospheric pressure is 13,500 in/s or 343 m/s. The velocity of sound at other temperatures can be readily calculated from the equation,

$$c = \left(\frac{\partial p}{\partial \rho}\right)_S = \sqrt{\frac{\gamma p}{\rho}} = \sqrt{\gamma G T} \propto \sqrt{T}$$

(12.4)

$$\gamma = C_p / C_v$$

where T is the absolute temperature in either deg. R or deg. K, depending on which unit system is used. At atmospheric pressure and 0 deg. C (32 deg. F) the velocity of sound in water is 55,288 in/s or 1,404 m/s. The velocity of sound in water, steam or water-steam mixture at any given temperature and pressure combination can be calculated from information given in the ASME Steam Table (1979), as outlined in Examples 12.2 and 12.3. A table of velocity of sound in water at selected values of temperatures and pressures are given in Chapter 2, Table 2.1.

In a heat exchanger, the velocity of sound in the direction transversal to the tube bundle axis is decreased by the presence of the tubes. If c_0 is the velocity without the tubes and c is the velocity in the presence of tubes, then

$$c = \frac{c_0}{\sqrt{1+\sigma}}$$

(12.9)

where σ is the ratio of the heat exchanger internal volume occupied by the tubes to the total volume.

Knowing the velocity of sound, the acoustic modal frequencies in ducts and cavities can be calculated. For slender pipes and ducts with lengths much larger than the diameters, the acoustic modal frequencies are given by

$$f_\alpha = \frac{\alpha c}{2L} \quad \text{Hz} \quad \alpha = 1, 2, 3, \ldots. \tag{12.12}$$

if both ends of the pipes are either open (to the ambient condition) or closed, and

$$f_\alpha = \frac{\alpha c}{4L} \quad \text{Hz} \quad \alpha = 1, 2, 3, \ldots. \tag{12.16}$$

if one end of the pipe or duct is open while the other end is closed. As long as the pipe is slender, the above equations are independent of the cross-sectional geometry. Likewise, the natural frequencies of the acoustic modes in a rectangular cavity are given by,

$$f_{\alpha\beta\gamma} = \frac{c}{2}\left(\frac{\alpha^2}{L_1^2} + \frac{\beta^2}{L_2^2} + \frac{\gamma^2}{L_3^2} \right)^{1/2} \quad \text{Hz} \quad \alpha, \beta, \gamma = 1, 2, 3, \ldots. \tag{12.18}$$

if opposite ends of the cavity are all either open or closed. However, if one end (for example in the "1" direction) is open while the opposite end is closed, the acoustic modal frequencies will be:

$$f_{\alpha\beta\gamma} = \frac{c}{2}\left(\frac{\alpha^2}{4L_1^2} + \frac{\beta^2}{L_2^2} + \frac{\gamma^2}{L_3^2} \right)^{1/2} \quad \text{Hz} \quad \alpha, \beta, \gamma = 1, 2, 3, \ldots. \tag{12.20}$$

The acoustic modal frequencies for other cases can be deduced similarly. To obtain the acoustic modal frequencies in cylindrical cavities, one has to solve numerically equations involving Bessel functions. However, this can be done easily with today's widely available spreadsheet, as shown in Section 12.4 and Examples 12.2 and 12.3.

To avoid potential vortex-induced excitation of acoustic modes, the most conservative way is to operate heat exchangers so that the vortex-shedding frequency is below the lowest acoustic modal frequency. However, experience with high-performance nuclear steam generators shows that, very often, acoustic resonance does not occur even theoretically one of the acoustic modal frequencies is within "lock-in" range of the vortex-shedding frequency. The resonance maps proposed by Ziada et al (1989) appear to predict the absence of acoustic resonance problems in nuclear steam generators correctly. These resonance maps are given in Section 12.5 and are far less conservative than the vortex "lock-in" rules.

The addition or removal of heat, such as in a furnace/duct system, can cause spontaneous acoustic oscillation with or without fluid flow in the duct. This is known as thermoacoustics. When the system has both ends open, it is known as a Rijke tube; when it has one open end and one closed end, it is know as a Sondhauss tube. A stability map based on the geometry of the system and the ratio of temperatures in the hot and cold portions of the system is given in Section 12.8. In general, one should avoid placing a furnace grid or a swirler in a furnace/duct system at the quarter point of a Rijke tube or at the mid-point of a Sondhauss tube.

Due to their small diameters compared with the wavelengths, slender structures are generally immune to normally impinging low-frequency acoustic waves.

Post-construction remedies of acoustic resonance problems usually depend on detuning—by changing either the frequencies of the excitation or, more commonly, by changing the frequencies of the acoustic modes. Because of the logarithmic nature of sound, any method based on energy dissipation would have to dissipate more than 90% of the acoustic energy before there is any noticeable effect on the sound level. Elimination or reduction of the excitation force is another, although usually not very practical, way.

Nomenclature

$a_{\alpha\beta}$	Amplitude function
a	Inner radius of annular cavity
A	Area of neck of Helmholtz resonator
b	Outer radius of annular cavity
B	Isothermal bulk modulus
c	Velocity of sound
c_0	Velocity of sound in the absence of tubes
C_p	Specific heat of gas at constant pressure
C_v	Specific heat of gas at constant volume
d	Diameter of tube or pipe
D	Diameter of cylindrical cavity
f	Frequency in Hz
$f_{\alpha\beta\gamma}$	Acoustic modal frequency
f_s	Vortex-shedding frequency
G	Universal gas constant
G_i	Resonance parameter for in-line tube array
G_s	Resonance parameter for staggered tube array
J	Bessel function of first kind
k	Wave number, $=2\pi/\lambda$
ℓ	Length of cold portion of a thermoacoustic system
L	Length of pipe, tube or characteristic length of flow channel discontinuity
L_i	$i=1,2,3$, lengths of rectangular cavity in x, y, z directions

p	Pressure
p'	Acoustic pressure
p_0	Ambient pressure or 0-peak acoustic pressure
p_i	Acoustic pressure of incident wave
p_s	Acoustic pressure of scattered wave
P	Tube pitch
P_L	Tube pitch, streamwise direction
P_T	Tube pitch, in direction perpendicular to flow
Q	Volume of Helmholtz resonator
r	Radial coordinate
$\vec{r}$	Position vector in cylindrical coordinate system
R	Radius of tube or cylindrical shell
R_a	Acoustic Reynolds number
S	Strouhal number
SPL	Sound pressure level
R_c	Reynolds number
T	Absolute temperature
T_h, T_c	Absolute temperatures in the hot and cold portions of a thermoacoustic system
V	Flow velocity
V_p	Pitch velocity
x	Length along "1" direction
X_L, X_T	Pitch-to-diameter ratio, streamwise direction and perpendicular to flow direction
Y	Bessel function of second kind

α	Critical temperature ratio (in thermoacoustics)
α, β, γ	Acoustic modal indices
ε	$=\alpha$ for pipes with acoustically open or closed ends
	$=(2\alpha-1)/2$ for pipes with one end acoustically open and the other closed
θ	Angular coordinate
ρ	Density of fluid
σ	Solidity ratio in a tube bundle
λ	Wavelength
$\lambda_{\alpha\beta\gamma}$	Roots of the radial function in the acoustic wave equation
ξ	Ratio of the length of the hot to the length of the cold portions of a gas column
Φ	Radial function of the acoustic wave equation
χ_α	$=\sqrt{(2f/c)^2 - (\varepsilon/L)^2}$

12.1   Introduction

Power and process plant components and piping systems form resonant cavities while fans, pumps, flow turbulence, cavitation, vortex-shedding, and the addition or removal of heat from the fluid are potential sources of excitation. When the conditions are right, the component or piping system can be set into acoustic resonance. Sometimes this results in nothing more than irritatingly loud noise. On other occasions, acoustic resonance can lead to rapid component failure. In Chapter 1, the kinematics of acoustics is briefly discussed; and in Chapters 8, 9 and 10, the responses of structures to flow turbulence are discussed in detail. This chapter is devoted to problems associated with noise and responses of structures to standing waves formed in enclosed flow channels associated with industrial piping systems and components. External flow problems associated with aerodynamic noises are considered beyond the scope of this book. Following our convention established in earlier chapters, we shall use the Greek alphabet α, β, etc., to denote acoustic modes and the English alphabet m, n, j, k to denote structural modes.

In the previous chapters, we have encountered numerous examples of sources of acoustic excitation. The most common and important ones are:

- Pressure pulses generated by fans and pumps. As shown in Figure 2.3 (Chapter 2), pressure pulses are generated by fans and pumps at the frequencies of the shaft rotation frequency, the blade passing frequency and its higher harmonics. These pulses are harmonic forcing functions with well-defined frequencies. The first blade passing frequency is equal to the number of blades multiplied by the shaft rotation frequency. In most pumps and fans, the most prominent pulse occurs at the first, second or the third blade passing frequency. The magnitudes of these pulses, unfortunately, usually have to be obtained from tests.

- Highly turbulent flow, cavitation and shock generated by valves of different designs and functions are powerful sources of acoustic excitation. Unlike pressure pulses generated by fans and pumps, pressure pulsation generated by valves can be a combination of discrete and continuous components. Usually, the higher the pressure drop, the higher the excitation force. One large flow opening also generates more turbulence than a multiple of small orifices that together pass the same volume of fluid flow.

- Vortices shed from the trailing edges of structures or from structural discontinuities are important sources of acoustic excitation. In Chapter 6, criteria are given for vortex-shedding from isolated tubes and tube bundles. The latter is especially important in designing heat exchangers to avoid acoustic resonance. More of this will be discussed later in this chapter. For flows over surface discontinuities like a step (Figure 12.1 (A)) or a cavity (Figure 12.1 (B)), the shedding frequencies are expressed as a generalization of Equation (6.1) for vortex-shedding off isolated cylinders,

$$f_s = SV / L \qquad\qquad (12.1)$$

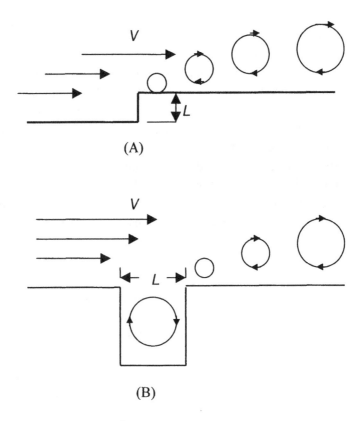

(A)

(B)

Figure 12.1 Vortices generated from flow over discontinuities

where L is the characteristic length of the discontinuity (Figure 12.1) and S is the Strouhal number discussed extensively in Chapter 6. Values of S can vary between 0.2 to 0.5. For flow over cavities such as in Figure 12.1 (B), the mechanism is probably more complicated than vortex-shedding. Alternate empirical equations for the frequencies associated with the shear wave instability had been suggested (Blevins, 1990):

$$f_\alpha = 0.33(\alpha - 1/4)V/L \qquad \text{for turbulent boundary layer} \qquad (12.2)$$

$$f_\alpha = 0.52V/L \qquad \text{for laminar boundary layer} \qquad (12.3)$$

where $\alpha = 1, 2, 3 \ldots$

For applications to power and process plant piping, Equation (12.2) holds.

• Flow turbulence can also excite standing waves associated with acoustic modes as will be discussed in the following sub-section. In Chapter 2, we discuss a case in which standing waves excited by the turbulence of leakage flow from a valve seat was used as a diagnostic tool to identify the location of the leak (Case Study 2.2).

Flow turbulence can also excite standing waves in pipes, sometimes with annoying, loud hums that can be heard hundreds of feet away.

- Cavitation can generate loud noises. In some piping systems, the cavitating venturi, briefly discussed in Chapter 10 under axial flow-induced vibration (Figure 10.8), is installed to "choke," or limit, the mass flow rate. The resulting high-velocity vapor shear flow can generate high-frequency noises that are much worse than the noise generated by cavitation itself. Cavitation can also generate low-frequency pressure pulses due to the unsteady exit flow. These pulses can excite low-frequency oscillations in close flow loops, which in turn can excite piping components such as valves nozzle-mounted on the main piping system. The high-frequency part of the force spectrum can shake loose threaded joints such as hand wheels and cause local shell-mode deformations of the pipes, resulting in rapid fatigue failures of welded joints.

- Since heat is a form of energy, the addition to or removal of heat from the fluid can excite acoustic standing waves. This phenomenon, called thermoacoustics, is especially important in furnace-duct systems.

We shall return to discuss in greater detail specific cases of acoustic modes in industrial piping systems and components excited by the above-mentioned forcing functions. Meanwhile, let us focus on the other condition for the acoustic problems: The formation of standing waves in pipes and cavities.

12.2   Velocities of Sound in Material Media

A pre-requisite for acoustic resonance is the formation of standing waves. Standing waves are formed when acoustic waves traveling in two opposite directions superimpose onto each other, resulting in alternate cancellations and reinforcement of the amplitudes. In order to find the frequencies of the standing waves, it is necessary to know the velocity of sound in the acoustic medium. In the following, the equation of sound in gases, liquids and solids are given. The derivation of these equations can be found in books on physics and chemistry. These equations are also widely available in physics, chemistry and mechanical engineering handbooks.

Velocity of Sound in Gases

Sound propagates in gases according to the adiabatic law, with a velocity equal to

$$c = \left(\frac{\partial p}{\partial \rho}\right)_S = \sqrt{\frac{\gamma p}{\rho}} = \sqrt{\gamma GT} \propto \sqrt{T}$$

$$\gamma = C_p / C_v$$

(12.4)

where the subscript S denotes the fact that the change in pressure p and the density ρ of the gas occurs isentropically (i.e., at constant entropy). C_p, C_v are the specific heats of the gas at constant pressure and constant volume respectively and T is the absolute temperature defined as,

$$T = (459.7 + \text{deg. in F}) \ \text{R(ankine)} \quad \text{or} \tag{12.5}$$
$$T = (273.2 + \text{deg. in C}) \ \text{K(elvin)} \tag{12.6}$$

G is the universal gas constant.

G=1.9865 calories/gm-mole per deg. K in cgs units,

G=8315 N-m/Kg-mole per deg. K in SI units

G=49720 ft-lb/slug-mole per deg. R in US foot units

=49720 in-lb/(lb-s^2/in)-mole per deg. R in US inch units

The values of C_p and C_v, as well as the densities and other physical constants of gases, can be obtained from physics and chemistry handbooks. In particular for air,

$\gamma = 1.4$

At 20 deg. C (293.2 deg. K, 68 deg. F or 527.7 deg. R), the velocity of sound in air is

c= 343 m/s or   13,504 in/s

Using Equation (12.4), the velocity of sound in air at other temperatures can be readily calculated without using the values for the universal gas constant.

Velocity of Sound in Liquids

Analogous to Equation (12.4), the velocity of sound in liquids is given by,

$$c = \sqrt{\frac{\gamma B}{\rho}} \tag{12.7}$$

where B is the isothermal bulk modulus of the liquid obtained from handbooks.

Velocity of Sound in Water, Steam or Water/Steam Mixture

At atmospheric pressure and 0 deg. C (32 deg. F), the velocity of sound in water is 1404 m/s or 55,288 in/s. At higher temperature and higher pressure, the velocity of sound in sub-cooled water is actually lower. An abridged table for the velocities of sound in water at different pressures and temperatures is given in Chapter 2 (Table 2.1). The velocity of sound in water, superheated steam or in two-phase water steam mixtures under any given ambient condition can be obtained from the ASME Steam Table (1979). At a given temperature T and pressure p, we first find the density and entropy of the water/steam at $p+\Delta p/2$, where $\Delta p/2$ is a small but convenient increment of pressure from the given pressure at which we want to calculate the velocity of sound. Keeping the entropy S constant, find the density of water/steam again at a pressure of $p+\Delta p/2$. Let the difference between the densities at these two slightly different pressures be $\Delta\rho$. From Equation (12.4), the velocity of sound in water/steam at T, p is equal to:

$$c = \sqrt{\frac{\Delta p}{\Delta\rho}}$$

Examples 12.2 and 12.3, given later in this chapter, illustrate the process numerically.

Velocity of Sound in Solids

The velocity of sound in slender solids such as pipes and rods depends on the mode of propagation (see also Chapter 2). For compressive waves,

$$c = \sqrt{\frac{B}{\rho}} \tag{12.8}$$

where B is the isothermal bulk modulus of the solid. In steel at room temperature, the velocity of compressive wave is about 198,800 in/s (5,049 m/s). This must not be confused with the velocity of bending waves (see Chapter 2) at which most noises propagate in piping systems.

Apparent Velocity of Sound through a Tube Bundle

When waves are scattered by a lattice grid, their velocities of propagation decrease. When light waves enter a material medium, they are scattered by the molecular lattice of the material, resulting in decreases in the velocities of propagation. The familiar sight of the apparent bending of a stick half-immersed in water is caused by a decrease in the velocity of light in water from that in air. Likewise, when acoustic waves impinge normally on a heat exchanger tube bundle, they are scattered by the tubes, resulting in decreases in the velocity of propagation.

Parker (1979), Blevins and Bressler (1987) found that if σ is the solidity ratio of the heat exchanger internal in the transversal direction, i.e., the ratio of the cross-section occupied by the tubes to the total cross-section, then the apparent velocity of sound in the transversal direction is given by the equation:

$$c = \frac{c_0}{\sqrt{1+\sigma}} \tag{12.9}$$

Example 12.4, given later in this chapter, shows the application of Equation (12.9) to calculate the acoustic resonant frequencies in a heat exchanger.

12.3   Standing Waves in Pipes and Ducts

Disturbances propagate in two opposite directions in the fluid media contained in pipes and ducts. When the wave fronts reach the ends, they are reflected back. The reflected wave then superimposes onto the original wave, forming interference patterns. The result is the formation of standing waves in the pipes or ducts. The name standing wave comes from the apparently non-propagating nature of the resultant wave, as one can readily see from surface waves reflected back from the edge of a pond. These standing waves are similar to the standing waves in a vibrating string. Like the guitar string, standing waves in a pipe possess their characteristic wavelengths and discrete frequencies, which are the principle behind the organ. In a power or process plant piping system, if these natural frequencies coincide with the frequency of an excitation force, a condition of resonant vibration would occur, very often with annoying or even damaging consequences. Thus, the first step in preventing acoustically induced vibration is to predict the natural frequencies of the standing waves in pipes, ducts and cavities. Some of these equations, particularly those for rectangular cross-sections, are derived in elementary books on physics while others may require more advanced techniques. Some of these formulas are given in the following sub-sections. The readers are referred to the literature for their derivations.

Following the convention established in Chapter 4, we shall in this Chapter use the Greek alphabet $\alpha,\ \beta,\ \gamma,\ \ldots$ to denote the acoustic modal indices (versus the English alphabet $m,\ n,\ k\ \ldots$ for the structural modal indices), with α for the axial mode and β for the transversal mode. We shall return to the third index later.

In the special case of one-dimensional systems such as slender pipes and ducts, the natural frequencies are governed by a single axial index. The acoustic pressure distribution over the cross-section of the pipe is assumed to be constant. Figure 12.2 gives the three common acoustic end conditions of a pipe. The acoustical pressure distribution p' and/or the modal frequencies are different in these three cases.

In the remaining part of this chapter, we shall use the term pipe to denote any conduits in which the characteristic length of the cross-section is much smaller than the length. This may include pipes, ducts, tubes and other flow conduits.

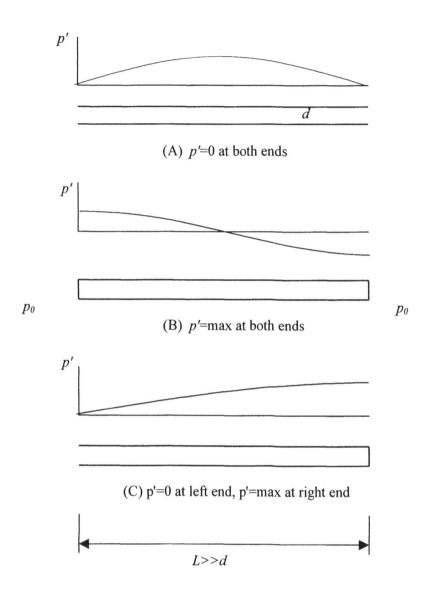

Figure 12.2 Slender pipe with different pressure boundary conditions

Slender Pipes with Both Ends Open (Figure 12.2(A))

This means the ends of the pipe are opened to the atmosphere or are connected to a large chamber with ambient pressure p_0. We are interested only in the perturbation of the pressure from the ambient pressure. If we denote,

$$p' = p - p_0 \qquad\qquad\qquad\qquad (12.10)$$

then the acoustic pressures at the ends of the pipe are the same as the ambient pressures, which are assumed to be the same at both ends. In this case,

$$p'_\alpha = p_{max} \sin \frac{\alpha \pi x}{L} \tag{12.11}$$

$$f_\alpha = \frac{\alpha c}{2L} \quad \text{Hz} \quad \alpha = 1, 2, 3, \ldots \tag{12.12}$$

Slender Pipes with Both Ends Closed (Figure 12.2(B))

In this case the maximum acoustic pressure occurs at the two ends. The pressure distribution and the frequencies are:

$$p'_\alpha = p_{max} \cos \frac{\alpha \pi x}{L} \tag{12.13}$$

$$f_\alpha = \frac{\alpha c}{2L} \quad \text{Hz} \quad \alpha = 1, 2, 3, \ldots \tag{12.14}$$

Slender Pipes with One End Open and the Other Closed (Figure 12.2(C))

In this case the maximum acoustic pressure occurs at the closed end and is zero at the open end. The pressure distribution and the frequencies are:

$$p'_\alpha = p_{max} \sin \frac{\alpha \pi x}{2L} \tag{12.15}$$

$$f_\alpha = \frac{\alpha c}{4L} \quad \text{Hz} \quad \alpha = 1, 2, 3, \ldots \tag{12.16}$$

Slender Annular Pipes and Pipes of Other Cross-Sections

As long as the pipe is slender, that is, as long as the length of the pipe is much larger than its hydraulic diameter, Equations (12.11) to (12.16) hold true. The cross-section of the pipe or duct has no effect on the acoustic mode shape or the frequencies.

Example 12.1

Figure 12.3 shows a square duct. At one end of the duct is a fan driving the air through the duct; at the other end is a right angle T. The dimensions of the duct and the ambient conditions of the air inside the duct are given in Table 12.1. The maximum available

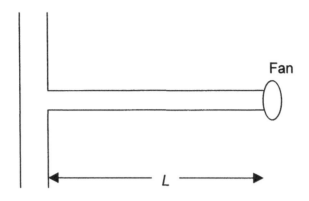

Figure 12.3 Standing waves in a duct

space to install this system is 10 ft (3.0 m). What should be the length of the duct L in order to avoid standing waves resonating with the blade passing frequency and its harmonics?

Table 12.1: Known Parameters for Example 12.1

	US Unit	SI Unit
Cross-section of duct	12 in x 12 in	0.305 m x 0.305 m
Air pressure	Standard atmospheric	Standard atmospheric
Temperature	80 deg. F	26.7 deg. C
Fan rotation speed	1200 rpm	1200 rpm
No. of blades	3	3
Velocity of sound at atmos. pressure	13504 in/s at 68 deg. F	343 m/s at 20 deg. C

Solution

We have to find the velocity of sound in the duct. Assuming the air pressure in the duct remains approximately at atmospheric pressure, the velocity of sound is proportional to the absolute temperatures, which are:

$T = 80 + 459.7 = 539.7$ deg. R or,
$T = 26.7 + 273.2 = 299.9$ deg. K

Knowing that the velocity of sound at 68 deg. F (527.7 deg. R or 593.2 deg. K) and atmospheric pressure is 13,504 in/s (343 m/s), we can readily find the velocity of sound in the duct using Equation (12.4). The results are given in Table 12.2. One end of the duct is attached to the fan, at which the sound pressure is generated by the blades. Therefore, this is a "closed" end. The other end is a T opened in both directions. As an approximation, we can treat this as an open end. The acoustic modal frequencies in the duct are given by Equation (12.16). Solving for L, we get

$$L = \frac{\alpha c}{4 f_n}$$

where f_n=60, 120, 180 . . . . Hz and α=1, 2, 3, ... Table 12.3 gives the duct lengths at which the air inside the duct would resonate with the blade passing frequency and its higher harmonics.

Table 12.2: Computed Parameters for Example 12.1

	US Unit	SI Unit
Fan blade passing f=1200*3/60	60 Hz	60 Hz
Absolute temperature, T	539.7 deg. R	299.9 deg. K
Velocity of sound in duct	13657 in/s	346.9 m/s
Acoustic wavelength at 60 Hz	113.8 in	2.89 m

Table 12.3: Duct Lengths in inch (meter in parentheses) at Which
Acoustic Modal Frequency=Blade Passing Frequency

Blade Passing Harmonics n	Acoustic Modal Number α			
	1	2	3	4
1, 60 Hz	56.9 (1.45)	113.8 (2.89)	171.7 (4.36)	227.6 (5.78)
2, 120 Hz	28.5 (0.72)	56.9 (1.45)	85.3 (2.17)	113.8 (2.89)
3, 180 Hz	19 (0.48)	37.9 (0.96)	56.9 (1.45)	75.8 (1.93)

The worst choice would be L approximately equal to 57 inch (1.5 m) because at this length, every acoustic mode in the duct would resonate with every harmonics of the blade passing frequency. Since our objective is to avoid resonance with the lower harmonics of the blade passing frequency, at which the pressure pulsations are highest, a good and convenient choice for L is around 70 in. (1.8 m).

12.4 Standing Waves in Cavities

In this chapter cavities are defined as enclosed or almost enclosed volumes in which the three dimensions are comparable to one another. The problem is therefore three-dimensional. In rectangular cavities, we shall encounter the familiar sine and cosine trigonometric functions. In cavities of circular cross-sections, we shall encounter the Bessel functions as we do in Chapter 4. This is because Bessel functions are the cylindrical coordinate counterpart of the sine and cosine functions in the Cartesian coordinate system. Table 12.4 gives the eigenfunctions of the wave equation in the Cartesian, cylindrical and spherical coordinates systems.

Table 12.4: Eigenfunctions of the Wave Equation

Geometry	Co-ordinate System	Eigenfunctions	Physical Term
Rectangular	Cartesian	Sine and Cosine	Harmonics
Cylinders	Cylindrical	Bessel functions	Cylindrical harmonics
Spheres	Spherical	Legendre functions	Spherical harmonics

With today's widely available spreadsheet, one can evaluate Bessel functions just as easily as trigonometric functions. Concentric spherical cavities are rare in power and process plants. We shall not discuss the Legendre functions and spherical harmonics in this book.

The following sub-sections give equations for the acoustic pressure distribution p', as well as modal frequencies for a few most commonly encountered geometry. More equations can be found in Blevins (1979).

Enclosed Rectangular Cavities (Figure 12.4)

$$p'_{\alpha\beta\gamma} = p_{\max} \cos\frac{\alpha\pi x}{L_1}\cos\frac{\beta\pi y}{L_2}\cos\frac{\gamma\pi z}{L_3} \tag{12.17}$$

$$f_{\alpha\beta\gamma} = \frac{c}{2}\left(\frac{\alpha^2}{L_1^2}+\frac{\beta^2}{L_2^2}+\frac{\gamma^2}{L_3^2}\right)^{1/2} \quad \text{Hz} \quad \alpha,\ \beta,\ \gamma = 1, 2, 3, \ldots. \tag{12.18}$$

Rectangular Cavity Opened at End $x=0$, Closed at All Others

$$p'_{\alpha\beta\gamma} = p_{\max} \sin\frac{\alpha\pi x}{2L_1}\cos\frac{\beta\pi y}{L_2}\cos\frac{\gamma\pi z}{L_3} \tag{12.19}$$

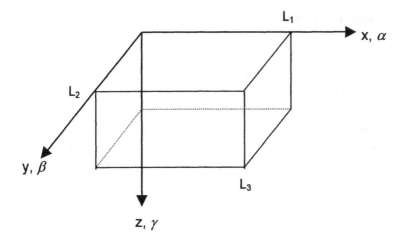

Figure 12.4 A rectangular cavity

$$f_{\alpha\beta\gamma} = \frac{c}{2}\left(\frac{\alpha^2}{4L_1^2} + \frac{\beta^2}{L_2^2} + \frac{\gamma^2}{L_3^2}\right)^{1/2} \quad \text{Hz} \quad \alpha,\ \beta,\ \gamma = 1, 2, 3, \ldots \qquad (12.20)$$

Rectangular Cavity Opened at End $x=0$, L_1, Closed at All Others

$$p'_{\alpha\beta\gamma} = p_{\max} \sin\frac{\alpha\pi x}{L_1}\cos\frac{\beta\pi y}{L_2}\cos\frac{\gamma\pi z}{L_3} \qquad (12.21)$$

$$f_{\alpha\beta\gamma} = \frac{c}{2}\left(\frac{\alpha^2}{L_1^2} + \frac{\beta^2}{L_2^2} + \frac{\gamma^2}{L_3^2}\right)^{1/2} \quad \text{Hz} \quad \alpha,\ \beta,\ \gamma = 1, 2, 3, \ldots \qquad (12.22)$$

The Annular Cavity

In Chapter 4, the equations for the hydrodynamic mass of a system of two finite coaxial cylindrical shells are derived. The readers should refer to Figures 4.1 and 4.2 in Chapter 4, as the geometry is the same in what follows. From Chapter 4, standing waves exist in an annular cavity only if the natural frequency is high enough so that,

$$f > \varepsilon c / 2L \qquad (12.23)$$

where

$\varepsilon = \alpha$ for fluid annulus with either both ends open ($p = 0$) or both ends closed ($p = p_{\max}$),
    $= (2\alpha - 1)/2$ if one end is open and the other end is closed.

When this condition is satisfied, the natural frequencies of the standing waves are given by Equation (4.31), which will be re-numbered as Equation (12.24) in this chapter:

$$J'_\beta (\chi_\alpha a) Y'_\beta (\chi_\beta b) = J'_\beta (\chi_\alpha b) Y'_\beta (\chi_\beta a) \tag{12.24}$$

where

$$\chi_\alpha^2 = (2f/c)^2 - (\varepsilon/L)^2 > 0 \tag{12.25}$$

In spite of its formidable appearance, Equation (12.24) can be easily solved numerically with today's spreadsheets and the following equations for the derivative of the Bessel functions (see Chapter 4, Equation (4.38), but with subscript n replaced by β to be consistent with our convention of using the Greek alphabet for the acoustic modes and the English alphabet for the structural modes):

$$
\begin{aligned}
J'_\beta (x) &= (\beta J_\beta(x) - x J_{\beta+1}(x))/x \\
Y'_\beta (x) &= (\beta Y_\beta(x) - x Y_{\beta+1}(x))/x
\end{aligned}
\tag{12.26}
$$

The pressure distribution in the cavity is given by (from Chapter 4, note α is the axial mode number and β is the circumferential mode number and $a_{\alpha\beta}$ are amplitude functions),

$$p'_{\alpha\beta} (\vec{r}) = a_{\alpha\beta}[J_\beta(\chi_\alpha r) - J'_\beta (\chi_\alpha a) Y_\beta(\chi_\alpha r)/Y'_\beta (\chi_\alpha a)]\cos\beta\theta \sin\frac{\alpha\pi x}{L} \tag{12.27}$$

both ends open

$$p'_{\alpha\beta} (\vec{r}) = a_{\alpha\beta}[J_\beta(\chi_\alpha r) - J'_\beta (\chi_\alpha a) Y_\beta(\chi_\alpha r)/Y'_\beta (\chi_\alpha a)]\cos\beta\theta \cos\frac{\alpha\pi x}{L} \tag{12.28}$$

both ends closed

$$p'_{\alpha\beta} (\vec{r}) = a_{\alpha\beta}[J_\beta(\chi_\alpha r) - J'_\beta (\chi_\alpha a) Y_\beta(\chi_\alpha r)/Y'_\beta (\chi_\alpha a)]\cos\beta\theta \sin\frac{\alpha\pi x}{2L} \tag{12.29}$$

or

$$p'_{\alpha\beta} (\vec{r}) = a_{\alpha\beta}[J_\beta(\chi_\alpha r) - J'_\beta (\chi_\alpha a) Y_\beta(\chi_\alpha r)/Y'_\beta (\chi_\alpha a)]\cos\beta\theta \cos\frac{\alpha\pi x}{2L} \tag{12.30}$$

one end is open and the other end is closed.

-

We shall return to the annular cavity later. Meanwhile, let us study the special case of a finite cylindrical cavity.

Finite Cylindrical Cavity

By letting the radius, a, of the inner cylinder go to zero and using the properties of the Bessel functions for small argument:

$$Y'_\beta(x) \to \infty \quad as \quad x \to 0 \qquad \text{for any integral values of } \beta \tag{12.31}$$

$$J'_\beta(x) \to 0 \quad as \quad x \to 0 \qquad \text{for any integral values of } \beta \text{ except } \beta=1 \tag{12.32}$$

$$J'_1(x) \to 1/2 \quad as \quad x \to 0 \tag{12.33}$$

We get from Equations (12.27) to (12.30), the limiting case of a single cylindrical cavity (see also Chapter 5):

$$p'_{\alpha\beta}(\vec{r}) = a_{\alpha\beta} J_\beta(\chi_\alpha r)\cos\beta\theta\sin\frac{\alpha\pi x}{L} \qquad \text{both ends open} \tag{12.34}$$

$$p'_{\alpha\beta}(\vec{r}) = a_{\alpha\beta} J_\beta(\chi_\alpha r)\cos\beta\theta\cos\frac{\alpha\pi x}{L} \qquad \text{both ends closed} \tag{12.35}$$

$$p'_{\alpha\beta}(\vec{r}) = a_{\alpha\beta} J_\beta(\chi_\alpha r)\cos\beta\theta\sin\frac{\alpha\pi x}{2L} \tag{12.36}$$

or

$$p'_{\alpha\beta}(\vec{r}) = a_{\alpha\beta} J_\beta(\chi_\alpha r)\cos\beta\theta\cos\frac{\alpha\pi x}{2L} \tag{12.37}$$

if one end is open and the other end is closed.

Analogous to Equation (12.24), the acoustic modal frequencies in this case are given by

$$J'_\beta(\chi_\alpha R) = 0 \tag{12.38}$$

where $R=D/2$ is the radius of the cylindrical cavity. Equation (12.38) can be easily solved numerically using a spreadsheet and with the recursive relationship for the derivative of the Bessel functions, Equation (12.26). Table 12.5 gives the roots of Equation (12.38) obtained from a spreadsheet. The numbers agree with those in Table 13.4 of Blevins (1979).

The natural frequencies of the acoustic modes are given by,

$$\chi_{\alpha\beta\gamma} R = \lambda_{\alpha\beta\gamma}$$

Table 12.5: Roots of the equation $J'_\beta(\lambda_{\alpha\beta\gamma}) = 0$

$\lambda_{\alpha\beta\gamma}$	β				
	0	1	2	3	4
$\gamma=0$	0	1.8412	3.0542	4.2012	5.3176
$\gamma=1$	3.8317	5.3314	6.7061	8.0152	9.2824
$\gamma=2$	7.0156	8.5363	9.9695	11.3459	12.6819
$\gamma=3$	10.1735	11.7060	13.1704	14.5859	15.9641

or

$$f_{\alpha\beta\gamma} = \frac{c}{2\pi}\sqrt{\left(\frac{\lambda_{\alpha\beta\gamma}}{R}\right)^2 + \left(\frac{\varepsilon\pi}{L}\right)^2} \tag{12.39}$$

For each pair of modal indices (α, β), there are infinitely many modes with different modal indices γ. These are the radial modes. At the modal frequencies, the pressure distributions are, from Equations (12.34) to (12.37), given by:

$$p'_{\alpha\beta\gamma}(r) = a_{\alpha\beta\gamma}J_\beta(\lambda_{\alpha\beta\gamma}r/R)\cos\beta\theta\sin\frac{\alpha\pi x}{L} \quad \text{if both ends are open} \tag{12.40}$$

$$p'_{\alpha\beta\gamma}(r) = a_{\alpha\beta\gamma}J_\beta(\lambda_{\alpha\beta\gamma}r/R)\cos\beta\theta\cos\frac{\alpha\pi x}{L} \quad \text{if both ends are closed} \tag{12.41}$$

$$p'_{\alpha\beta}(r) = a_{\alpha\beta\gamma}J_\beta(\lambda_{\alpha\beta\gamma}r/R)\cos\beta\theta\sin\frac{\alpha\pi x}{2L} \tag{12.42}$$

or

$$p'_{\alpha\beta}(r) = a_{\alpha\beta\gamma}J_\beta(\lambda_{\alpha\beta\gamma}r/R)\cos\beta\theta\cos\frac{\alpha\pi x}{2L} \tag{12.43}$$

if one end is open and the other is closed

where $\lambda_{\alpha\beta\gamma}$ are given in Table 12.5. The three factors on the right-hand side of the above equation represent the radial, tangential and axial distribution of the acoustic pressures respectively. Figures 12.5 show the first four radial modes for $\beta=0$, 1 and 2.

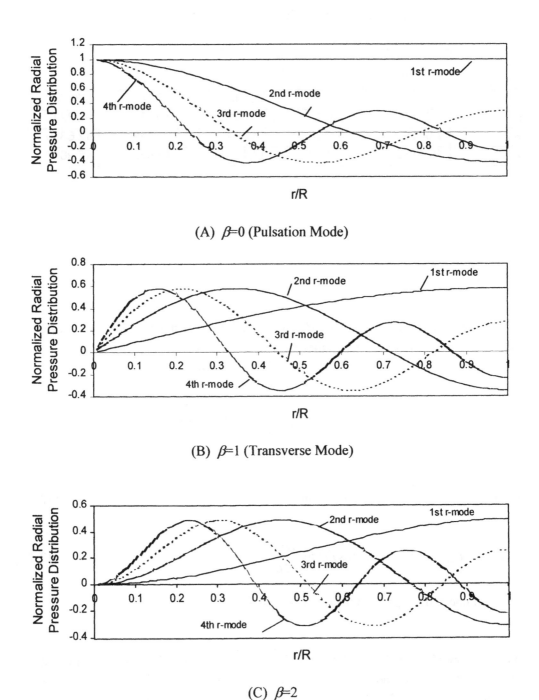

(A) $\beta=0$ (Pulsation Mode)

(B) $\beta=1$ (Transverse Mode)

(C) $\beta=2$

Figure 12.5 The first four radial modes of each of the first three circumferential acoustic modes in a finite cylindrical cavity

The Helmholtz Resonator

A

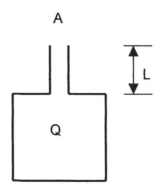

Figure 12.6 The Helmholtz resonator

When an acoustic cavity ends with a long, narrow neck, it is know as a Helmholtz resonator. When the area A of the neck is much smaller than the other dimensions of the cavity, the natural frequency of the Helmholtz resonator is given by the simple equation,

$$f \approx \frac{c}{2\pi}\sqrt{\frac{A}{QL}} \qquad\qquad (12.44)$$

where Q is the internal volume of the resonator.

This equation is only approximate. A more accurate and complicated equation, together with equations for the frequencies of coupled Helmholtz resonators, are given in Blevins (1979).

Because of its narrow neck, considerable energy is dissipated for driving the gas through the Helmholtz resonator. Thus the Helmholtz resonator is often used as a muffler to subdue noises. The Helmholtz resonator principle can also be used to de-tune the natural frequencies of resonant cavities from the frequencies of potential excitation forces, such as blade passing frequencies from pumps or vortex-shedding frequencies.

Example 12.2

Figure 12.7 shows a cylindrical tank containing water at atmospheric pressure and 80 deg. F (26.7 deg. C). A pump rotating at 1,200 rpm and with five blades is used to transfer water to the tank. Assess the concern of acoustic resonance of the water column due to the pump blade passing frequencies and the resultant effects on the structural integrity of the water tank. (This is the way this kind of problems is usually posed to practicing engineers).

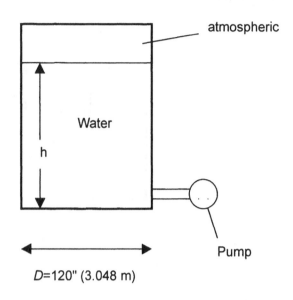

Figure 12.7 Water pumped into a cylindrical tank

Solution

Even though the velocity of sound in water at room temperature and standard pressure is readily available in many handbooks, here we want to illustrate a general method of calculating the velocity of sound in water, steam or water/steam two-phase mixtures using data from the ASME Steam Table (1979). The following data are actually obtained from an electronic version of the Steam Table, which gives the entropies and densities for arbitrary input values of pressure and temperature. Hard copies of the Steam Table will require inputting the pressure and temperatures at the discrete values listed. This may give results that differ slightly but not significantly. First, we find the density ρ and entropy S of water at $p-\Delta p/2$ (Δp is an arbitrary small but convenient increment of p) and temperature T. Next, keeping the entropy constant at S, we find, from the ASME Steam Table again, the density of water at $p+\Delta p/2$. The results are given in Table 12.6. From this table, at the constant entropy line of $S=9.33149$ Btu/lbm-R and around the thermodynamic state of $p=14.7$ psi and $T=80$ deg. F, a change in pressure $\Delta p=6$ psi would cause a change of density of water $\Delta\rho=62.2216-62.2204=0.0012$ lbm/ ft^3. Before we carry out the remainder of the calculation, we must change the units into a consistent unit set in the US unit system. That is, we must change the 0.0012 lbm/ ft^3 into 1.797E-9 (lb-s^2/in)/in^3 (see Chapter 1). We find the velocity of sound at 80 deg. F and 14.7 psi to be approximately $c=57,778$ in/s. If we carry out the entire calculation in SI units, we would find $c=1,476$ m/s. Due to round-off errors in the Steam Table, the velocities computed in US and SI units do not agree with each other exactly. A more exact method of computing the velocity of sound in water or steam is to program the equations of state directly.

Table 12.6: Water Properties in the Neighborhood of the Given Ambient Conditions

	US Unit	SI Unit
Pressure = $p-\Delta p/2$	11.7 psi	80668.7 Pa
Temperature T=	80 deg. F	26.7 deg. C
From Steam Table:		
Density, $\rho=$	62.2204 lb/ft^3	996.675 Kg/m^3
Entropy, $S=$	9.33149 Btu/lbm-R	390.691 J/Kg-K
Pressure = $p+\Delta p/2$	17.7 psi	122037 Pa
Entropy, $S=$	9.33149 Btu/lbm-R	390.691 J/Kg-K
Density, $\rho=$	62.2212 lb/ft^3	996.694 Kg/m^3
Therefore,		
$\Delta p=$	17.7-11.7=6 psi	41368 Pa
$\Delta\rho=$	0.0012 lbm/ ft^3	0.019 Kg/m^3
	(1.797E-9 (lb-s^2/in)/in^3)	
$c = \sqrt{\dfrac{\Delta p}{\Delta\rho}}$	57778 in/s	1476 m/s

The first blade passing frequency of the pump is 100 Hz, with its higher harmonics equal to 200, 300, 400 . . . Hz. The acoustic modal frequencies are given by Equation (12.39) with the eigenvalues given in Table 12.5. From Equation (12.39), we can express the water level h in terms of the blade passing frequencies f_n, the axial modal index α and the eigenvalues $\lambda_{\alpha\beta\gamma}$:

$$h = \frac{(2\alpha-1)\pi/2}{\sqrt{\left(\dfrac{2\pi f_n}{c}\right)^2 - \left(\dfrac{2\lambda_{\alpha\beta\gamma}}{D}\right)^2}}, \quad f_n = 100, 200, 300, \ldots \tag{12.45}$$

For a given set of modal numbers, α, β, γ, its associated eigenvalues $\lambda_{\alpha\beta\gamma}$ and the blade passing frequency f_n, Equation (12.45) gives real values for the water level only if the quantity inside the square root is positive. This means that not all of the acoustic modes can exist at all blade-passing frequencies. Table 12.7 gives a summary of all the permissible acoustic modes at the lowest 10 harmonics of the pump blade passing frequencies. Obviously, the tank should be designed for the modes with the highest water levels and the lowest frequencies, such as the (1, 0, 0) mode at 100 Hz and a water level of 144 inches (3.67 m); the (1, 0, 1) mode at 600 Hz and a water level of 117 inches (2.97 m); and the (1, 1, 0) mode at 300 Hz and a water level of 142 inches (3.61 m). The first

Table 12.7: Permissible Acoustic Modes for
Different Blade Passing Frequencies (*h* in inches)

α=1 β=0					
Frequency (Hz)	γ=	0	1	2	3
	λ=	0	3.8317	7.0156	10.1735
100	h=	144.4	--	--	--
200	h=	72.2	--	--	--
300	h=	48.1	--	--	--
400	h=	36.1	--	--	--
500	h=	28.9	--	--	--
600	h=	24.1	117.4	--	--
700	h=	20.6	37.9	--	--
800	h=	18.1	26.6	--	--
1000	h=	14.4	17.8	--	--

α=1 β=1					
Frequency (Hz)	γ=	0	1	2	3
	λ=	1.8412	5.3314	8.5363	11.706
100	h=	--	--	--	--
200	h=	--	--	--	--
300	h=	141.8	--	--	--
400	h=	51.0	--	--	--
500	h=	35.0	--	--	--
600	h=	27.3	--	--	--
700	h=	22.5	--	--	--
800	h=	19.3	--	--	--
1000	h=	15.1	25.1	--	--

α=1 β=2					
Frequency (Hz)	γ=	0	1	2	3
	λ=	3.0542	6.7061	9.9695	13.1704
100	h=	--	--	--	--
200	h=	--	--	--	--
300	h=	--	--	--	--
400	h=	--	--	--	--
500	h=	82.2	--	--	--
600	h=	38.5	--	--	--
700	h=	27.8	--	--	--
800	h=	22.3	--	--	--
1000	h=	16.3	--	--	--

mode is a plane wave mode with uniform pressure distribution across the cross-section of the tank. This is the same mode as that in a long pipe. It will cause the tank to vibrate in the symmetric, or breathing mode. The (1, 0, 1) mode is also a symmetric mode but with a nodal circle. The pressure distribution along a radius is not constant (see Figure 12.5). The (1, 1, 0) mode again has constant pressure distribution across the cross-section of the tank. However, it is a transverse mode and will cause the tank to vibrate laterally. In this particular example, the first 10 harmonics of the pump blade passing frequency will not resonate with any shell modes ($\beta \geq 2$). Since in a short cylindrical shell, the lowest modes are usually shell modes with $n > 2$, this tank can be easily designed against any pump-induced vibration.

Example 12.3

Return to Example 4.2 in Chapter 4, where we calculate the *structural* frequencies of the coupled fluid-shell system. Use the data given in Chapter 4, calculate the *acoustic* modal frequencies in the water annulus. Refer to Figure 4.5 in Chapter 4 for the schematics of the system. The relevant input data from Table 4.6 are reproduced in Table 12.8 for easy reference in this example.

Table 12.8: Data for Finite Water Annular Cavity

	US Unit	SI Unit
Length L	300 in	7.620 m
Inner radius of water annulus, a	78.5 in	1.994 m
Outer radius of water annulus, b	88.5 in	2.248 m
Boundary condition of water annular gap:	Closed at the top and pressure-released at the bottom	
Water temperature	575 deg. F	301.7 deg. C
Water pressure	2200 psi	1.517E+7 Pa

The velocity of sound was calculated with data from the ASME Steam Table (1979) following the method outlined in Example 12.2. Table 12.9 shows the pressures at which we find the densities.

Knowing the length L, inner and outer radii, a, b, of the annular gap and the velocity of sound, Equation (12.24) can be readily solved numerically for the modal frequencies using a spreadsheet. For any assigned frequency f, we calculate the parameter

Table 12.9: Water Properties in the Neighborhood of the Given Ambient Conditions

	US Unit	SI Unit
Pressure = $p-\Delta p/2$	2100 psi	1.4479E+7 Pa
Temperature, T=	575 deg. F	301.7 deg. C
From Steam Table:		
Density ρ=	45.0506 lb/ft^3	721.64 Kg/m^3
Entropy S=	0.774821 Btu/lbm-R	3244.02 J/Kg-K
At this S and		
Pressure = $p+\Delta p/2$	2300 psi	1.58579E+7 Pa
Density ρ=	45.1474 lb/ft^3	723.19 Kg/m^3
Therefore,		
Δp=	2300-2100=200 psi	1.3789E+6 Pa
$\Delta \rho$=	9.68E-2 lbm/ ft^3	1.55 Kg/m^3
	(1.44975E-7 (lb-s^2/in)/in^3)	
$c = \sqrt{\dfrac{\Delta p}{\Delta \rho}}$	37142 in/s (3095 ft/s)	943 m/s

$$\chi_\alpha = \sqrt{(2f/c)^2 - (2\alpha - 1)/2L}$$

Using Equations (12.26) for the derivatives of the Bessel functions and the BESSELJ and BESSELY mathematical functions in a spreadsheet, we can calculate and plot the functions;

$$\Phi(f) = J'_\beta (\chi_\alpha a)Y'_\beta (\chi_\beta b) - J'_\beta (\chi_\alpha b)Y'_\beta (\chi_\beta a)$$

Table 12.10 reproduces a few rows from the spreadsheet, as well as the plots $\Phi(f)=0$ for the first few modes. From Figure 12.8, we get the acoustic modal frequencies. These are summarized in Table 12.11. Of these modes, the plane wave modes ($\gamma=0$) for the first transversal mode ($\beta=1$) and that for the first shell mode ($\beta=2$) are the ones we have to pay attention to because of their low modal frequencies.

It took this author, who is computer-illiterate by today's standard, less than one hour to program this entire problem.

Table 12.10: Roots (in Hz) of Equation (12.24)

β	γ		
	0	1	2
0	1860	3720	5380
1	77	1860	3720
2	145	1860	3726

Table 12.11: Partial Listing of the Spreadsheet Used to Solve Example 12.3

This   sheet   for beta =2

	1 m=	39.37 in
	US Unit	SI Unit
Cylinder a	78.5 in	1.994 m
Cylinder b	88.5 in	2.248 m
L=	300 in	7.620 m
p=	2200 psi	1.52E+07 PA
T=	575 F	301.7 C
C=	37142 in/s	944.5 m/s

alpha =   1   axial   mode
epsilon=   0.5
cut on f=   30.95 Hz   30.99 Hz

					---------- Beta =		2	------------	
delta f=	10 Hz								Function (f)
Freq f	Chi	Chi*a	Chi*b	J'(chi*a)	Y'(chi*b)	J'(chi*b)	Y'(chi*a)		J'aY'b - J'bY'a
31	2.93E-04	2.30E-02	2.59E-02	5.74E-03	1.46E+05	6.48E-03	2.10E+05		
41	4.55E-03	3.57E-01	4.03E-01	8.74E-02	3.90E+01	9.79E-02	5.59E+01		
51	6.86E-03	5.38E-01	6.07E-01	1.28E-01	1.13E+01	1.43E-01	1.63E+01		
61	8.89E-03	6.98E-01	7.87E-01	1.61E-01	5.17E+00	1.77E-01	7.42E+00		
71	1.08E-02	8.49E-01	9.57E-01	1.88E-01	2.88E+00	2.04E-01	4.12E+00		-3.02E-01
81	1.27E-02	9.94E-01	1.12E+00	2.09E-01	1.80E+00	2.25E-01	2.57E+00		-1.99E-01
91	1.45E-02	1.14E+00	1.28E+00	2.27E-01	1.24E+00	2.39E-01	1.73E+00		-1.34E-01
101	1.63E-02	1.28E+00	1.44E+00	2.39E-01	9.16E-01	2.47E-01	1.25E+00		-9.01E-02
111	1.80E-02	1.42E+00	1.60E+00	2.46E-01	7.25E-01	2.49E-01	9.55E-01		-5.89E-02
121	1.98E-02	1.55E+00	1.75E+00	2.49E-01	6.09E-01	2.44E-01	7.68E-01		-3.58E-02
131	2.15E-02	1.69E+00	1.91E+00	2.47E-01	5.39E-01	2.33E-01	6.48E-01		-1.83E-02
141	2.33E-02	1.83E+00	2.06E+00	2.40E-01	4.97E-01	2.17E-01	5.70E-01		-4.64E-03
145	2.40E-02	1.88E+00	2.12E+00	2.35E-01	4.85E-01	2.09E-01	5.48E-01		-1.48E-05

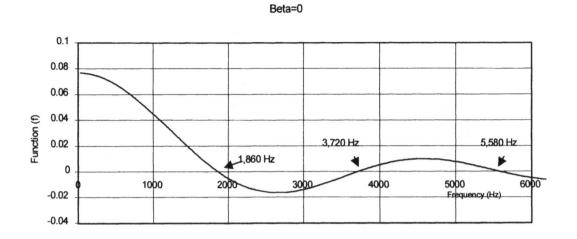

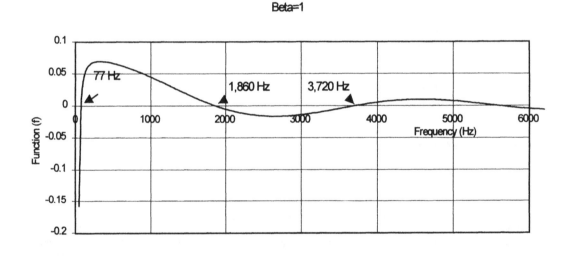

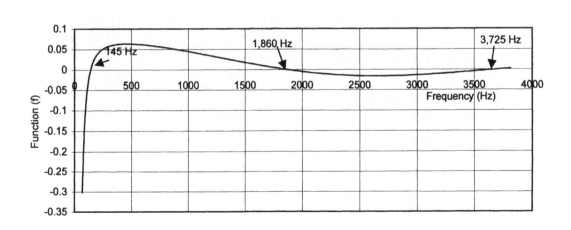

Figure 12.8 Plots of the function $\Phi(f)=0$

Example 12.4

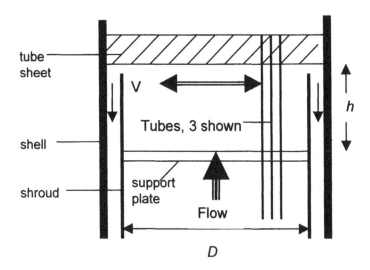

Figure 12.9 Top span of a heat exchanger

Figure 12.9 shows the top span of a large heat exchanger. Superheated steam flows up through the tube bundle to the top span, where it turns 90 deg. and flows across the tube bundle and down the annular gap between the shroud and the shell. The dimensions of the components as well as the ambient conditions of the steam are given in Table 12.12. Assuming the gap between the tubesheet and the top of the shroud is negligibly small compared with the other dimensions, find the lowest few acoustic modal frequencies in the top span of this heat exchanger, and a safe cross-flow velocity range for the steam.

Table 12.12: Dimensions of Heat Exchanger and Condition of Shell-Side Fluid

	US Unit	SI Unit
Diameter of shroud, D	120 in	3.048 m
Height of top span, h	48 in	1.219 m
Tube diameter, d	0.625 in	0.01588 m
Tube pitch to diameter ratio, P/d	1.4	1.4
Tube pattern	Equilateral triangle	Equilateral triangle
Shell-side pressure, p	1000 psi	6.895E+6 Pa
Shell-side temperature, T	575 deg. F	301.7 deg. C

Solution

Table 12.13: Computed Results for Examples 12.4 and 12.5

	US Unit	SI Unit
From ASME Steam Table (1979), density, ρ	2.06 lb/ft^3 (6.4E-2 slug/ft^3) (3.085E-6 (lb-s^2/in))	33 Kg/m^3
Following method of Ex.12.2, 12.3 $c_0=$	2.0256E+4 in/s	514.6 m/s
Solidity ratio, σ	0.463	0.463
Effective velocity of sound, c	1.7035E+4 in/s	432.8 m/s
Dynamic viscosity, μ	4.144 lbf-s/ft^2	1.984E-5 Pa-s
Kinematic viscosity, $\nu=\mu/\rho$	6.475E-6 ft^2/s (9.32E-4 in^2/s)	6.012E-7 m^2/s
$R_a=$	1.15E+7	1.15E+7
At $V_p=$	480 in/s	12.2 m/s
$R_c=$	3.2E+5	3.2E+5

Given the pressure and temperature of steam, one can find its density from the ASME Steam Table (1979). It is left as an exercise for the readers to calculate the velocity of sound in the steam at the given ambient conditions, following the method outlined in Examples 12.2 and 12.3. The results are given in the second row of Table 12.13. Since the steel tubesheet and support plate bound the top span of the heat exchanger, the end condition of the cylindrical cavity is "closed-closed." Using Equation (12.39) with the given $R=D/2$, h, $\varepsilon=\alpha=1, 2, 3 \ldots$ and the eigenvalues λ given in Table 12.5, we can calculate the acoustic modal frequencies provided we know the velocity of sound in the steam. For that we must calculate the solidity ratio and then the apparent velocity of sound through the tube bundle. For an equilateral triangular array, we get from Figure 12.10:

Total volume for each of the rectangular pitch $= hPP\sin 60 = 0.866hP^2$

Volume occupied by the tubes$= h\pi d^2 / 4$

Solidity ratio,

$$\sigma = \frac{h\pi d^2 / 4}{0.866hP^2} = 0.9069(d/P)^2 = 0.9069/1.4^2 = 0.463$$

From Equation (12.9), the velocity of sound through the tube bundle is:

$$c = \frac{c_0}{\sqrt{1+0.643^2}} = 0.841c_0 = 17035 \text{ in/s} \ (432.8 \text{ m/s})$$

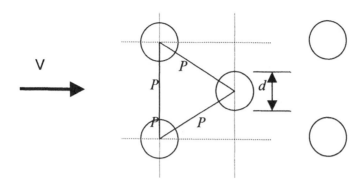

Figure 12.10 Solidity ratio in an equilateral triangular pitch

Table 12.14 Natural Frequencies in the Top Span (α=1)

β	γ			
	0	1	2	3
0	177	248	363	493
1	196	299	425	558
2	225	351	484	621
3	260	403	543	683
4	299	455	600	743

With this, we can proceed to calculate the acoustic modal frequencies using Equation (12.30) and Table 12.5. The results are tabulated in Table 12.14.

The most detrimental acoustic problem in a heat exchanger occurs when one or more of the lower acoustic modes is excited by vortex-shedding in the tube bundle. The best way to prevent this is to operate the heat exchanger at a cross-flow velocity so that the vortex-shedding frequency is below the lowest acoustic modal frequency. In this example, this is 177 Hz. At a cross-flow velocity V, the vortex-shedding frequency is, from Equation (6.1), given by,

$$f_s = SV/d$$

with the Strouhal number given by Equation (6.16) in Chapter 6 for a normal triangular array,

$$S = \frac{1}{1.73(P/d - 1)} = \frac{1}{1.73(1.4 - 1)} = 1.45$$

With this, we get,

$V = 76.3$ in/s or 1.9 m/s

We would avoid vortex-induced acoustic resonance in this heat exchanger if we keep the cross-flow velocity 30% below the above values (see Chapter 6), or 53 in/s (1.35 m/s). Note that in common with vortex-shedding analysis, the velocity is the free stream velocity, not the pitch velocity V_p commonly used in tube bundle fluid-elastic stability analysis. In terms of the pitch velocity,

$$V_p = \frac{P}{P-d} = 184 \ \text{in/s} \quad (15.3 \ \text{ft/s or 4.7 m/s})$$

Considerable research work has been published in the literature on heat exchanger acoustics. The following section gives a short review of this special topic.

12.5   Heat Exchanger Acoustics

The above examples show that once the acoustic modes are "cut on," that is, once the minimum frequency for the formation of standing waves is exceeded, acoustic modes are "dense" in the frequency axis. This, together with the uncertainties in predicting the vortex-shedding frequency, especially in tube bundles, and the ability of vortices to shift their shedding frequencies to "lock-into" the structural modes (see Chapter 6), makes it very difficult to avoid vortex-induced acoustic resonance in heat exchangers once the shedding frequency exceeds the acoustic "cut-on" frequency. However, experience with operating heat exchangers shows that even when the theoretical vortex-shedding frequency coincides with one of the acoustic modal frequencies, very often there is no noise problem. There are two reasons for this: First, probably due to its extremely chaotic nature, no organized vortex-shedding has ever been observed in two-phase flows. As a result, there probably will be no vortex-induced acoustic vibration in heat exchangers with two-phase shell-side fluids. Second, in order to maintain acoustic resonance, there must be a steady supply of energy to drive the acoustic standing waves. Inside a heat exchanger, the flow is highly inhomogeneous. It is not likely that all, or even most, of the tubes would shed vortices at the same frequency. Most likely only a small percentage of the tubes may shed vortices at a certain acoustic modal frequency. If structural, squeeze film and viscous damping of the entire tube bundle dissipate more energy than what a few vortices can generate, steady acoustic modal resonance cannot be established.

At least some of the tubes in some high-performance nuclear steam generators often have to operate above the acoustic cut-on frequency. Thus, more realistic guidelines to avoid heat exchanger acoustic resonance are needed. Before 1970, heat exchanger tube vibration research was much influenced by Chen (1968), who believed heat exchanger tube bundle dynamics was governed by vortex-shedding. After Connors published his

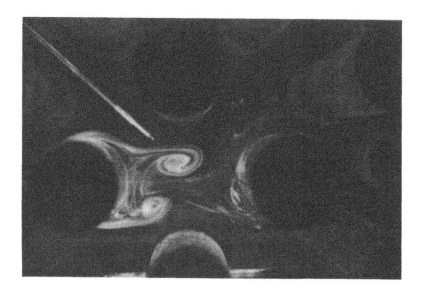

Figure 12.11 Vortex-shedding in a tube bundle (Weaver, 1993)

landmark paper on tube bundle fluid-elastic instability (see Chapter 7), the question of vortex-induced vibration and acoustic resonance was largely ignored. From 1970-1990, most of the research activities on tube bundle dynamics focused on fluid-elastic instability.   Indeed, the very existence of vortex-shedding in a tube bundle was questioned.

However, noise problems do occur in heat exchangers, though only in those with shell-side gases. Very often the resulting low-frequency noises, which were undoubtedly the result of acoustic resonance in the heat exchanger internal compartments, could be heard even a mile away and were expensive to correct. Some forces are exciting these acoustic modes.

In the late 1980s, there was renewed interest in heat exchanger acoustics by researchers such as Blevins and Bressler (1987), Weaver (1993) and Ziada et al (1989), to mention a few. The readers are referred to an excellent review article by Weaver (1993) for a summary of this effort. This paper also contains an extensive bibliography on related publications.   Figure 12.11, reproduced from the above-mentioned paper, conclusively shows the existence of vortex-shedding in a tube bundle. Guidelines for designing heat exchangers to avoid acoustic resonance with vortex-shedding were published.

Resonance Maps (Ziada et al, 1989)

Ziada et al (1989) proposed to classify the susceptibility of a tube bundle to acoustic resonance based on a "resonance parameter" G, defined as (refer to Figure 12.12 and Chapter 7 for array geometry definitions),

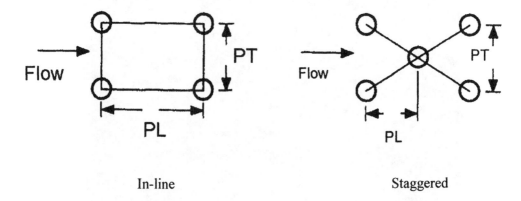

In-line   Staggered

Figure 12.12 Definition of P_T and P_L (Note $P_L=L/2$ in Ziada's definition)

$$G_i = \frac{\sqrt{R_c}}{R_a} X_T \qquad \text{for in-line arrays} \qquad (12.46)$$

$$G_s = \frac{\sqrt{R_c}}{R_a} \frac{[2X_L(X_T-1)]^{1/2}}{(2X_L-1)} \qquad \text{for staggered arrays} \qquad (12.47)$$

where

$$R_c = \frac{V_p d}{v} \qquad (12.48)$$

is the Reynolds number (see Chapter 6) at the onset of acoustic resonance and V_p is the pitch velocity (see Chapter 7),

$$R_a = \frac{cd}{v} \qquad (12.49)$$

with c = velocity of sound in the tube bundle (Equation (12.9)), is the "acoustic" Reynolds number,

$$X_L = P_L/d \qquad (12.50)$$
$$X_T = P_T/d \qquad (12.51)$$

Based on experimental data, the authors gave acoustic resonance maps as functions of tube bundle geometry and the resonance parameter G. These are shown in Figures 12.13 and 12.14.

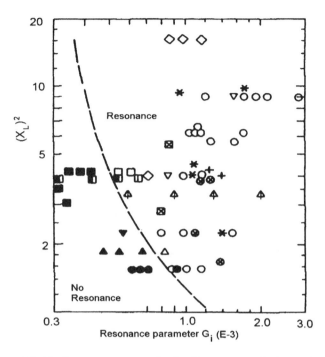

Figure 12.13 Resonance map for in-line array (Ziada et al, 1989)

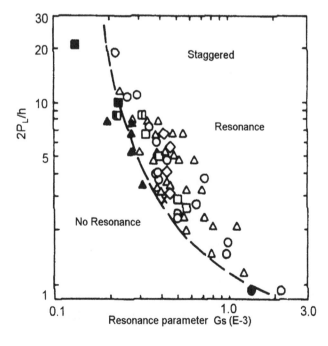

Figure 13.14 Resonance map for staggered array (Ziada et al, 1989)
(*h=(PT-d)/2* or *g* (minimum gap between tubes), whichever is smaller)

Sound Pressure Level

If resonance cannot be avoided, Blevins (1990) suggested the following equation to estimate the resulting acoustic pressure:

$$P_{rms} = \frac{12V_p}{c}\Delta p \qquad (12.52)$$

Example 12.5

In the heat exchanger in Example 12.4, the maximum pitch velocity is $V_p=40$ ft/s (12.2 m/s). Would there be any acoustic resonance according to Ziada's resonance maps? If the pressure drop across the tube bundle is 3 psi (20,700 Pa), what would be the sound pressure level if acoustic resonance does occur?

Solution

From Figures 12.10 and 12.12, we get,

$P_T=P=1.4d$
$P_L=P\sin60°=1.212d$
$X_T=P_T/d=1.4$
$X_L=P_L/d=1.212$

The gap between adjacent tubes is $P-d=0.4d$, while $(P_T-d)/2=0.2d$. Thus,

$h=0.2d$ and

$2P_l/h=12.1$

To find the resonance parameter G_s at the given pitch velocity, we must calculate the acoustic Reynolds number from Equation (12.49) and the Reynolds number R_c from Equation (12.48). As seen from Example 1.1 in Chapter 1, this can be tricky in the US unit system. At the ambient conditions given in Table 12.12, one can find the dynamic viscosity From the ASME Steam Table (1979). This is given in both the US and SI units in Table 12.13, together with the densities. However, the density given in the ASME Steam Table is in lb/ft³. To form a consistent unit set with the other units used in the US unit system, we must convert the mass density into slug/ft³, then use the equation,

$v = \mu/\rho$ =4.144E-7/6.4E-2=6.475E-6 ft²/s=9.32E-4 in²/s

This value is given in Table 12.13. The corresponding calculation in the SI unit system is much more straightforward. Knowing the kinematic viscosity and the velocity of sound in the tube bundle from Example 12.4, we now proceed to calculate the Reynolds number from Equations (12.48) and (12.49),

$$R_a = cd/v = 1.15\text{E+7}$$
$$R_c = 3.2\text{E+5}$$

As expected, the Reynolds numbers are the same in both unit systems. We can now substitute R_a, R_c, X_L=1.212, X_T=1.4 into Equation (12.47) to compute,

$$G_s = 3.4\text{E-5}$$

From Figure 13.14, we see that the point $2P_l/h$=12.1, G_s=3.4E-5 is well in the "no resonance" zone. Thus, there should be no acoustic resonance problem at this pitch velocity.

If acoustic resonance cannot be avoided, from Equation (12.52),

$$p_{rms} = \frac{12V_g}{c}\Delta p = \frac{12*480*3}{17035} = 1.01 \text{ psi} \quad (6900 \text{ Pa})$$

The sound intensity generated is given by Equation (2.10),

$$I = c\, p_{rms} = 1.01*17035 = 17280 \text{ (lb-in/s)/in}^2 = 302 \text{ w/cm}^2$$

Using the reference sound intensity, I_0=1E-16 w/cm^2 at 0 dB (see Chapter 2), we come to the conclusion that this heat exchanger will generate acoustic noise in excess of,

$$SPL = 10\log_{10}(301/1.\text{E-16}) = 185 \text{ dB}$$

Nuclear steam generators with comparable dimensions have been operating under similar conditions for years without emitting any noticeable acoustic noises. This example clearly shows that the mechanism of acoustic resonance in heat exchangers is much more complicated than just the coincidence of vortex-shedding and acoustic modal frequency, and that the conventional "separation" rule of preventing vortex-induced vibration will lead to unreasonably overconservative results when applied to a heat exchanger. However, this example also shows that acoustic resonance in heat exchangers is to be avoided at all costs. An *SPL* of 185 dB is not acceptable by any standard. Furthermore, once the resonance condition is established, any means of reducing the excitation energy or attenuating the acoustic energy will not by itself significantly change the sound intensity much. In the above example, even if we reduce the pressure differential by a factor of 2.0, the reduction in *SPL* will only be 3 dB—a barely noticeable result.

12.8   Thermoacoustics

The addition to or removal of heat from a column of gas can cause pressure oscillations in the gas. This leads to the very important subject of thermoacoustics. This section briefly describes the physics of the phenomenon and simple ways to avoid thermoacoustic instability, which may lead to detrimental results. The readers are referred to a review article by Eisinger (1999) and references cited there for further studies.

Figure 12.16 (A) shows a column of gas with one end opened (pressure-released) and the other end closed (pressure reaches maximum). If heat is added either internally or externally to the closed end, or if heat is removed either internally or externally from the open end, spontaneous acoustic pressure oscillation in the column of gas may be initiated. The condition for spontaneous pressure oscillation depends on the temperature gradient between the hot end and the cold end. If there is insufficient temperature gradient between these two ends, spontaneous pressure oscillation cannot be established. This thermoacoustic system is called a Sondhauss tube after the person who first studied it.

Figure 12.16 (B) shows a column of gas with both ends open (pressure-released). If the gas column is positioned vertically and heat is added either internally or externally to the lower part of the column, again spontaneous acoustic pressure oscillation may be initiated. Again, in order to sustain this spontaneous pressure oscillation, there must be sufficient temperature gradient between the cold end and the hot end of the gas column. This thermoacoustic system is called a Rijke tube after the person who first investigated it. Note that in either system, fluid flow is not a requirement for thermoacoustic oscillation, although it may contribute to its initiation or its severity.

Many components in the power and process industries have the Sondhauss or Rijke tube arrangement. Gas furnaces, for example, are Rijke tubes while some nuclear steam generators are Sondhauss tubes. Sometimes thermoacoustic pressure oscillations is designed into the system to help the combustion process, very often unwanted thermoacoustic oscillation may be harmful to the environment or the equipment.

The temperature ratio $\alpha_c = T_h / T_c$, with T_h and T_c in degrees K, between the hot and the cold ends at which spontaneous pressure oscillation is initiated is governed by the ratio of the lengths of the "hot" column to the "cold" column (see Figure 12.16),

$$\xi = (L - \ell) / \ell$$

where ℓ is the length of the "cold" column, usually separated from the "hot" column by a furnace grid, a swirler or some other physical object (see Figures 12.16 and 12.17). Or ℓ can be the length of the burner while $L - \ell$ is the length of the furnace. Thus, ℓ is usually well defined. The relationship between the critical temperature ratio α and the geometry parameter ξ was obtained based on experimental data,

$$(\log_{10} \xi)^2 = 1.52(\log_{10} \alpha - \log_{10} \alpha_{min}) \tag{12.53}$$

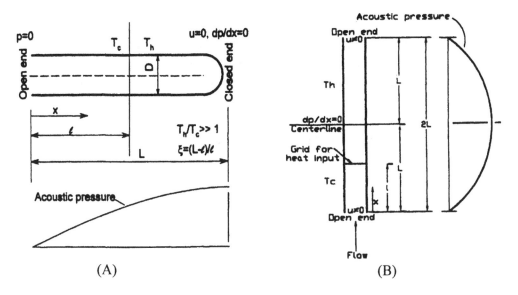

Figure 12.16 Thermoacoustic systems: (A) Sondhauss tube; (B) Rijke tube
(Eisinger, 1999)

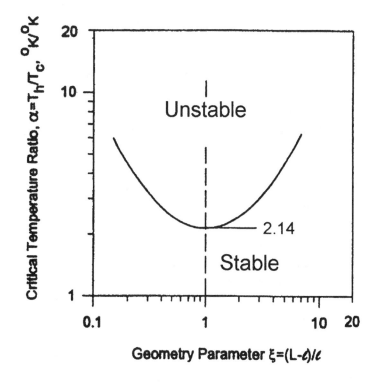

Figure 12.17 Stability diagram for thermoacoustic systems.
(Eisinger, 1999; note temperatures must be in deg. Kelvin)

Figure 12.17 is a plot of Equation (12.53). From the diagram, it can be seen that:

- Decreasing the temperature ratio suppresses thermoacoustic vibration.
- Decreasing the length of the "hot" column (by, e.g., moving the furnace grid or swirler) or increasing the length of the cold column suppresses thermoacoustic vibration.

The stability diagram also shows that the worst situation occurs when the swirler or grid is placed at the mid-point of a Sondhauss-type (closed-open) system, or at the quarter point of a Rijke-type (open-open) system. When the condition of instability is established, the gas column will vibrate spontaneously with acoustic frequencies given by Equation (12.12) for the Rijke tube, and Equation (12.16) for the Sondhauss tube.

12.9   Suppression of Acoustic Noise

Noise and acoustic forces can excite structural components, causing sonic fatigue failure. Very often, the noise level itself requires corrective action, even if it does not compromise the structural integrity of any components. According to the US Occupational Safety and Health Adminstration (OSHA) Guide, if the ambient noise level exceeds 105 dB (see Chapter 2 for definition of dB), ear protection is required for long-term (more than one hour) exposure to the noise. Even if the noise is below this level, complains from the neighbors or just the feeling that something is not right often requires corrective action. As discussed in Chapter 2, due to the logarithmic nature of the human ear, a tenfold decrease in acoustic energy is often perceived as "half the loudness." Thus, once the condition of excessive noise is established, often it is very difficult and expensive to correct. Post-construction correction to noise problems centers around the following methods:

- Suppression by filtering. This may involve something as simple as enclosing the noise source with sound-absorbing material or protecting the ears with ear plugs or ear cups. This method has limited success in suppressing high-frequency noise such as that generated by cavitation or turbulent shear flow. Low-frequency noises from acoustic resonance are much more difficult to attenuate. Enclosing a heat exchanger suffering from acoustic resonance usually cannot solve the problem. Because human beings "hear" low-frequency noise through their body (face, chest, abdomen and back) rather than through their ears, even ear protection may not help.
- Suppression by energy dissipation. A prime example of this is the automobile muffler. The Helmholtz resonator, because of its energy dissipation ability, is often employed to muffle acoustic noise. This has limited success if the noise level is not excessive and if the component that generates the noise is relatively small. Due to the logarithmic nature of the human ears, at least 90% of the acoustic energy has to

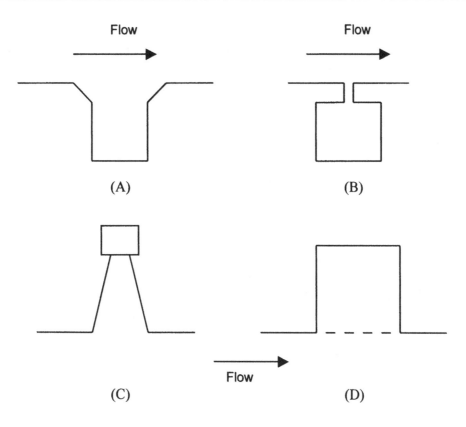

Figure 12.18 Four ways to reduce acoustic resonance due to flow over a cavity: (A) Double ramp to smooth the flow; (B) Narrow neck to change the resonant frequency and increase damping; (C) Conical nozzle; (D) Grid covering cavity opening.

be dissipated before there is any perceivable effect. This means the size of the Helmholtz resonator has to be comparable to the volume of the component that generates the noise. For a large heat exchanger, as an example, this is obviously not practical.

- Removing or reducing the source of excitation. If there is no external force to excite the acoustic mode, no noise would be generated. This can be as simple as not using a certain device. Lowering the operational fluid flow velocity so that cavitation does not occur, or reducing the turbulence level, is another way of eliminating the excitation source. The noise level of a resonant cavity excited by flow over it (Figure 12.18(A)) can be significantly reduced by rounding off the corners or using a ramp to reduce the flow turbulence and eliminate the formation of vortices. Two common ways to reduce noise generation by a valve is to divide a single-stage pressure drop into several smaller stages, and to replace a large flow window with many smaller orifices.

- Detuning at the source. The most effective method of post-construction correction of acoustic problems is to separate the excitation frequency from the acoustic modal frequency, so that resonance does not occur. There are two alternate approaches to de-tuning. One is to alter the frequency of the source; the other is to alter the natural frequency of the acoustic mode. The latter will be addressed in the following paragraph. By reducing the operation flow velocity, for example, one can decrease the vortex-shedding frequency below the acoustic cut-on frequency so that no acoustic modes can resonate with the vortex-shedding frequency. Installing a pump or fan with a different blade passing frequency is another example, though not a very practical one.

- Detuning the acoustic modal frequencies. This is probably the most common way of correcting acoustic resonance problems. Installing baffles in a heat exchanger, as an example, effectively alter the dimensions of the resonant cavity and, as a result, their natural frequencies. Likewise, removing tubes in a heat exchanger reduces the "solidity ratio" and increases the effectively velocity of sound in the heat exchanger. As a result, the acoustic modal frequencies increase (Equation (12.9)). Installing a "neck" between a cavity and the flow surface changes the cavity into a Helmholtz resonator with usually higher resonant frequencies (Equation (12.44)). For years, designers of high-fidelity loudspeaker systems know that non-parallel walls tend to suppress acoustic resonance in loudspeaker cabinets. This principle has been used in at least one instance to correct the acoustic problems in the cavity formed by a nozzle and the valve cavity (Case Study No. 12.3).

12.10   Response of Structures to Acoustic Waves

The examples as well as the case studies in this chapter all point to one fact: even at a very loud noise level of 160 dB, the acoustic pressure is not very high. A 0.1 psi (700 Pa) acoustic pressure would produce a deafening noise. This, together with the relatively long wavelength associated with acoustic noise, makes slender structures relatively immune to normally impinging acoustic waves. This can be intuitively seen in Figure 12.19. Suppose an acoustic wave with wavelength length λ and acoustic pressure $\pm p_0$ impinges normally onto a cylinder with diameter d, the pressure differential across the cylinder is only in the order,

$$\Delta p = \frac{2 p_0 d}{\lambda}$$

with the force on the cylinder equal to,

$$|F| = d\Delta p = \left(\frac{2d}{\lambda}\right) p_0 d \quad \text{per unit length} \tag{12.54}$$

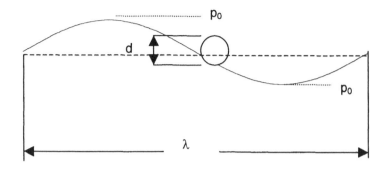

Figure 12.19 Acoustic wave impinging normally on a slender cylinder

Since acoustic wavelengths are relatively long compared with the diameter of tubes and pipes, thin structures are usually not very responsive to normally impinging acoustic waves. Only large-scale structures, such as buildings, large storage tanks or similar objects, are easily excited by normally incident acoustic waves.

In the Appendix, the force on a cylinder due to a normally incident acoustic wave is derived more rigorously based on wave scattering theory. The exact force on the cylinder is shown to be (Equation (12.60) in the Appendix),

$$| F | = \frac{\pi^3}{\sqrt{32}} \left(\frac{d}{\lambda} \right)^{3/2} p_0 d$$

12.11   Case Studies

Acoustic problems are among the most common vibration problems in the power and process industries. Furthermore, acoustic problems distinguish themselves from fluid-elastic instability and turbulence-induced vibration in that, very often, the exact cause of the problem is not known even after the problem is corrected. Since so much is dependent on experience in the diagnosis of acoustically induced vibration problems, this section is devoted to more than the usual number of case studies that are likely to be encountered in power and process plants.

Case Study 12.1: Noise and Vibration Caused by Cavitating Venturi

The cavitating venturi is briefly discussed in Chapter 10 under axial and leakage-flow-induced vibration. Figure 10.8 shows a schematic of this device, which is installed in piping systems to limit the water mass flow by "choking" it. The bubble collapses combined with the exit turbulent shear flow generate intense acoustic energy, which accounts for the high noise levels (in excess of 100 dB) normally associated with cavitating venturis. The fluctuating pressure spectra due to cavitating flow resemble those due to turbulence in that they are continuous. However, unlike turbulence, the power spectra of which normally decrease rapidly with frequencies, the energies in cavitating flows do not drop off with frequency until well beyond the range of human audibility. As a result, cavitating flow tends to excite the higher coupled fluid-shell modes in piping systems. As we have seen from previous examples, higher-order shell and acoustic modes are closely spaced in the frequency axis. This further intensifies the noise generated from cavitation and the shear flow. This high-frequency force tends to shake loose treaded joints and excite shell modes in the pipes, sometimes causing fatigue failures of welded joints in hours.

During a pre-operational test of a cavitating venturi in a nuclear plant, the noise level generated was so loud that it could be heard hundreds of feet away from the building. Valves mounted on top of the main piping system by nozzles were seen to vibrate with "alarming" amplitudes. Within minutes, accelerometers with design limits of over 50-g used to monitor the vibration of the venturi were overloaded and damaged.

The cavitating venturi was then removed and installed in a laboratory piping system for testing, with the venturi mounted in a Schedule 80 pipe (about 10", or 0.25 m diameter). With supply pressure at 160 psig (1.1 MPa), back-pressure at 35 psig (0.24 MPa) and flow rate at 2,900 gpm (0.18 m^3/s), "pounding frequency" in the axial direction was felt, noise level in excess of 115 dB was recorded three feet (one meter) away from the venturi and acceleration in excess of 300 g was recorded. The accelerometers were eventually damaged, but not before some useful data were recorded. Figures 12.20 and 12.21 show typical acceleration spectra at the venturi in the axial and transversal directions. In the transversal direction, sharp spectral spikes characteristic of shell-mode vibrations persist well beyond 1,000 Hz and did not fall off in amplitude even beyond 12 KHz. In the axial direction, the spectral peaks were more characteristic of piping vibration. After several hours of testing, cracked welds were observed in the attached piping tens of feet (10 meters) from the venturi.

The test continued without instrumentation for a total of 100 accumulated hours. After that, the venturi was removed and its internal as well the internal of the pipe immediately downstream of the venturi were inspected with a magnifying glass for cavitation damage. There was no trace of cavitation damage on the pipe immediately downstream of the venturi. However, there was considerable pitting at the throat of the venturi—characteristic of cavitation damage (Figure 12.18). The pitting became progressively worse as the cumulative test duration went from 50 to 100 hours.

This cavitating venturi was not used in this particular nuclear plant. However, cavitating venturis have been used in other power plants without causing any damage to

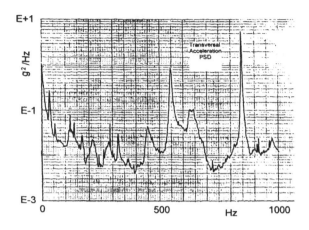

Figure 12.20 Acceleration PSD in the direction perpendicular to the venturi axis

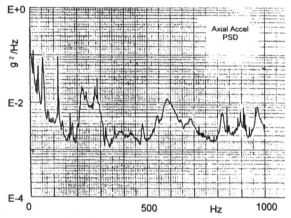

Figure 12.21 Acceleration PSD in the axial direction

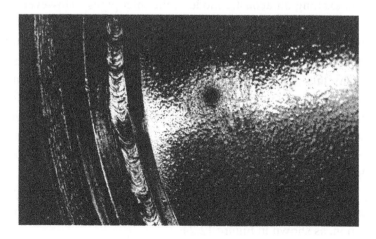

Figure 12.22 Cavitation damage of venturi throat

the attached piping or components. Carefully designed, it is a viable alternative to valves to control the flow. Since cavitation damage at the throat of the venturi will occur, this device is not for continuous operation. In addition, the following precautions should be taken:

- Personnel working in the neighborhood of the cavitating venturi must wear hearing protection as the noise level from cavitating venturis normally exceeds 100 dB.
- Avoid using long nozzles or standoff pipes to mount components such as valves on the pipes near the venturi.
- Avoid welded joints in the neighborhood of the venturi if possible. If weld joints must be used, they should be designed to withstand the stresses caused by the expected shell-mode deformation of the pipes.
- All threaded joints should either be removed or locked.
- The affected piping system should be properly anchored.

Case Study 12.2: Acoustic Noise Generated by a Spherical Elbow (Ziada et al, 1999)

As often happens in the industry, a problem occurs and demands immediate corrective action. The problem is fixed, most likely by trial and error, but the root cause of the problem is never understood. This case study is such an example.

Figure 12.23 is a schematic drawing of the high-pressure steam bypass piping system in a coal-fired plant. During the commissioning of this plant, severe piping vibration and intense noise level were observed. Field measurement showed that the noise spectrum contained spikes, one of which agreed quite closely with a transversal acoustic modal frequency in the inlet pipe upstream of the bypass valve. Thus, it appeared that something was exciting a standing wave in this segment of the piping system. Since valves are common noise-generating sources, the first suspicion was that the turbine bypass valve was exciting an acoustic mode in the inlet pipes. However, the same field measurement also showed that the vibration amplitude of the pipe upstream of the bypass valve was higher than that in the downstream pipe. Furthermore, a separate test on the valve showed that it generated a broadband noise spectrum without any spikes. It was very unlikely that this valve would preferentially excite the upstream inlet pipes. Nonetheless, the valve was modified with the objective of reducing flow excitation. When the modified valve was installed into the otherwise un-modified piping system, the problem persisted.

A model test of the piping system was then conducted with the suspected components removed one by one to see their individual effect on the piping vibration. It was found that removing the valve or the orifice plate had little effect on the piping vibration. However, when the spherical elbow was eliminated, there was a significant reduction in the piping vibration, as shown in Figure 12.24.

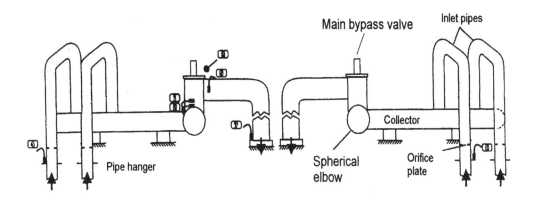

Figure 12.23 Schematic of piping system (Ziada et al, 1999)

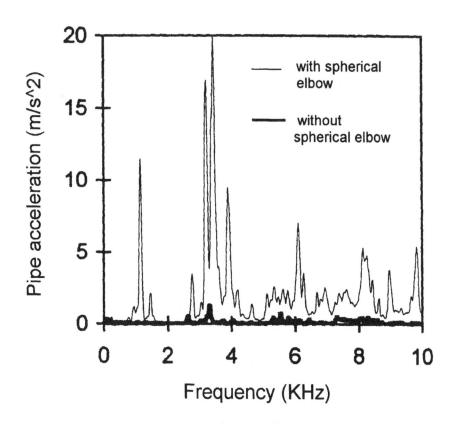

Figure 12.24 Acceleration spectra of model piping with and without
the spherical elbow (Ziada et al, 1999)

Thus, it was obvious that the piping vibration was caused not by the bypass valve or the orifice plate, but by the spherical elbow—an apparently much less efficient excitation force generating component. The spherical elbow was eliminated and excessive vibration has not been experienced since. The question why the spherical elbow would generate a force that preferentially excites the upstream piping system was never answered. However, elbows are known to be turbulence-generating components, although usually not as efficiently as orifice plates or valves. Still, it is possible that the spherical elbow had generated a turbulence spectrum with a peak that happened to be near one of the acoustic modes in the upstream piping system.

This case study shows real lives under the industrial environment. If there is a problem, it has to be fixed at all cost. However, once it is fixed, there will be no pressure to do a post-incident root cause analysis, no matter how small the cost.

Case Study 12.3: Acoustic Resonance in Valve Cavity (Coffman and Bernstein, 1979)

Figure 12.25 shows the safety relief valves mounted on top of a 1-1/2 inch (3.8 cm) diameter pipe carrying steam at 614 psig (4,233 KPag) and 635 deg. F (335 deg. C) at a mass flow rate of 3,360,000 lb (1,527,000 Kg) per hour. The piping system is part of the reheater header in Oklahoma Gas and Electric's 550 MW gas-fired plant. As originally designed, the valves were mounted on the pipe by straight nozzles (Figure 12.26). In 1971, shortly after the initial startup, unusual noise and vibration from these valves were noticed. Within a few months of operation, several of the valve drop levers fell off the valves after their pivot pins had worn through. The first annual inspection of these valves showed that all of them had suffered severe wear. In three of them, the internal components were virtually completely destroyed.

After a series of vibration amplitude and pressure measurements, it became obvious that standing waves in the valve cavities (formed by the valve internal and the nozzle) were excited by the steam flow over them (Sub-section 12.1). Corrective action was taken by shortening the nozzles to increase the frequencies of the acoustic modes in the valve cavity-nozzle combination. The unit was returned to service. Similar noise and vibration were noticed, with the exception that the prominent noise component was at a higher frequency, consistent with shortening of the nozzle. At the second annual inspection, again severe wear of the safety valve internal parts was noticed.

After judging several post-construction corrective measures, it appeared that the most economical one was to replace the cylindrical nozzle with a conical reducer with a base diameter almost twice that of the original straight nozzle (Figure 12.26). With this modification, the unit was returned to service. The loud noise stopped. Measurement showed that the vibration amplitudes of the valves were negligibly small. No unacceptable wear was observed in subsequent inspections.

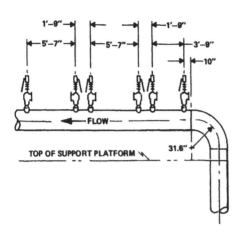
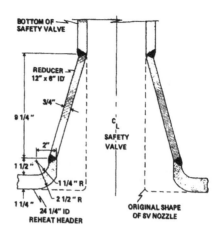

Figure 12.25 Safety valves mounted   Figure 12.26 Modified reducer versus
          on reheat header   original cylindrical nozzle
                                        (Coffman and Bernstein, 1979)

Although the problem was solved, again the reason why the reducer nozzle would eliminate the acoustic resonance problem was never satisfactorily explained. For years, however, designers of high-fidelity loudspeaker systems know that non-parallel walls tend to suppress cabinet resonance, which tends to color the musical sound. Thus it should not be surprising that the conical cavity suppressed the formation of standing waves in the valve-nozzle combination.

Case Study 12.4: Acoustic Resonance in Heat Exchanger (Tanaka et al, 1999)

Figure 12.27 is a simplified schematic drawing of a heat exchanger in a power plant in Japan. It has four tube bundles arranged in a normal square pitch with a pitch-to-diameter ratio of 1.57. The width of the duct is 7.5 m while the diameter of the tube, d=0.0254 m. During pre-operational tests with air at 32 deg. C flowing in the shell side, a loud noise suddenly started at a reduced velocity V/fd=2.9, and continued until this reduced velocity reached 8.0. The measured acoustic pressure at the center of the duct (point A in Figure 12.27) reached 1,600 Pa, while the measured frequency was almost constant at 45 Hz. Figure 12.28 shows the measured pressure distribution across the duct. It had the classic first transversal acoustic mode with closed ends.

The problem was solved by inserting four straight plates into the tube bundle, as shown in Figure 12.27(B). These four plates divided the duct into five cavities transversally, with a substantial increase in the first acoustical modal frequency. Assuming the center compartment has a dimension 1/3 of the duct width, the first acoustic modal frequency is now three times higher than before. Theoretically, the loud noise should not appear until a reduced velocity V/fd=8.7 is reached.

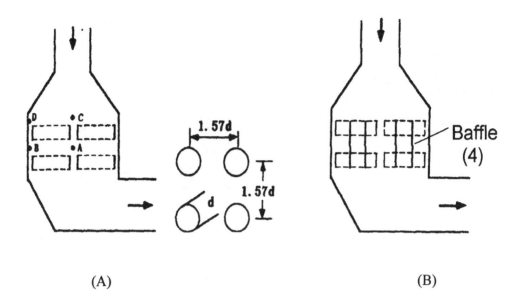

(A)   (B)

Figure 12.27 Schematic of heat exchanger—(A) Original design;
(B) Redesign (Tanaka et al, 1999)

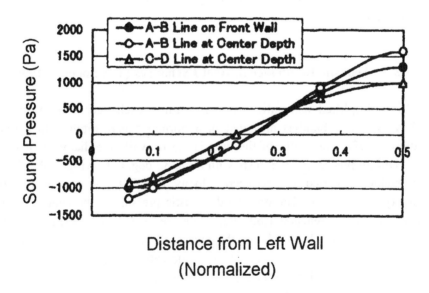

Figure 12.28 Measured acoustic pressure distribution across the duct
(Tanaka et al, 1999)

Case Study 12.5 Valve-Generated Acoustic Waves in a Pipe (Pastorel et al, 2000)

Steam dump piping systems are used in nuclear plants during start-up, shutdown, turbine trip and other un-scheduled events. The purpose of this system is to divert the steam from the nuclear steam generators to the condenser, thus bypassing the steam turbine. The steam flow is controlled by multiple control valves. The function of these valves is to reduce the pressure of the fresh steam from that of the steam generator outlet to that of the condenser.

At the Gentilly Nuclear Power Station, the steam flow is controlled by four control valves each with a 255-mm (10 in.) seat diameter and a 100 mm (3.9 in.) lift. The steam pressure and temperature at the valve inlets are 45 bar (652 psi) and 260 deg. C (500 deg. F), reduced to 8.0 bars (116 psi) at the outlets. Shortly after commercial operation, the pipes downstream of the control valves experienced cracks. Metallurgical examination indicated that the main cause of failure was fatigue. A subsequent dynamic analysis of the piping system ruled out transient thermal-hydraulic loads as the cause of fatigue crack in these pipes. An in-situ test was conducted using high-temperature accelerometers and strain gauges installed at several key locations of the piping system, including one at the elbow immediately downstream of one of the control valves. The measured dynamic strain spectra showed a sharp spike at 600 Hz and a broadband around 2,000 Hz (Figure 12.29). The total rms strain was 44.5E-6, corresponding to a maximum peak stress of 175 MPa (25,380 psi) for the piping material. The endurance limit of the material is 80 MPa (11,600 psi). Thus, a minimum of a twofold decrease of the dynamic strain was required to prevent future fatigue failure of the pipes.

The very narrow peak at 600 Hz indicates that the strain was mostly likely acoustically induced. Apparently the large, single pressure drop through the control valves had excited standing acoustic waves in the downstream pipe, which in turn excited the pipe, causing large stresses at critical points of the piping system. To correct this, the control valves were re-designed. The single-stage pressure drop was replaced by two stages. In addition, the large flow window of the original design was replaced by 12 narrow slits in the first stage and 408 small orifices through a cone in the second stage. The length of the first-stage slits was designed in such a way that for small valve lifts, most of the pressure drop would occur in the first stage, while for large valve lifts, the pressure drop would be evenly divided between the two stages. Figure 12.30 shows the spectra of the noise generated in the downstream pipe by the original and re-designed valves and shows there is a significant decrease in the sound pressure level (SPL) throughout the frequency range.

After the re-designed valves were installed, another in-situ measurement showed that the dynamic strains were reduced by a factor of almost four. The maximum peak dynamic stress was 40 MPa (5,800 psi), which is half the endurance limit of the pipe material.

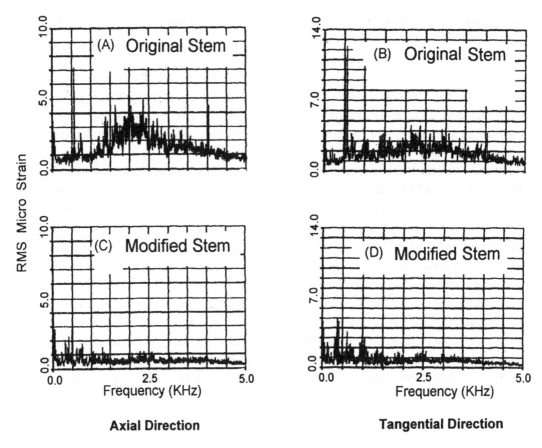

Figure 12.29 Typical measured dynamic strain PSD: (A) Axial strain, with original valve; (B) Tangential strain, original valve; (C) Axial strain, re-designed valve; (D) Tangential strain, re-designed valve (Pastorel et al, 2000).

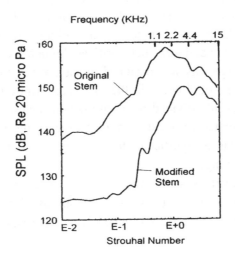

Figure 12.30 Measured sound pressure level PSD in pipe downstream of valve
(Pastorel et al, 2000)

Case Study 12.6 Thermoacoustic Vibration of a Gas Turbine Recuperator (Eisinger, 1992)

Figure 12.31 (A) is a schematic drawing of a gas turbine recuperator heat exchanger system (actual arrangement was vertical). During cold start-up, low-frequency vibrations were noted. Large pressure pulsation in the discharge duct/recuperator caused severe vibration of the entire system, from the recuperator to the turbine and the duct in between. The vibration stopped after the system was fully warmed up. In-situ measurement showed distinct quarter-wave length pressure pulsation at 3.5 Hz at a turbine speed of 3,000 RPM. At 4,000 RPM, the pressure pulsation was so severe that the turbine had to be tripped to prevent damage to the equipment. It was estimated that at this point, the mean gas flow velocity entering the recuperator tube bundle was about the same as the acoustic particle velocity. This coupling mechanism further amplified the resonant vibration, leading to an acoustic pressure of over 12 KPa (1.8 psi), or about 176 dB at 3.5 Hz (Figure 12.31 (C)).

Root cause analysis showed that the vibration was thermoacoustic in nature. The gas turbine exhaust and the recuperator acted as a Sondhauss tube (Section 12.9), with the exhaust gas acting as the hot portion of the tube, and the tube side of the recuperator acting as the cold portion of the tube. The location of the hot/cold interface in this particular case happened to be near the center of the system, making it especially sensitive to spontaneous thermoacoustic oscillation. During cold start-up, the temperatures in the hot and cold portions formed a step function, as shown in Figure 12.31 (B), giving the system sufficient temperature gradient to initiate thermoacoustic oscillation (see Figure 12.31). At this point, the flowing gas velocity happened to coincide with the acoustic particle velocity. This velocity coupling probably had either initiated or worsened the acoustic vibration. After it was fully warmed up, the temperature gradient became less steep (Figure 12.31 (B), dotted line). As a result, spontaneous thermoacoustic oscillation could no longer be sustained.

In new designs, the prevention of thermoacoustic oscillation was achieved by choosing the correct relative lengths of the cold and hot portions of the duct. Post-construction remedial action consisted of modified operational procedures, such as opening the turbine exhaust gas discharge duct or pre-heating the system externally prior to start-up. Other remedies such as partially blocking the recuperator discharge or use of acoustic dampers inside the duct are less desirable because of the increase in pressure loss.

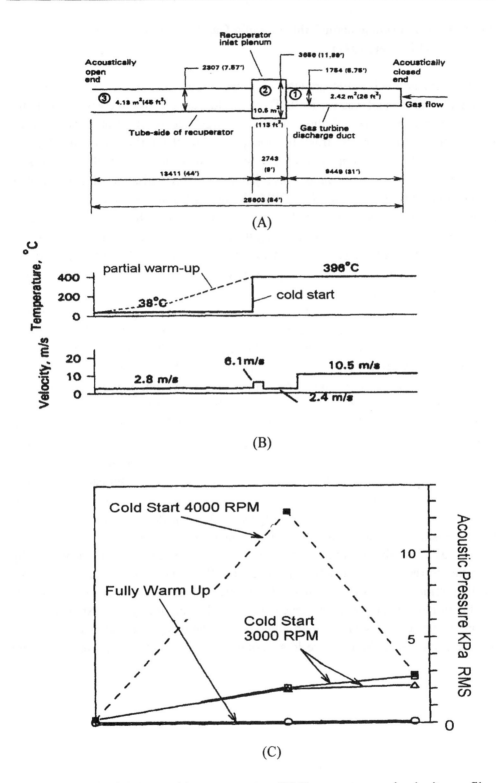

Figure 12.31 (A) Gas turbine recuperator, (B) Temperature and velocity profiles;
(C) Measured acoustic pressure (Eisinger, 1992).

Appendix 12A: Scattering of Normal Incident Acoustic Wave by a Cylinder

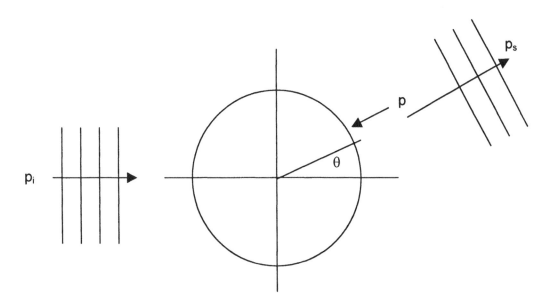

Figure 12.32 Scattering of a plane wave by an infinite cylinder

The pressure distribution on the surface of a rigid circular cylinder of infinite length due to a normally incident acoustic wave (Figure 12.32) can be obtained by following a procedure outlined by Skudrzyk (1971) or Moorse and Feshback (1953). First we expand the incident plane wave in terms of a superposition of cylindrical waves:

$$p_i = p_0 e^{ikr\cos\theta} = p_0[2\sum_{\alpha=1}^{\infty} i^{\alpha} J_{\alpha}(kr)\cos\alpha\theta + J_0(kr)] \tag{12.55}$$

where p_0 is the 0-peak acoustic pressure amplitude and k is the "wave number", defined as,

$$k = 2\pi f / c = 2\pi / \lambda$$

and c is the velocity of sound and λ is the wavelength at frequency f. This incident wave is scattered by the cylinder. The scattered wave is an outgoing cylindrical wave. Hence, it can be expressed as a linear combination of Hankel function of the second kind:

$$p_s = \sum_{\alpha=1}^{\infty} B_{\alpha} H_{\alpha}^{(2)}(kr)\cos\alpha\theta \tag{12.56}$$

On the surface of the stationary cylinder, the normal component of the velocity is zero. Therefore,

$$\frac{\partial p_i}{\partial r} + \frac{\partial p_s}{\partial r} = 0 \quad \text{at } r=R, \text{ the radius of the cylinder.}$$

This boundary condition enables us to find the constants B_α, and Equation (12.56) for the scattered wave reduces to:

$$p_s = p_0 \sum_{\alpha=1}^{\infty} -\varepsilon_\alpha i^\alpha \frac{J_{\alpha-1}(kR) - J_{\alpha+1}(kR)}{H_{\alpha-1}^{(2)}(kR) - H_{\alpha+1}^{(2)}(kR)} \frac{r}{R} H_\alpha^{(2)}(kR) \cos\alpha\theta \tag{12.57}$$

where $\varepsilon_0 = 1$ $\quad$ and $\quad$ $\varepsilon_{\alpha \geq 1} = 2$

For small values of kR, the term containing J_0 (kR) dominates. Therefore, we can keep only this term in the sum. Furthermore, we can use the series expansion of the Bessel function for small argument and reduce Equation (12.57) to the simple form:

$$p_s = -p_0 \sqrt{\frac{\pi}{2}} \frac{k^{3/2} R^2}{\sqrt{r}} \left[\frac{1}{2} + \cos\theta\right] e^{-i(kr+\pi/4)} \tag{12.58}$$

On the surface of the cylinder, $r=R=d/2$,

$$p_i \approx p_0 J_0(kR) \approx p_0$$

$$p_s \approx -p_0 \sqrt{\frac{\pi}{2}} (kR)^{3/2} \left[\frac{1}{2} + \cos\theta\right] e^{-i(kR+\pi/4)}$$

The 0-peak pressure distribution on the surface of the cylinder is the sum of the above two,

$$|p| = p_0 \left[1 - \sqrt{\frac{\pi}{2}} (kR)^{3/2} (\frac{1}{2} + \cos\theta)\right] \tag{12.59}$$

which is no longer symmetric. Projecting the force in the $\theta=0$ direction and integrating, we get a net force on the cylinder given by,

$$| F |= \int_{0}^{2\pi} p\cos\theta R d\theta \quad \text{per unit length of the cylinder.}$$

The only term that contributes to the net force is the term containing $\cos\theta$. Upon integrating, we get,

$$| F |= p_0 d \sqrt{\frac{\pi}{2}} \frac{\pi}{4}(kR)^{3/2} = \frac{\pi^3}{\sqrt{32}}\left(\frac{d}{\lambda}\right)^{3/2} p_0 d \qquad (12.60)$$

As an example, at $f=130$ Hz, velocity of sound $c=3,430$ ft/s, $d=9$ in (22.86 cm),

$kR=0.09$ so the assumption $kR<<1.0$ is justified.

Reference:

ASME, 1979, Steam Tables, Fourth Edition.

Blevins, R. D. 1979, Formulas for Natural Frequencies and Mode Shapes, Van Nostand Reinhold, New York.

Blevins, R. D. and Bressler, M. M., 1987, "Acoustic Resonance in Heat Exchanger Tube Bundles, Part I and Part II," Journal of Pressure Vessel Technology, Vol. 109, pp. 275-288.

Blevins, R. D., 1990, Flow-Induced Vibration, Second Edition, Van Nostrand Reinhold, New York.

Chen, Y. N., 1968, "Flow-Induced Vibration and Noise in Tube Bank Heat Exchangers due to Von Karman Streets," Journal of Engineering for Industry, Vol. 90. pp. 134-146.

Coffman, J. T. and Bernstein, M. D., 1979, "Failure of Safety Valves due to Flow-Induced Vibration," Flow-Induced Vibrations, edited by S. S. Chen, ASME Press, New York, pp. 115-128.

Eisinger, F. L., 1992, "Fluid-Thermoacoustic Vibration of a Gas Turbine Recuperator Tubular Heat Exchanger System," in Symposium on Flow-Induced Vibration and Noise, Vol. 4, edited by M. P. Paidoussis and J. B. Sandifer, ASME Press, New York, pp. 97-121.

Eisinger, F. L., 1999, "Eliminating Thermoacoustic Oscillations in Heat Exchangers and Steam Generator Systems," in Flow-Induced Vibration—1999, edited by M. J. Pettigrew, ASME Press, New York, pp. 153-163.

Morse, P. M. and Feshbach, H., 1953, Method of Theoretical Physics, McGraw Hill, New York.

Parker, R., 1978, "Acoustic Resonance in Passages Containing Banks of Heat Exchanger Tubes," Journal Sound and Vibration, Vol. 57, pp. 245-260.

Pastorel, H., Michaud, S. and Ziada, S., 2000, "Acoustic Fatigue of a Steam Dump Pipe System Excited by Valve Noise," in Flow-Induced Vibration, edited by S. Ziada and T. Staubli, Balkema, Rotterdam, pp. 661-668.

Skudrzyk, E., 1971, The Fundamentals of Acoustics, Springer-Verlag, New York.

Weaver, D. S., 1993, "Vortex-Shedding and Acoustic Resonance in Heat Exchanger Tube Arrays," in Technology for the '90s, edited by M. K. Au-Yang, ASME Press, New York, pp. 777-810.

Tanaka, H., Tanaka, K. and Shimizu, F., 1999, "Analysis of Acoustic Resonant Vibration Having a Mutual Exciting Mechanism," in Flow-Induced Vibration—1999, edited by M. J. Pettigrew, ASME Press, New York, pp. 145-152.

Ziada, S., Oengoren, A. and Buhlmann, E. T., 1989, "On Acoustic Resonance in Tube Arrays-I, Experiments," Journal of Fluids and Structures, Vol. 3, pp. 293-314.

Ziada, S., Oengoren, A. and Buhlmann, E. T., 1989, "On Acoustic Resonance in Tube Arrays-II, Damping Criteria," Journal of Fluids and Structures, Vol. 3, pp. 325-324.

Ziada, S., Sperling, H., and Fisker, H., 1999, "Flow-Induced Vibration of a Spherical Elbow Conveying Steam at High Temperature," Symposium on Flow-Induced Vibration, PVP-Vol. 389, edited by M. J. Pettigrew, ASME Press, New York, pp. 349-357.

CHAPTER 13

SIGNAL ANALYSIS AND DIAGNOSTIC TECHNIQUES

Summary

Although present-day frequency-domain signal analysis (spectral analysis) usually involves the use of the Fast Fourier Transform together with digital computers, neither of these are absolute requirements for spectral analysis of vibration data. However, the digital format offers much higher dynamic ranges with great reduction in the weight of the data acquisition and analysis equipment. Care must be exercised when one converts the continuous time domain data into a discrete time series. If an insufficient number of data points is selected, that is, if the sampling rate is not high enough, a phenomenon called aliasing will occur, whereby data higher than one half of the sampling frequency will fold back into the lower frequency range and contaminate the useful data. The following table outlines the inter-relationship between the sampling frequency, the maximum frequency of interest, the resolution in the time and frequency domains, the number of data blocks to analyze to ensure a certain statistical accuracy and the total length of time record required. This applies to steady vibration data. In analyzing single-event transient data, such as in impact tests, one should not include data outside of the transient in averaging the data.

Relationships between Sampling Rate and Analysis Results

Quantity	Relationship
Sampling Interval	ΔT
Sampling Rate	$f_s = 1/\Delta T$
Maximum (Nyquist) Frequency (Hz)	$1/(2\Delta T)$
FFT Sample Block Size (Number of data points per block)	n (must be 2^k where k is an integer)
FFT Spectrum Lines	$n/2 + 1$ (including $f=0$)
FFT Frequency Resolution (Hz)	$\Delta f = 1/(n\Delta T)$
Number of Data Blocks	N (64 - 100 blocks recommended)
Total Length of Time Record Needed	$T = Nn\Delta T$
Normalized Error in PSD Estimate	$\varepsilon = 1/\sqrt{N}$

While too low a sampling rate will cause aliasing, too high a sampling rate will cause loss of resolution in the frequency domain. This is because the frequency and time domain resolutions are related by the equation,

$$\Delta f * \Delta t = 1/n \qquad\qquad\qquad (13.12)$$

a high sampling rate (small Δt) will result in insufficient frequency resolution.

Because of its low cost, compactness and versatility, the digital filter is a very powerful tool for the diagnostic engineer to eliminate noise contamination from test data, even after the test is completed. Because of the great increase in the dynamic range of digital equipment, sensors that yield signals that are too weak for analog equipment are perfectly useful for digital equipment. In the last ten years, some new sensors had been developed specifically to work with digital equipment.

Acronyms

BPF   Blade passing frequency
FFT   Fast Fourier transform
MOV   Motor-operated valve
PSD   Power spectral density
RAM   Random access memory

Nomenclature

f Frequency, in Hz
f_s Sampling frequency
f_N Nyquist frequency
Δf Resolution in the frequency domain
$G(f)$ Single-sided power spectral density
i An integer number
j $= \sqrt{-1}$
k An integer number
n Number of data points in one block of data to be Fourier-transformed
N Total number of data blocks used to average the transformed data
$S(\omega)$ Two-sided power spectral density
t Time
T Length of time record in one block of data Fourier-transformed
Δt Resolution in the time domain
$x(t_i)$ Time domain data at discrete time point t_i
$X(f_i)$ Fourier transform of time domain data at the discrete frequency point f_i

ε Normalized error
ω Frequency, in radian/s

13.1   Introduction

Readers who have gone through the work examples and read through the case studies in the previous chapters probably realize by now that, being a very complex subject, flow-induced vibration problems often cannot be predicted or diagnosed by analysis alone. In designing a new product, the analysis is only a starting point. Laboratory tests are usually used to verify that the design is free of the most serious flow-induced vibration problems before it is put into production. Large, complicated and expensive equipment such as airplanes and nuclear reactors probably will need further pre-operational tests before the equipment is put into commercial operation. When a vibration problem occurs in the field, almost without exception some measurement will be made. Based on the measured data, engineers try to diagnose the problems and find ways to correct them. Thus, testing is an integral part of flow-induced vibration analysis. To be successful, the engineer who plans the test and analyzes the data should be thoroughly familiar with the fundamentals of flow-induced vibration and signal analysis. Other than the simplest on-line monitoring in which one just watches the traces on the oscilloscope, modern data acquisition and analysis are, without exception, carried out digitally with the help of computers and spectrum analyzers. In order to understand the phenomenal success of the digital data recording format, one must first understand the deficiency of the analog equipment it replaced. Before the introduction of the digital format, the most common medium for recording and storing data was the magnetic tape. In this record/playback medium, the weakest signal is one that can be distinguished from the background noise by the recorder tape head. The strongest signal the tape can store is when all the magnetic dipoles in it are lined up—a phenomenon known as saturation. The dynamic range, which is the log to the base 10, of the square of the ratio between these two signals (Section 2.13), is typically between 50 and 60 dB for an analog scientific recorder. To achieve the highest signal-to-noise ratio, thick and wide magnetic tapes were used and were recorded at high speeds so the volume of magnetic particles that passed under the record/playback head per second was high. These factors translated into bulky and expensive data acquisition, storage and playback systems. By contrast, the dynamic range of the digital format is governed by the word length of the data acquisition and analysis software. If a k-bit word is used in the software, the largest number the word can represent is $2^k - 1$ while the smallest, non-trivial number that word can represent is unity. The dynamic range of the software is equal to $20\log_{10}(2^k - 1)$ dB. For a 16-bit word, this is equal to 96 dB. Even software with a 12-bit word structure has a dynamic range of 72 dB, compared with about 60 dB for an audiophile quality analog vinyl record playback system of the 1970s, 62 dB for an audiophile quality analog cassette tape machine with Dolby B noise reduction, and 68 dB for a professional, studio analog master tape.

With increasingly higher-speed portable computers, larger, inexpensive mass storage systems and less expensive analog-to-digital conversion circuit boards available, present day digital signal processing software uses at least a 16-bit word structure; some are available with 20 or even 24-bit words. The compact disc, as an example, is based on 16-bit words. With its 96-dB dynamic range, the background noise in a compact disc virtually disappears. This, in short, is the main reason for the compact disc's success in the home sound reproduction market.

In signal analysis, this 40 dB additional dynamic range the digital equipment has over its analog predecessor means that sensors that generate signals that are 10,000 times too weak to be useful for analog equipment, can now be developed to provide useful information that can often shed new light on the conditions of the component being monitored. In Section 13.8, we shall briefly discussed some recent developments in the applications of these new sensors to plant component condition monitoring and diagnosis.

To the engineers whose primary interest is problem-solving, spectral and digital signal analysis often cause some level of discomfort or even confusion because of the unfamiliar terms used (especially before the compact disc and digital video disc became popular). Worse, without some knowledge about the fundamentals of digital signal processing, the diagnostic engineers may misuse the equipment or mis-interpret the data.

The objective of this chapter is to explain, in non-mathematical terms, the basic concept of spectral and digital signal analysis, with the objective of eliminating the mysteries behind these modern technologies and preventing future mistakes that could arise from lack of understandings of these concepts. For this reason, this chapter is purposely written in a "light" and non-sophisticated way. The mathematically inclined readers should refer to the many books in digital signal analysis, Fourier transform and wavelet analysis.

13.2   Representation of a Continuous Wave by a Series of Discrete Points

The first step in digital signal analysis is to convert the continuous analog signal to a series of discrete numbers. In the following discussion, it is assumed that the signals from the sensors (example: accelerometers) are either digitally recorded onto a tape or directly acquired into the hard disk of a data acquisition system. If the signal is recorded analogly, then it must be digitized. This digital information is very much like the musical notes coded on a compact disc. However, while on the compact disc, the signals are sampled at 44.1 K per channel per second (i.e., 44,100 points in each of the two stereo channels) to represent one second of music, commercial data acquisition systems usually offer user-selectable sampling rates. The sampling rate, or frequency, f_s, selected determines the upper frequency limit the data can represent. Theoretically, the upper frequency limit of the data is equal to one half the sampling rate, and is called the Nyquist frequency,

$$f_N = f_s / 2 \tag{13.1}$$

In the compact disc format in which the sampling rate is fixed at 44,100 per second; the highest musical note a compact disc can reproduce, irrespective of the reproduction hardware, is 22,050 Hz, which is beyond the audible range of human ears. In practice, some margin of safety must be allowed.

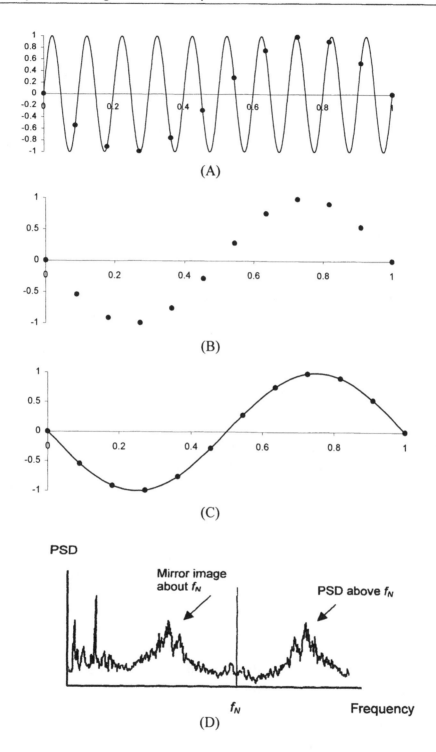

Figure 13.1 Aliasing of cyclic data due to an insufficiently high sampling rate: (A) Digitization (A/D conversion) of original continuous wave; (B) Resulting time series; (C) Reconstruction (D/A conversion) of the continuous wave from the time series; (D) Foldback of PSD above the Nyquist frequency into PSD at lower frequencies

Aliasing

Ideally, we should represent a continuous wave by an infinite number of points. In practice, the number of points is restricted by the speed and memory of the computing equipment. However, if the number of points representing a continuous wave, i.e., the sampling rate, is not sufficiently high, a phenomenon called aliasing will occur, thereby misrepresenting the original vibration form. To understand the origin of aliasing, refer to Figure 13.1(A) in which the original continuous wave is digitized (a process called A/D conversion). The resulting time series is shown in Figure 13.1(B). Figure 13.2(C) shows how this time series is reconstructed by a process called D/A conversion. When the number of discrete points representing the original continuous wave is not high enough as is in this case, the reconstructed continuous wave will have a much lower frequency than the original one. This is the origin of aliasing. If the original spectrum contains information above the Nyquist frequency, as shown in Figure 13.1(D), the spectrum obtained from the digitized time series will not contain that information. Instead, spectral content above the Nyquist frequency will appear as its mirror image about the line defined by $f=f_N$. All of us have witnessed the problem of aliasing. In a movie, the picture frames are sampled at a fixed rate, usually 18 per second. This is not high enough to, e.g., represent the rotation of a helicopter rotor or a stagecoach wagon wheel. As a result, the rotor or the stagecoach wagon wheels appear to rotate backward at a slower speed.

Theoretically, one needs at least two points to represent one cycle of a wave. (This will be clear in Section 13.3 on the Fourier transform.) This means that, for example, the theoretical minimum sampling rate of a 100 Hz wave is 200 samples per second, 200/s in short. In practice a number higher than 2.0 times is recommended to avoid aliasing. Various experts in the field of signal processing have recommended a minimum of 2.2 to 3.0/s. Exactly how much margin one should allow has been a subject of debate for many years among experimental engineers and digital audio purists alike. In the case of the compact disc, it was decided years ago that a sampling rate of 44.1 K per second will have sufficient margin to cover an upper frequency limit of 20 KHz for a digitization-frequency ratio of 2.2.

Apparent Frequencies

According to the Nyquist theorem, after digitization, one cannot see any frequencies that are higher than half the sampling frequency (the Nyquist frequency). The next question is what happens to the frequencies higher than the Nyquist frequency. While this can be derived mathematically from the properties of the trigonometric functions, some physical insight can be obtained by close study of Figure 13.1(A). First, we notice that frequencies that are $(1+x)$, $(2+x)$, $(3+x)$ etc., times that of the sampling frequency will all have the same frequency after sampling. Second, we notice that if the frequency is a fraction higher than half the sampling frequency, the phase angles of the sampling points are opposite to those of the original wave, as shown in Figure 13.1(A). If the frequency is a fraction higher than an integral multiple of the sampling frequency, the phase angles of the sampling points are the same as those of the actual wave. In a motion picture, when the engine of the helicopter first starts, the blade passing frequency (BPF) is less than the Nyquist frequency, i.e., less than 9/s. We see the actual rotation of the blades (Figure 13.2(A)). As the engine revs up, the BPF increases to, say 10/s. The sampling points are now opposite in phase to those of the actual rotation—we see the mirror reflection of the rotation about the Nyquist frequency of 9/s and the rotor rotates backward at an apparent BPF of 8/s (Figure 13.2(B)). As the engine continues to rev up, the BPF increases to, e.g., 20/s, the amplitude and phase at the sampling points are exactly the same as those at 20-18=2 per second. The rotor appears to rotate forward, but at a BPF of 2/s. As the engine continues to rev up, we see the blades keep moving forward, become stationary, then rotate backward, become stationary, then forward again, until the picture becomes blurry (Figures 13.2(C) and (D)).

In processing discretized time series data, the digital signal will not become blurry as in the motion picture. Instead, signals with frequencies many times higher than the Nyquist frequency will continue to fold back and contaminate the data in the range of interest. For this reason, signals above the Nyquist frequency must be analogly filtered off before digitization. It happens more than once that the 60Hz AC hum and its second harmonic at 120 Hz, both of which are common in electronic equipment not properly grounded, folded back into the lower frequency range and caused the diagnostic engineer to mis-interpret the results.

If the signal above the Nyquiest frequency is not filtered off before digitization, the following simple rule can be used to find their apparent frequencies:

If Remainder of $(f_{actual} - if_s) < f_s/2$, $f_{app} = f_{actual} - if_s$

If Remainder of $(f_{actual} - if_s) > f_s/2$, $f_{app} = $ *mirror image of* $(f_{actual} - if_s)$ *about* $f_s/2$ (13.2)

where i = an integer and f_s = sampling frequency

Example 13.1: Motor Gear Meshing Frequency

Equipment designed for diagnosis of motor-operated valves (MOV) based on the motor current signature analysis (see Section 13.8) often has low sampling rates of 1,000 to 3,000 per second. These sampling rates are high enough to reveal spectral peaks around the frequencies of motor rotation speed, the drive sleeve rotation and the worm shaft rotation frequency. Suppose the sampling frequency is 1,000 per second. If the motor rotates at 2,400 RPM and the first motor gear has 88 teeth, what would be the apparent gear meshing frequency if the signal is not filtered off before digitization?

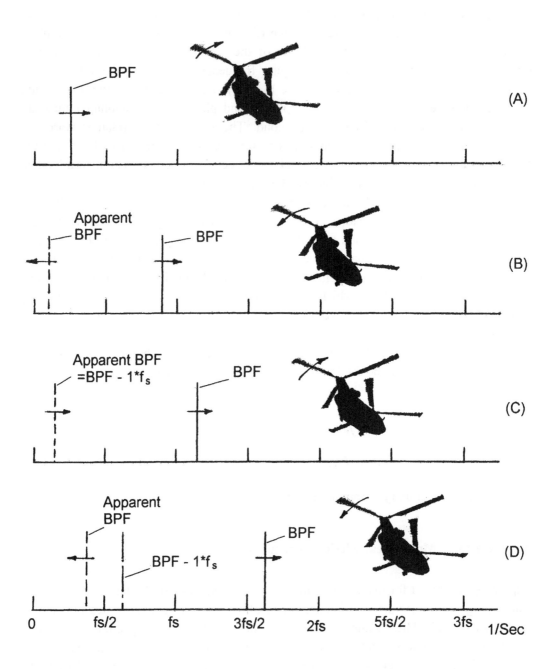

Figure 13.2 Apparent blade passing frequencies of a helicopter rotor in a motion picture due to aliasing. (A) BPF<f_N; (B) f_N<BPF<f_s; (C) f_s<BPF<$3f_s/2$; (D) $3f_s/2$<BPF<$2f_s$.

Solution

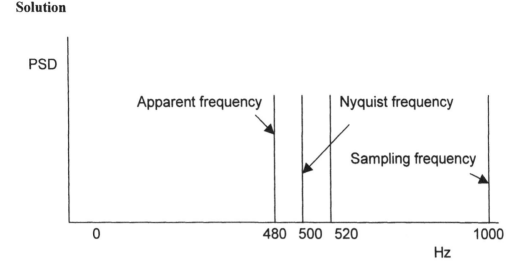

Figure 13.3 Aliasing of gear meshing frequency

The motor gear meshing frequency is 88*40 = 3520 Hz. This is much higher than the Nyquist frequency of 500 Hz. The real spectral peak corresponding to the motor gear meshing frequency will be folded back several times and appear as an alias below 500 Hz. To find out what is the apparent frequency, we must subtract integral multiples of 1000 from 3520, until the remainder is below the sampling frequency 1000 Hz. If, after the subtraction, the remainder is less than the Nyquist frequency (500 Hz in this case), this is what we shall see. If the remainder is higher than the Nyquist frequency, what we shall see is the mirror image about the Nyquist frequency. Since,

3520 - 3*1000 = 520 Hz

which is still higher than the Nyquist frequency (500 Hz). We shall see the mirror reflection of the signal about 500 Hz, or 480 Hz, as shown in Figure 13.3.

13.3   Spectral Analysis

The concept of time domain and frequency domain representation of vibration data is briefly discussed in Chapter 2, where it is shown that very often, a vibration signal in the time domain representation does not convey much information to the diagnostic engineers, who then resolve to analyze the data in the frequency domain. Figures 2.6(B), 2.7(B), 2.8(B), 2.10(B), 2.11(B), 2.12(B) and 2.13(B) in Chapter 2 are PSD plots for the vibration data, the time history plots of which are shown in the corresponding Part (A) of these figures. The decomposition of a total quantity into its frequency bands is called spectral analysis.

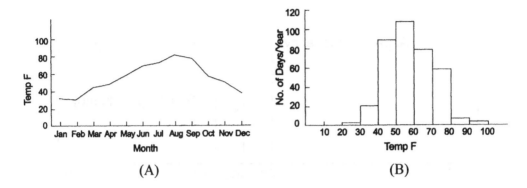

Figure 13.4 Two different ways of showing the temperature in a city:
(A) Time domain presentation and (B) Frequency domain presentation

We live in a world in which time is the prime parameter. We go to work on certain days of the week and at certain times of the day and carry out projects according to pre-set schedules. Our customers order products and demand delivery by certain days. Our customers pay after they receive the products and the invoices. Not surprisingly, each of us is very aware of the flow of sequence of events in our everyday lives and the most common form of presenting data, whether financial, biological or engineering, is to display the data as functions of time.   As an example, we can plot the average temperatures of a city as a function of the months of the year (Figure 13.4 (A)). However, time histories are not always the only, or even the most preferable, form of presenting data. Instead of plotting the average daily temperature of a city in a certain year, one can, for example, plot the number of days in which the average temperatures are between 0 to 20, 20 to 40, 40 to 60, 60 to 80, 80 to 100 degrees for the same year (Figure 13.4 (B)). The former is a time domain representation of the temperature, while the latter is a form of "spectral" representation of the temperature in that city. Instead of studying the temperature variation as a function of time, we are now studying the "frequency" distribution of the temperatures, which may make more sense in some applications, such as to the utility company trying to forecast the demand of electricity in that city. The word "spectral" originally meant "as a function of color," i.e., as a function of frequency since the colors of light are dependent upon the frequency of the light waves. However, its applications are far more reaching than its original terminology implies.   Spectral analysis is a powerful tool for analyzing data, including social, biological, medical and engineering data. To be specific and to simplify the writing, we shall assume, in the following discussions, that the data are vibration and acoustic data, even though the general methods are equally applicable to other kinds of data, such as fluctuations in electric motor currents, as we shall discuss in Section 13.8.

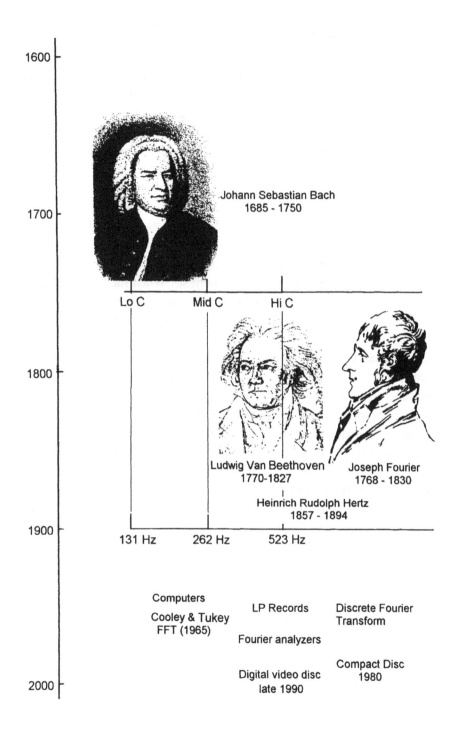

Figure 13.5 History's great composers studied the frequency composition of sound long before digital signal analysis and the Fast Fourier Transform were introduced

Although spectral analyses are commonly performed with main frame, workstation or microcomputers together with their software, spectral analyses had been carried out long before digital computers and Fourier analyzers became common analysis tools. Indeed, history's great composers, such as Bach and Beethoven were very familiar with the frequency content of musical sound, even though they did not have the benefit of the basic unit of frequency, the Hz, which did not came until almost one hundred years later, as shown in Figure 13.5. Most of us have performed spectral analysis of sunlight in the following high school science experiment: Place a prism under the sun and a sheet of white paper below the prism—a spectrum of color ranging from red to violet similar to what is seen in a rainbow will appear on the paper. What we did was to decompose the sun's white light into its constituents—the various colors ranging from infra-red (which we cannot see) to red, orange, yellow, green, blue, purple, violet to ultra-violet (which we cannot see). Like random vibration, the sun's light consists of a combination of infinitely many light and other electromagnetic waves of different frequencies, and hence different wavelengths. It is the different wavelengths that give rise to different colors, with red having the longest wavelength and violet the shortest. When the sunlight hits the prism, the angles of refraction depend on the wavelengths. As a result, the different colors are separated after passing through the prism. In short, the prism "decomposes" the sunlight into its constituents.

In digital signal analysis, spectral decomposition is usually carried out with a mathematical process called Fourier transform. However, this is not the only way of spectral analysis. The spectral analysis of the sunlight in the above high school science experiment, for example, is achieved without the use of the Fourier transform. Before the digital age, spectral analysis of vibration data was performed with the analog auto-correlators to calculate the auto-correlation function first. By a well-known theorem in mathematical physics, the power spectral density was obtained by the Fourier transform of the correlation function. Conversely, the auto-correlation function is obtained by Fourier transform of the power spectral density. In mathematical language, the time and frequency domains form a set of dual spaces. Transformation from one space to the other can be carried out by the Fourier transform.

Another method of analog spectral analysis is to use a bank of band-pass filters to separate the energy contained within each frequency band. The power spectral density (PSD) was obtained by dividing the measured total energy within each frequency band by the bandwidths (Figure 13.6). However, it was more customary in the analog days to plot the amplitude, represented by the voltage output, in each frequency band. Some experimentalists still prefer this form of presenting spectral analysis results even with today's digital spectrum analyzers (Figure 13.7).

It should be pointed out that while the frequency-domain representation of the data sometimes enables us to see things better, it does not contain any information that is not in the original time signal. Fourier transform is just a mathematical manipulation of the data. It does not create any new phenomena.

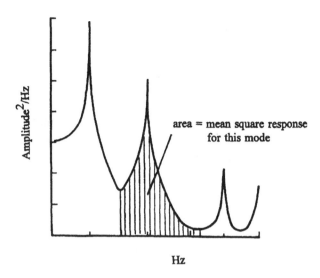

Figure 13.6   The area under the power spectral density (PSD) curve should be equal to the mean square response

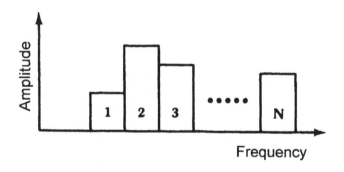

Figure 13.7 A histogram or frequency distribution plot

Unit of Power Spectral Density (PSD)

As discussed in the above section, PSD, or power spectral density, is a representation of the energy density distribution over the frequency band. Indeed, the term was actually derived from the prism experiment discussed above. The colors we see in a rainbow or after the sunlight passes through the prism depend on the frequencies of the electromagnetic waves sent out by the sun. The intensity of each color depends not on the total energy contained in the color band, but on the total energy contained in the color band, divided by the frequency bandwidth of the color. Because the color bandwidths are not the same, equal energy in two color bands does not necessarily mean their energy densities are the same. In other words, the intensity of each color band depends on the energy density per unit frequency bandwidth, and thus the term spectral (for color) density. Finally, since we are concerned with the flow of energy, rather than energy itself, the term power spectral density (PSD) was created.

In vibration and acoustics, the main objective of spectral analysis is to see the energy content in each frequency band. From the above discussion, it is obvious that the natural way of presenting PSD data is to tabulate or plot the power density as a function of frequency. Since energy is proportional to the square of amplitudes, the unit in the ordinate of a PSD plot should be the quantity squared per Hz, and the abscissa should be Hz. Examples of the correct units for the ordinate are:

in^2/Hz, m^2/Hz	displacement PSD
$(in/sec)^2/Hz$, $(m/sec)^2/Hz$	velocity PSD
$(in/sec^2)^2/Hz$, $(m/sec^2)^2/Hz$	acceleration PSD
lb^2/Hz, N^2/Hz	force PSD
psi^2/Hz, Pa^2/Hz	fluctuating pressure PSD

In this representation, the area under the PSD curve represents the mean square quantity (Figure 13.6). Examples are mean square displacement, mean square acceleration and mean square pressure fluctuation. Indeed, this is the standard method of obtaining the rms values in a frequency band—by numerical integration of the area under the PSD curve between the two frequency limits of the frequency band, and taking the square root of the result. The total rms value is obtained by integrating the area under the PSD curve from zero frequency to an upper frequency at which the PSD curve asymptotically approaches zero, and then taking the square root of the result. While the above example, or its variants, is the only correct way of presenting PSD data, there are several ways of presenting results of spectral analysis. Tabulating or plotting the total amplitude in different frequency bands as a function of the center frequencies of the band (Figure 13.7) is another way. As discussed in the previous sub-section, this method of presenting spectral analysis results has roots that can be traced back to the days when a bank of analog filters was used to separate the spectral components of vibration data. This graphic or tabular display is still preferred by some test engineers, and the resulting function is a spectrum, frequency distribution or histogram. It should not be referred to as a power spectral density.

13.4   The Fourier Transform

The Fourier transform was originally formulated by Joseph Fourier, in the days of Napoleon, purely as a mathematical curiosity. At that time quantitative description of vibration or acoustics was not yet established. Even the fundamental unit of vibration, the Hz, was not introduced until 100 years later (see Figure 13.5). Vibration analysis was not in Fourier's mind when he published his famous transform. It was not until many years later that mathematical physicists discovered that the Fourier transform is the vehicle to convert time domain events into frequency distributions, and hence became a powerful tool in the quantitative study of vibration and sound.

As originally formulated, the Fourier transform is an integral. With the advent of high-speed digital computers, this integration is carried out numerically, that is, by summation. The resulting mathematical process is known as the discrete Fourier transform.

Discrete Fourier Transform

To carry out the Fourier transform digitally by numerical integration, the original continuous time domain data $x(t)$ must first be discretized as described in Section 13.2:

$$x(t) \xrightarrow{\text{digitization}} x_i(t_i), \quad i = 0,1,2,3,4,......n\text{-}1$$

Analogous to the Fourier integral, the discrete Fourier transform is defined as

$$X(k\Delta f) = \Delta t \sum_{i=0}^{n-1} x(i\Delta t)e^{-j(2\pi/n)ik}, \quad j = \sqrt{-1} \tag{13.3}$$

Here the time step, Δt, is the inverse of the sampling rate, f_s,

$$\Delta t = 1/f_s$$

and determines the resolution in the time domain. Note: We use $x(t)$ to represent a variable in the time domain and $X(f)$ its corresponding Fourier transform. In general, $X(f)$ is a complex number. The power spectral density (PSD) is not the Fourier transform itself, but is related to the Fourier transform of the time domain data by,

$$PSD(k\Delta f) = \frac{2|X(k\Delta f)|^2}{n\Delta t}, \quad k = 0,1,2,....n-1 \tag{13.4}$$

The inverse discrete Fourier transform is, consistent with the above definition,

$$x(i\Delta t) = \Delta f \sum_{k=0}^{n-1} X(k\Delta f)e^{+j(2\pi/n)ik}, \quad i = 0,1,2,....n-1 \tag{13.5}$$

Here the frequency resolution, Δf, is the inverse of the total duration of time record contained in n data points. That is,

$$\Delta f = 1/n\Delta t = 1/T \tag{13.6}$$

where T is the total time record in one block of n data points. There is confusions in the literature due to variations in the definitions of the discrete and inverse discrete Fourier transforms, and of the power spectral density function. The above definitions can be regarded as "engineers'" definitions of the Fourier transform and PSD. The resultant frequency and time series X_i and x_i have the proper dimensions of inverse time (i.e., frequency) and time, respectively, and the resultant PSD has the proper dimension of amplitude squared per Hz, and is defined only for positive values of frequency. The mathematicians, on the other hand, are interested only in numbers. They often define the discrete Fourier transform $X(f)$ with 1 instead of Δt in front of the sum. In this case the definition of the inverse discrete Fourier transform must have $1/n$ instead of Δf in front of the sum, and the constant in the denominator of the definition of the PSD will be n instead of $n\Delta t$. To make the situation even more confusing, mathematicians usually define the PSD as amplitude squared per radian, with radian extending from $-\infty$ to $+\infty$. The engineers' definition is called single-sided PSD and is denoted by the notation G throughout this book; the mathematicians' definition is called double-sided PSD and is denoted by the notation S. In general,

$$G(f) = 2S(f), \qquad f \geq 0$$
$$ = 0, \qquad f < 0 \tag{13.7}$$

Since $f = \omega/2\pi$, as a function of Hz, the PSD is equal to 2π times the PSD as a function of ω:

$$G(f) = 2\pi G(\omega) \tag{13.8}$$

Very often digital signal analyses are carried out with personal computers, workstations, or Fourier analyzers (which are microcomputers). To conserve memory space, the time series is read into the RAM in blocks of finite sizes, such as blocks of $n = 1,024, 2,048$ etc., data points. This block of n data points are then Fourier-transformed into n complex numbers in the frequency domain.

Fourier Transformed

Time Series -----------------------------------> Frequency Series
x_i, $i=0,1,...n$ $X_i + jY_i$, $i = -n/2 ... 0 ... +n/2$

At first it looks like we start with n real data points in the time domain and end up with $n+1$ complex data points in the frequency domain. Careful examination of the equation in the discrete Fourier transform, however, reveals that the real part of the Fourier transform, X, is always symmetrical about $i = 0$, while the imaginary part Y is always antisymmetric, so that the modulus $|X|$ of the Fourier transform is always symmetrical about the point $i = 0$. Furthermore, the imaginary part Y at $i=0$ and $i=+n/2$ and $i=-n/2$ are always zero, so that $X(i=-n/2) = X(i=n/2)$. In short, starting with n real data points in the time domain, we end up with $n/2+1$ independent real data points and $n/2-1$ independent and non-zero imaginary data points in the frequency domain, for a total of n data points as in the time domain. Finally,

$$PSD(f_k) = G(f_k) = \frac{2(X_k^2 + Y_k^2)}{n\Delta t}, \quad k \geq 0 \tag{13.9}$$

The total number of data points representing the PSD is $n/2 + 1$. Since

$$\Delta f = 1/(n\Delta t) = f_s/n \tag{13.10}$$

it follows that the maximum frequency attained after Fourier transform is,

$$f_{max} = (n/2)\Delta f = f_s/2 \tag{13.11}$$

That is, the maximum frequency attained after Fourier transform of the discrete time series is half of the sampling frequency. This is the well-known Nyquist theorem mentioned in Section 13.2. Table 13.1 summarizes the relationship between the sampling rate, the maximum frequency, and the resolutions in the time and frequency domains.

Table 13.1 Relationships between Sampling Rate and Analysis Results

Quantity	Relationship
Sampling Interval	ΔT
Sampling Rate	$f_s = 1/\Delta T$
Maximum (Nyquist) Frequency (Hz)	$1/(2\Delta T)$
Sample Block Size	n
Spectrum Lines	$n/2 + 1$ (including $f=0$)
Frequency Resolution (Hz)	$\Delta f = 1/(n\Delta T) = 1/T$

After one block of time data is transformed into the frequency domain, another block of n data points is read into the memory and the above process is repeated. The resulting blocks of PSDs is then averaged to give better statistical accuracy (see Section. 13.7). This process is repeated until either the entire time record is exhausted or when sufficient statistical accuracy is attained. For a long time history involving tens of thousands of time steps, these computational procedures can be extremely time-consuming. This is why as late as the mid-sixties, even after the digital computer had been introduced as a common research tool, digital signal analysis seldom involved more than a few hundred data points and were used more in bio-medical and social research than in engineering, which usually involved much more data points.

The Fast Fourier Transform

In 1965, Cooley and Tukey published a special algorithm to calculate the Fourier transform of a discrete time series with digital computers. For n data points, where n must be equal to 2^k with k an integer, this algorithm is a ratio of $n\log_2 n/n^2$ times faster than the classical method of computing the discrete Fourier transform and hence earned the name Fast Fourier Transform (FFT). Using the equation $\log_a X = (\log_a b)*(\log_b X)$, we get as an example, that for $n=1,024$, the FFT is 102 times faster than the classical discrete Fourier transform. For $n=1,000,000$, the ratio is 30 seconds versus two weeks of computation time.

The FFT algorithm revolutionized the technique of signal analysis, including, but not limited to, vibration and acoustic data analysis, and gave birth to a whole new product line of compact Fourier analyzers, which quickly knocked the old analog auto-correlator completely off the market. Today power spectral densities are computed directly from the FFT of a time series, as is discussed in the previous sub-sections, and the auto-correlation function is computed by FFT of the PSD function, exactly the reverse of what engineers did a generation ago. Without the FFT algorithm, digital signal processing probably would not be as widely accepted as it is today.

Resolution in Time and Frequency Domains

Equipment catalogs are full of spectrum (or Fourier) analyzers with various degrees of capabilities and sophistication. One of the features to judge an analyzer is its maximum sampling rate. Usually the higher the maximum sampling rate, the more expensive the analyzer. An engineer who has made the justification to acquire the state-of-the-art analyzer with a high sampling rate will, therefore, have the tendency to select the highest sampling rate in all of his tests. Very often, this compromises the resolution in the frequency domain. The resolutions in the time domain, Δt, and the resolution in the frequency domain, Δf, are related to each other by the "Heisenberg uncertainty principle," which is one of the fundamental principles in physics:

$$\Delta f * \Delta t = 1/n \qquad (13.12)$$

For the same n, a high resolution in the time domain, i.e., a high sampling rate in digitizing the original continuous time data, will result in a lack of resolution in the frequency domain. Sometimes this will lead to inaccurate results from spectral analyses. Figure 13.8, for example, shows a series of spectral peaks from testing a heat exchanger tube, the dominant frequencies of which are less than 200 Hz. The apparent triangular shape of the resonance peak is an indication of insufficient resolution in the frequency domain (Δf too large), usually caused by selecting too high a sampling rate. Not only is there an insufficient number of points in the frequency space to show the true resonance shape of the response, but it may also miss the true peak in the PSD plot, as shown in Figure 13.9. Since the area under the resonance curve represents the mean square

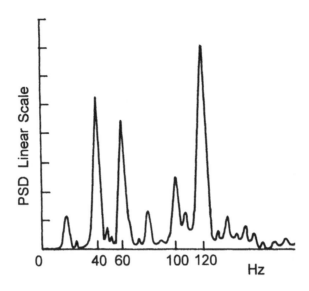

Figure 13.8 The triangular spectral peaks in a PSD plot usually indicate that
there is insufficient resolution in the frequency domain

amplitude, if the rms response is computed by integrating the area under the PSD curve,
lack of frequency resolution will lead to errors in the estimate of the rms responses.

One way to increase the frequency resolution is to increase the number of data points,
n, in each block of data to be transformed. This will increase the computational time as
well as the size of the data files. A better way to obtain finer resolution in frequency is to
decrease the sampling rate, as shown in the following example.

Example 13.3

The maximum total sampling rate of a four-channel spectrum analyzer is 96,000 per
second. This analyzer will be used to acquire piping vibration data in which the highest
frequency of interest is expected to be less than 50 Hz, with a damping ratio of about 5%.
If a data block size n=1,024 is chosen, would the PSD peak be properly resolved if the
maximum sampling rate is chosen? What sampling rate would you choose?

Solution

If we choose the maximum sampling frequency f_s=96000/4= 24000/s per channel and
block size n = 1024, the resolutions in time and frequency are, from Table 13.1,

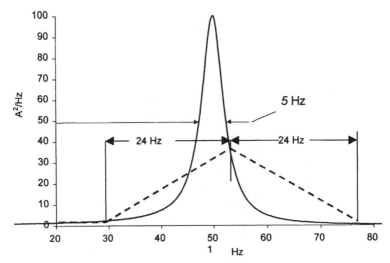

Figure 13.9 True (solid line) and apparent resonance
peak (dotted line) at the maximum sampling rate

$$\Delta t = 1/f_s = 1/24000$$

$$\Delta f = 1/n\Delta t = 24000/1024 \cong 24 \quad \text{Hz}$$

From Equation (3.3), the half bandwidth of the resonance peak is approximately,

$$\Delta f = 2*50*0.05 = 5 \quad \text{Hz}$$

As shown in Figure 13.9, the resolution is too coarse to reveal the true shape of the resonance curve. What we would get will be a triangular spectral peak. If we are interested in measuring the vibration of a piping system with frequencies no higher than 50 Hz, we can choose a sampling rate of 200/s. The theoretical maximum frequency we can attain after FFT will be 100 Hz, which is sufficient to envelop the 50 Hz of interest and with sufficient margin. At

$$f_s = 200/s$$

while keeping n=1040 as before, the time and frequency resolutions will be,

$$\Delta t = 1/200 = 5.0 \text{ ms}$$

$$\Delta f = 1/n\Delta t = 1/(1024 \times 0.005) = 0.20 \text{ Hz}$$

There will be 40/0.20=200 points to define the resonance peak between 30 and 70 Hz.

13.5   Windows

One requirement for a perfect Fourier transform of a time record is that the function must be periodic and an integer number of waves included in the transform. Otherwise, an infinitely long time record will be required. In practice, the time series must be truncated and the data points read in and transformed in finite block sizes of n data points (see Section 13.4). Unless the truncation occurs at the exact point in time so that an exact number of complete waves is included in the data, the truncated time history will no longer be a sinusoidal wave. Since a non-sinusoidal wave is a linear combination of many sine waves of different frequencies (harmonics), this results in spurious signals in the form of side bands in the Fourier transform of the time data—a phenomenon commonly called leakage, which causes the smearing of energy contained in a spectral peak throughout the entire spectrum (Figure 13.10). To minimize this problem, the time series is truncated, not abruptly by multiplication with a "box car" function (Figure 13.11), but by a function with more gentle tapers. These functions are called "windows" as they effectively permit one to look at the time series through a window opening. Although windows of many different types that serve different purposes exist (Bendat and Piersol, 1971), the most commonly used, other than the box car function, is the Hanning window. The two functions are shown in Figure 13.11. For most applications, the differences between the Hanning window and other non-box car window functions are not significant.

The Hanning window reduces, but does not eliminate the leakage problem (Figure 13.12); it causes distortions of the PSD curve by widening the spectral peaks. Since the area under a spectral peak represents the modal contribution to the vibration, the area under the spectral peak must not be changed by the window function. This means that if the spectral peak is widened, its amplitude must be decreased to preserve the area under the curve. From the above discussions, we can see that: (1) If the modal damping ratio is to be determined by the half-power-point method, as in Example 13.3, corrections must be made to account for the distortion caused by windowing. More direct estimates of the damping ratio can be obtained from time domain analysis using, e.g., the logarithmic decrement method, or in the frequency domain, using the sine sweep method; (2) The amplitudes in a PSD function have no absolute physical meaning. If the amplitudes of two PSD plots are to be compared, we must make sure the same window function is used to obtain the spectra in both cases.

While the Hanning window reduces the leakage problem in the frequency domain in the case of stationary data, it actually smears out the energy contained in a transient event, such as an impact, and thus causes a loss of resolution in the waveform in the time domain. Therefore, window functions other than the box car function should not be used for impact studies in the time domain. Further cautions on the analysis of transient data will be discussed in Section 13.7 on statistical accuracy.

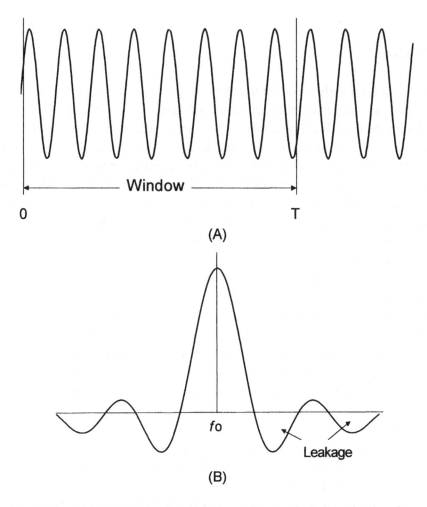

(A)

(B)

Figure 13.10 (A) Truncation of a continuous time domain signal;
(B) Resulting leakage in the frequency domain

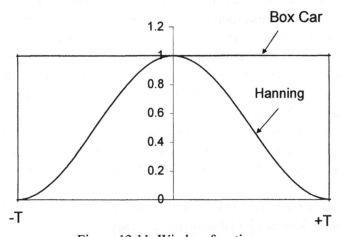

Figure 13.11 Window functions

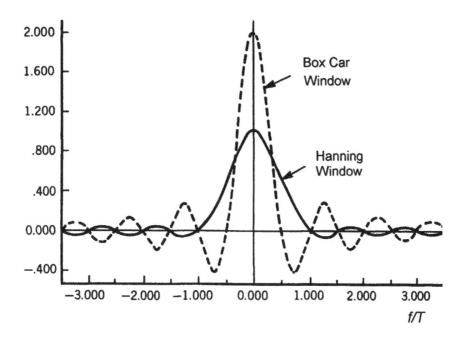

Figure 13.12 Fourier transform of Hanning and box car windows

13.6   Digital Filtering

It frequently occurs that while analyzing vibration data, some noise masks the useful data, making interpretation of the results difficult. Filtering can eliminate these unwanted data. Like analog filters, digital filters can be of the low-pass, high-pass, band-pass, or band-reject type. But unlike analog filters that are bulky pieces of hardware, digital filters can be implemented in the form of computer software. The filtering action is based on mathematical manipulation of the digital data, and can be done after the test is completed. Numerical algorithms for high-pass, low-pass, band-pass, and band-reject filters have been developed. Figure 13.13(A) shows an example of accelerometer data contaminated by electrical hum because of a faulty ground. Digital high-pass filtering of the data at 150 Hz, which eliminated both the 60-Hz hum and its 120-Hz second harmonic, resulted in the data given in Figure 13.13(B). In Figure 13.14(A), data obtained from an ultrasonic transducer mounted on the bottom of check valve with the objective of detecting the flutter motion of the disk assembly (see Chapter 2, Section 2.4 and also Section 13.8) was completely masked by AC hum. Since disk flutter typically has frequencies below 5 Hz, this data was low-passed filtered at 10 Hz to eliminate the AC hum. The resultant signal is shown in Figure 13.14(B) and clearly shows the motion of the disk. The following example shows a more unusual application of the digital filter in plant component diagnosis.

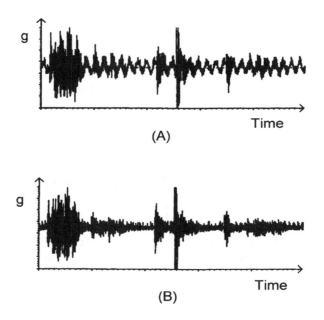

Figure 13.13 (A) Accelerometer signal contaminated by AC hum;
(B) Same signal after digital high-pass filtering at 150 Hz

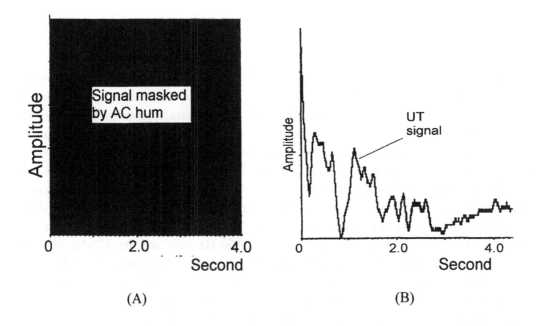

Figure 13.14 (A) Ultrasonic dynamic data completely masked by AC hum;
(B) Digital low-pass filtering at 10 Hz reveals the UT data, showing flutter of the disk

Example 13.4

Figure 13.15(A) shows the time data from an accelerometer mounted on a check valve. The frequent spikes show that there was a continuous tapping or popping noise inside the check valve, which may be backstop tapping of the disk assembly, or cavitation inside the check valve, or a combination of both. Eliminate or confirm the presence of one or the other.

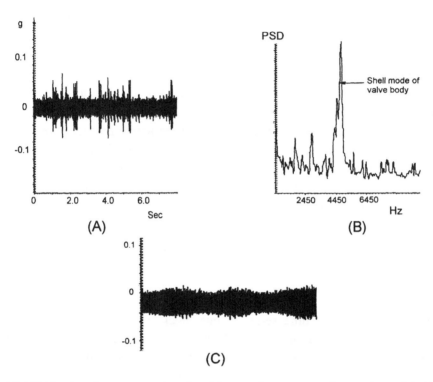

(A)

(B)

(C)

Figure 13.15(A) Accelerometer signal with continuous tapping noise; (B) PSD of vibration data shows prominent body mode at 4500 HZ; (C) Band-reject filtering the raw data between 4,000 and 5,000 Hz eliminated the tapping noise

Solution

The time signal was first Fast Fourier-transformed to obtain a PSD function, Figure 13.15(B). The prominent spectral peak at about 4,500 Hz is most likely a valve body mode excited by the repeated tapping of the disk assembly against the backstop of the valve. The original time series is then band-reject-filtered from 4,000 to 5,000 Hz. The resulting time trace is shown in Figure 13.15(C). The absence of any spikes in the filtered data shows that the original spikes in (A) were all caused by backstop tapping. Cavitation would have caused a continuous high-frequency spectrum, which will not be removed by the band-reject filter. Thus, it can be concluded that there was no cavitation inside this valve.

13.7   Statistical Accuracy

It is well known that when we make an estimate based on a sample of a much larger ensemble, then the larger the sample, the more accurate the estimate will be. In a time series, each block of n data points is one sample of a larger ensemble. Thus it is intuitively apparent that when deriving the PSD from the time history, the more data points we include in the Fourier transform, the more accurate the resulting PSD will be. However, there is a catch in this intuition, and that is related to the resolution of the PSD.

When spectral analyses are carried out with microcomputers and the FFT algorithm, blocks of data containing n data points are read into the computer one at a time. This block of data is then Fourier-transformed and the next block of data is read in. The two PSD samples are then averaged and the resulting averaged PSD is stored in the memory. A third block of n data points is then read in and the above process is repeated until a total of N data blocks are transformed. Our intuition is correct in this case—as more data blocks are transformed and the resultant PSDs averaged, the statistical accuracy for the computed PSD increases. A simple way to quantify the statistical scatter of data points is through the use of a single parameter called the normalized error ε. In the above computation of the PSD, the normalized error is proportional to the inverse square root of the total number of data blocks transformed and averaged. That is,

$$\varepsilon^2 = 1/N \tag{13.13}$$

If we transform the entire time series in one block, we would get a high-frequency resolution according to Equation (13.12). However, the normalized error is unity, which means that the probable error in our PSD estimate is as large as the standard deviation, which is obviously not very good. On the other hand, if we divide the time series into 100 blocks, the normalized error will be 0.1.

In the more rigorous language of statisticians, the normalized error, ε is related to the confidence level and confidence limits of our PSD estimate. Without going into the details of statistics, we give in Figure 13.16, the effect of the number of averages N on the scatter of data, while Figure 13.17 plots the confidence levels and confidence limits in the PSD estimates as a function of number of averages. From these figures, it is apparent that while less than four averages are obviously not enough, one starts to get into the domain of diminishing returns when N exceeds 100.

The above discussions are valid as long as the data is stationary or quasi-stationary. Examples of stationary random events are the pressure fluctuation in a turbulent boundary layer, as shown in Figure 13.16(A) and the structural response to it, as shown in Figure 13.16(B). Block averaging in both cases reduces the statistical scatter. Figure 13.15(A), which we refer to earlier, shows the time history of continuous tapping of the disk assembly of a check valve against the backstop. This is an example of a quasi-

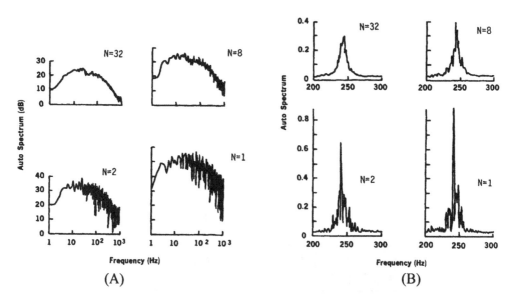

Figure 13.16 (A) Effect of number of averages, N, on the statistical scatter in the PSD of the fluctuating pressure beneath a turbulent boundary layer; (B) Effect of number of averages, N, on the statistical scatter in the PSD of the structural response.

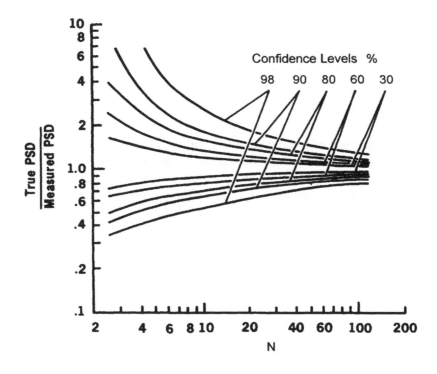

Figure 13.17 Confidence level and confidence limits as functions of number of averages, N.

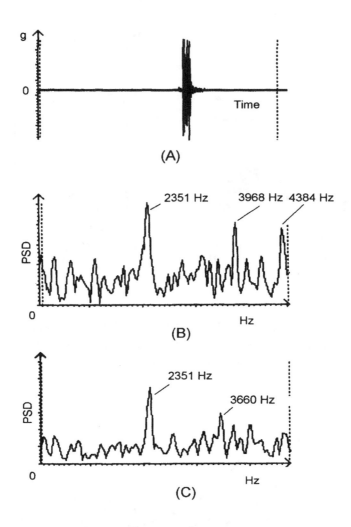

Figure 13.18 Impact test of a component in a piping system: (A) Time history of single impact; (B) FFT of one data block of time record covering just the transient brings out the lowest three modal frequencies of the component; (C) Averaging over 50 blocks of data extending beyond the transient tends to dilute the second and third modal frequencies, bringing out the background piping frequency of 3,660 Hz instead.

stationary event. Block averaging in this case also gives a better-defined structural response spectrum, as shown in Figure 13.15(B). Figure 13.18(A) shows the time history of an impact test of a component mounted in a piping system. This is an example of a transient event in which non-trivial data existed only during a fraction of a second. If only this short time interval is transformed, the resulting PSD plot will bring out the component's modal frequencies, as shown in Figure 13.18(B). On the other hand if block averaging over a long time record beyond the duration of the impact is used, the information contained in the impact will be spread out over the entire time record and the background structural vibration will mask the modal frequencies of the component, as shown in Figure 13.18(C). Thus block averaging beyond the duration of the transient is not recommended.

Example 13.5   Sampling Rate and Length of Data Record Requirement to Resolve a Spectral Peak

The fundamental shell mode of a large vessel is expected to be around 10 Hz. It is necessary to accurately measure the modal-damping ratio of this shell, using the half power point method with a steady white noise excitation. What sampling frequency should we use and how long should we record the steady vibration data?

Solution

Modal damping ratios for metallic shells are usually small, less than 1%. Assume the expected damping ratio is 0.005. To achieve reasonable statistical accuracy, we need 100 averages for a normalized error of 0.1. From Equation (3.13), the half power bandwidth of the spectral peak is expected to be about,

$$\Delta f = 2 f_0 \zeta = 2*0.005*10 = 0.1 \text{ Hz}$$

To accurately represent this resonance peak, we need about 10 points. A frequency resolution of

$$\Delta f = 0.02 \text{ Hz}$$

will give about 10 points to represent this peak. To represent a frequency of 10 Hz, we need a minimum sampling rate of 20 per second according to the Nyquist theorem. However, in practice some margin is necessary. We shall use a sampling rate of

$$f_s = 30 \text{ /s}$$

To take advantage of the Fast Fourier Transform (FFT) algorithm, the number of data points to be transformed each time must be $n = 2^k$, where k is an integer. Let us start with a block size

$$n = 2^{10} = 1024 \text{ points}$$

The length of time record per block is,

$$\delta T = n. \Delta t = 1024*1 / 30 = 34.13 \text{ sec}$$

After fast Fourier transform, we get 512 points on the $+f$ side and 512 points on the $-f$ side of the PSD curve. The PSD is symmetrical about $f=0$. Of these only the $+f$ side is useful to us. The maximum frequency we get is the Nyquist frequency, which is equal to one half the sampling frequency (15 Hz in the present case), and we have 512+1=513 points to represent it. The frequency resolution is

$\Delta f = 15/512 = 0.029$ Hz

Alternatively, from Table 13.1,

$\Delta f = 1/n\Delta t = 1/34.13 = 0.029$ Hz

However, we have already concluded earlier that we need a resolution of 0.02 Hz. One way to decrease Δf is, from equation (13.12), to increase the number of data point, n, per block. The next step up is to choose

$n = 2^{11} = 2048$

Keeping the same sampling rate $\Delta t = 1/30$ sec, we have now after FFT,

$\Delta f = 1/n.\Delta t = 1/(2048 \times 1/30) = 0.0146$ Hz

This is fine enough to define the resonance peak. But the time record per block is now

$\delta t = 2048*1/30$ seconds

and we need 100 blocks of this to achieve the required statistical accuracy. The total length of time record we need is

$T = Nn\Delta t = 100*2048/30 = 6827$ seconds = 1.9 hours!

If we use 16-bit words, which will give us a dynamic range of 96 dB, the total number of bits per test record is

$204800*16 = 3.28$M bit = 1.64M byte

This example shows that low-frequency tests involve very long time records. One way to economize is to sacrifice statistical accuracy. One hundred averages correspond to a normalized error of 0.1. In practice, more than 100 averages would not buy us too much additional accuracy. Since normalized error $\varepsilon = 1/\sqrt{N}$, we can drop the number of averages to 64 without sacrificing much statistical accuracy while cutting the test time by more than one-third.

13.8   Beyond the Accelerometer and Fast Fourier Transform

As mentioned in the introductory section, the 10,000 times increase in the dynamic range inherent in digital data acquisition systems enable new sensors that yield signals below the noise floor of analog equipment to be developed to yield useful signals. In the last ten years, some of these new sensors were developed for nuclear plant component monitoring. However, there is no reason why these sensors cannot be used in other power and process plants. The following are only a few of these examples.

<u>The Ultrasonic Transducer as a Dynamic Sensor</u>

The ultrasonic instrument was originally developed to locate flaws in solids. In the early 1990s its application was extended into vibration measurement (Au-Yang, 1992). This measurement is accomplished in several steps. The instrument first generates a series of up to 1000-per-second high-voltage pulses to be applied to the UT transducer, which converts the pulses into ultrasonic waves aimed at the target. At the same instant, the timing circuitry begins a counting sequence. When the sound waves impinge upon and reflect from the target, they are picked up by the transducer, which is now in a passive "listening" mode. The timing circuitry then determines the elapsed time. The UT signals are then recorded directly into the computer, with the peak voltage proportional to the lapse time between the emission and reception of the ultrasonic pulses. Knowing the speed of sound in water, the target positions can be computed as a function of time in the form of a time series, from which the vibration frequency and amplitude of the target can be derived.

With the proper data acquisition and analysis software, the commercial ultrasonic flaw detector has been used to accurately and quantitatively determine the disk opening angles and flutter frequencies and amplitudes of check valves without disassembling the valves or interfering with their normal operations. It has also been used to non-intrusively measure the vibration of nuclear fuel bundles when other types of sensors cannot be used. Two examples of its applications are given in Figures 2.4 and 2.5 in Chapter 2. However, ultrasonics works only if the component to be tested is carrying an ultrasonic conducting medium such as water. Furthermore, coarse-grained stainless steel tends to scatter ultrasonic waves so that well-defined echoes cannot be received through component casings built with this material.

Although in theory, the resolution of the UT sensor is limited by the wavelength of the ultrasound (0.1 to 0.2 mm typical), this is true only for unsteady flutter. For steady-state vibration, amplitudes as small as 0.001 mm have been measured and correlated with the accelerometer by phase averaging techniques (Au-Yang, 1992).

The Eddy Current Sensor

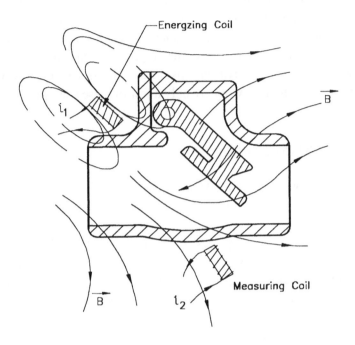

Figure 13.19 One example of using the eddy current sensor
to detect the motion of disk inside a check valve

The eddy current sensor was developed to supplement the UT vibration sensor for components with stainless steel casings and with water, steam or gases inside. In this application, one or two AC current-carrying elements are mounted externally on the valve body (Figure 13.19). When the disk starts to move, eddy currents induced in the elements perturb the inductance and thereby the total impedance in the elements. A current then passes through the circuit and the voltage generated is related to the position of the disk assembly. Because eddy current is highly non-linear, only qualitative information on the disk-opening angle can be derived from this voltage. Also because magnetic fluxes cannot penetrate strongly magnetic materials like carbon steel, the eddy current sensor works only on components with stainless steel housings. On the other hand, it can detect motion of components built even with non-ferrous metal, as shown in Figure 13.20. Figure 13.21 presents an example of field data from the application of the eddy current sensor to monitor the motion of the disk assembly inside a check valve.

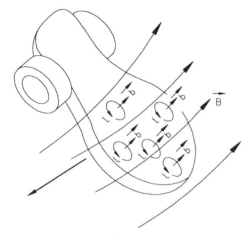

Figure 13.20 How the eddy current sensor works with moving parts built with non-ferrous metals: As a metallic object moves in a magnetic filed $\vec{B}$, eddy current loops i are generated in the surface. These eddy current loops in turn generate magnetic fields $\vec{b}$, which perturb the original magnetic field. Since this is a second order effect, the perturbation is much smaller than that caused by a moving part built of ferrous material.

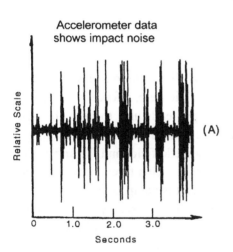

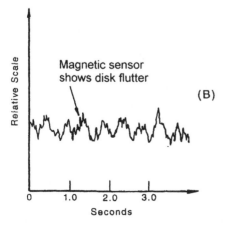

Figure 13.21 (A) Accelerometer signal showing frequent tapping noise in a check valve; (B) Eddy current sensor signal confirms disk flutter. Approximate line up of the wave crests indicates that the disk was hitting the backstop.

The Hall Effect Sensor

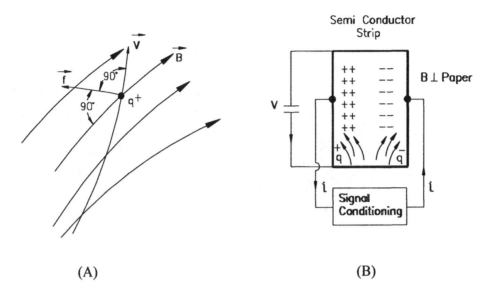

(A)   (B)

Figure 13.22 (A) The Lorentz force; (B) the Hall effect sensor

The Hall effect sensor was developed to monitor the disk position in valves carrying water, steam or gases. When a current flows in a magnetic field, a force is induced in the electrons. This force, called the Lorentz force, is perpendicular to both the velocity vector of the electron and the magnetic flux line (Figure 13.22(A)), and its magnitude depends on the applied current, the strength of the magnetic field, and the angle between the directions of current flow and the magnetic flux lines. This force induces an electric potential and thus a current flow when the external circuit is completed. In application to check valve monitoring, two powerful permanent magnets are externally mounted on the valve body, thereby inducing a magnetic field inside the valve. A Hall effect sensor is placed on the outside of the valve body, usually on top of the bonnet. When the circuit is completed, an induced current flows through it. By measuring the current, the position of the valve disk can be monitored. Again, since magnetic fields are highly non-linear, only qualitative information on the disk-opening position can be obtained.

The Hall effect sensor has the advantage that it works on components with both carbon and stainless steel housings and with any kind of fluid inside. Aside from being only a qualitative position sensor, its other major disadvantage is that its sensitivity decreases rapidly with increase of temperature and cannot operate at temperatures higher than 90 deg. C. To overcome this disadvantage and to increase the sensitivity of the Hall effect sensor, a combination cooler/concentrator/insulator unit is usually used to focus the magnetic flux lines onto the tiny Hall effect sensor, which for high temperature applications is usually cooled by some kind of forced air device, such as a vortex tube.

Figure 13.23 shows an example of Hall effect sensor data from the test of a large carbon steel check valve.

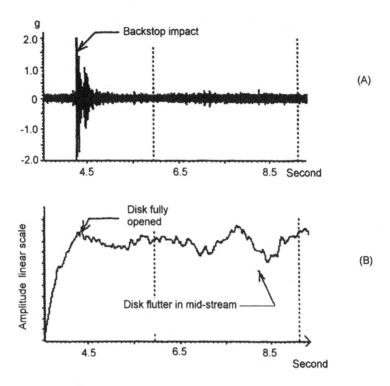

Figure 13.23 (A) Accelerometer data from a check valve showing disk fully opened to hit the backstop; (B) DC magnetic sensor trace also shows the disk fully opened to the backstop. However, the disk was not pressed firmly against the backstop; it was fluttering mid-stream.

Motor Current Signature Analysis

Kryter and Haynes (1989) found that as the mechanical load on an electric motor varies, so does the current through the motor power cables. This motor current variation can be large-step changes such as the in-rush current, or those induced by the limit and torque switches; they can also be minute variations caused by small changes in the mechanical loads. It is the latter, which can be regarded as "noises" riding on top of the motor current, that is particularly rich in information on the conditions of the motor, the gear train, as well as other equipment, such as valve actuators driven by the motor. This motor current "noise," or signature as it is more technically called, contains virtually all the information normally detected by mechanical sensors, such as the motor rotational speed; worm gear teeth meshing frequency; "hammer blow"; drive sleeve rotation speed; valve stem motion; valve disk seating and un-seating. In addition, it also contains information that is electrical in nature and thus cannot be detected by mechanical sensors, such as slip frequency, which is a function of the difference between the actual motor speed and the synchronous speed; no load, running and peak currents during disk seating and unseating.

Motor current signature analysis offers important advantages over mechanical signature analysis in the diagnosis of motor-driven equipment. First, the motor current can be picked up by clamped-on inductive current probes without cutting or splicing into the motor power cable. Second, since the same current flows through any section of the motor power cable, the current probe can be clamped onto the motor power cable at any convenient point. The most convenient location is usually the motor control center, which can be several hundred feet away from the motors. Not only can the motor-driven equipment be monitored non-intrusively, but they can also be monitored remotely at a convenient central location.

By acquiring motor current data regularly at the motor control center, the conditions of individual motor-driven equipment can be trended so that degradation of the motor itself, its drive train, or the equipment downstream of the motor and its drive train can be detected and remedial action taken before it becomes serious enough to jeopardize plant safety and operations. Figures 13.24 to 13.26 show examples of motor current signature and their interpretation to determine the equipment conditions.

Beyond the Fast Fourier Transform

Even more powerful analytical tools are currently being developed to replace or supplement the Fast Fourier Transform for analyzing digital data. It is widely believed that wavelet analysis may one day replace the FFT. With faster computers and larger and less expensive mass storage media, artificial intelligence and neural network, once limited only to military applications, are becoming more common tools in signal analysis and condition monitoring of plant components.

References

Au-Yang, M. K., 1992, "Application of Ultrasonics to Non-Intrusive Vibration Measurement," Proceedings, 1992 International Symposium on Flow-Induced Vibration and Noise, Vol. 4, edited by M. P. Paidoussis, ASME Press, New York, pp. 45-57.

Bendat, J. S. and Piersol, A. G., 1971, Random Data: Analysis and Measurement Procedures, Wiley-Interscience, New York.

Cooley, J. W. and Tukey, J. W., 1965, "An Algorithm for the Machine Calculation of Complex Fourier Series," Mathematical Computation, Vol. 19, pp. 297.

Kryter, R. C. and Haynes, H. D., 1989, "Condition Monitoring of Machinery Using Motor Current Signature Analysis", Sound and Vibration, September 1989, pp. 14-21.

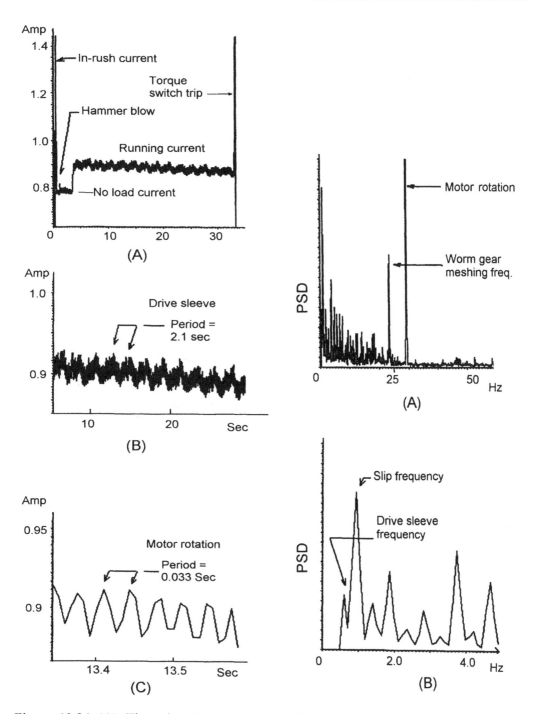

Figure 13.24 (A) Time domain motor current signature; (B) Zoom-in shows drive sleeve frequency; (C) Extreme zoom-in shows motor rotation and gear meshing frequency. Worn gear, if any, will show up in the waveforms.

Figure 13.25 Frequency domain analysis of motor current signature shown in Figure 13.24. (Note triangular spectral peak in (B) because of lack of resolution at this very low frequency)

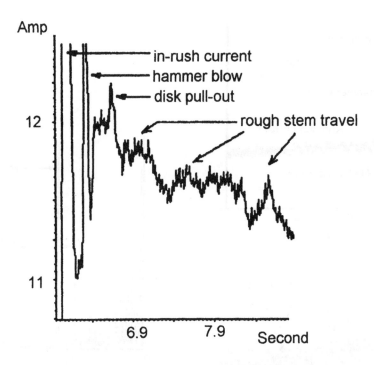

Figure 13.26 Motor current signature showing very rough
valve stem travel due to a bent stem.

AUTHOR INDEX

SUBJECT INDEX

About the Author

M.K. Au-Yang received his B.Sc. from the University of Hong Kong with the highest prize award in the School of Engineering and Architecture, the MS and the Ph.D. from the University of California at Berkeley. Before his retirement in 2001, he worked for 27 years in the nuclear industry at Framatome ANP and the Commercial Nuclear Power Division of the Babcock & Wilcox Company. Before that, he worked for six years at the NASA Ames Research Center. Dr. Au-Yang has 33 years of experience in vibration measurement, analysis, plant component diagnosis and troubleshooting. He is a registered professional engineer in the Commonwealth of Virginia, a Fellow of ASME, a Fellow of the Institution of Diagnostic Engineers of Britain and a member of the Executive Committee of the ASME Pressure Vessel and Piping Division. He was a former associate editor of the Journal of Pressure Vessel Technology and a former member of ASME's Committee on Operations and Maintenance of Nuclear Plants. He was granted two U.S. patents, has published over 40 technical papers, edited 24 ASME Special Publications and has authored and co-authored several ASME Sec. III and Operation and Maintenance Guides on flow-induced vibration and coupled fluid-structural dynamic analysis. He is presently a co-instructor of an ASME professional development course on Flow-Induced Vibration.

Printed and bound by CPI Group (UK) Ltd, Croydon, CR0 4YY

Printed and bound by CPI Group (UK) Ltd, Croydon, CR0 4YY

16/04/2025

14658832-0002